ZEBRAFISH

ZEBRAFISH

A practical guide to husbandry, welfare and research methodology

Edited by Claire Allen and Jean-Philippe Mocho

CABI

CABI is a trading name of CAB International

CABI
Nosworthy Way
Wallingford
Oxfordshire OX10 8DE
UK

Tel: +44 (0)1491 832111
E-mail: info@cabi.org
Website: www.cabi.org

CABI
200 Portland Street
Boston
MA 02114
USA

T: +1 (617)682-9015
E-mail: cabi-nao@cabi.org

A catalogue record for this book is available from the British Library, London, UK.
Library of Congress Control Number: 2024946596.

ISBN-13: 9781800629424 (hardback)
9781800629448 (ePDF)
9781800629431 (ePub)

DOI: 10.1079/9781800629431.0000

Typeset by Cheshire Typesetting Ltd, Cuddington, Cheshire and Exeter Premedia Services Ltd, Chennai, India
Printed and bound by CPI Group (UK) Ltd, Croydon, CR0 4YY
Photos by the authors unless otherwise indicated

Contents

Part II Practical research methodology

Contributors

Claire Allen Biological Services Aquarium, The University of Sheffield, Sheffield, UK

Rita Almeida formerly at Champalimaud Foundation, Lisbon, Portugal

Kamar E. Ameen-Ali School of Health and Life Sciences, Teesside University, Middlesbrough, UK

James Bull Department of Biosciences, Swansea University, Swansea, UK

Carolina Cabrera formerly at Champalimaud Foundation, Lisbon, Portugal

Joanna Cambray-Young unaffiliated veterinary surgeon, UK

Ana Catarina Certal Champalimaud Foundation, Lisbon, Portugal

Karishma Chhabria University of California San Diego, San Diego, California, USA

Pam S. Ellis The Bateson Centre, MRC-Arthritis Research UK Centre for Integrated Research Into Musculoskeletal Ageing (CIMA)

and Healthy Lifespan Institute (HELSI), School of Medicine and Population Health, University of Sheffield, Sheffield, UK

Catarina M. Henriques The Bateson Centre, MRC-Arthritis Research UK Centre for Integrated Research Into Musculoskeletal Ageing (CIMA) and Healthy Lifespan Institute (HELSI), School of Medicine and Population Health, University of Sheffield, Sheffield, UK

Sara Hilditch Statistical Services Unit, School of Mathematics and Statistics, University of Sheffield, Sheffield, UK

Michael L. Kent Zebrafish International Resource Center, University of Oregon, Eugene, Oregon, USA; Departments of Microbiology and Biomedical Sciences, Oregon State University, Corvallis, Oregon, USA

Peter Laud Statistical Services Unit, School of Mathematics and Statistics, University of Sheffield, Sheffield, UK

Christian Lawrence SmartLabs, Boston, Massachusetts, USA

Raquel R. Martins Sheffield Institute for Translational Neuroscience (SITraN), Faculty of Health, University of Sheffield, Sheffield, UK (formerly, The Bateson Centre, University of Sheffield, Sheffield, UK)

Sandra Martins formerly at Fish Platform, Champalimaud Foundation, Lisbon, Portugal, now at Marine and Environmental Sciences Centre / Aquatic Research Network, Laboratório Marítimo da Guia, Faculdade de Ciências, Universidade de Lisboa, Portugal, Cascais, Portugal and at Comparative Molecular and Integrative Biology, Centro de Ciências do Mar, Universidade do Algarve, Faro, Portugal

Mollie E. Millington Francis Crick Institute, London, UK

Jean-Philippe Mocho Joint Production System Ltd, Potters Bar, UK

Joana Monteiro Champalimaud Foundation, Lisbon, Portugal

Katrina N. Murray Zebrafish International Resource Center, University of Oregon, Eugene, Oregon, USA

Paul G. Schroeder Department of Biomedical Services, Oxford University, Oxford, UK; Red Kite Veterinary Consultants Ltd, Thame, UK
Lynne U. Sneddon University of Gothenburg, Department of Biological and Environmental Sciences, Gothenburg, Sweden

Amy Spencer Exploristics, Belfast, UK (formerly, Statistical Services Unit, School of Mathematics and Statistics, University of Sheffield, Sheffield, UK)

Eleanor Stillman School of Mathematics and Statistics, University of Sheffield, Sheffield, UK

Part I

Husbandry and care in practice

Chapter 1

The zebrafish model

An Introduction

Claire Allen and Jean-Philippe Mocho

Some species are global, invading every corner of our planet. For instance, the human and the rat, if we consider an urban setting, are always thought to be within 50 metres of each other. It can be surmised that you are often similarly distanced from a zebrafish (*Danio rerio*), since they are thriving in the wild (localized), your local pet shop, private and commercial aquaria and many research facilities globally. Originally native to South Asia, its popularity in aquarist tanks has remained constant and, interestingly, captive escapees are now known to be thriving in the freshwaters of America and Australia. The zebrafish is a small shoaling cyprinid fish, frequently sold under the name Zebra Danio, with several advantageous characteristics: natural hardiness; tolerance to measured fluctuations in water quality parameters (e.g. temperature, pH, conductivity); easy oviparous breeding; and a non-selective omnivorous diet (Chapter 3). These characteristics remove challenging husbandry demands that might be experienced with other less tolerant species. One problem to consider, nevertheless, is their preference for filial cannibalism, but separating young fish from adults is a simple solution in the research environment.

In the laboratory setting, George Streisinger pioneered molecular biology research with *D. rerio* about 50 years ago. Predictions were made then that large-scale screens on zebrafish were an important route into answering fundamental biological and medical questions, and that genetic tools would become research options, and so it has proved. The advent of genetic engineering and a fully sequenced

Claire Allen and Jean-Philippe Mocho (eds) **Zebrafish**
DOI: 10.1079/9781800629431.0001,

genome have now propelled this small tropical fish even further into the forefront of medical research and discovery; so much so that this species has become the third most commonly used for research in the UK and the European Union. This is explained by several key benefits to using the zebrafish for research. They are photoperiodic breeders and thus breeding begins at sunrise or when the lights are switched on, with parents either selected and paired or prepared for mass breeding the previous afternoon. At the start of the day, the newly born embryos can be harvested and manipulated easily. Removal of embryos from the parents can occur weekly providing their fecundity is maintained through optimized husbandry practices, providing researchers with a regular and ample supply of embryos. Their transparency is convenient for genetic manipulation by micro-injection of the single-celled embryo and allows invasive microscopic visualization of cellular and molecular processes, particularly in the first few days post fertilization. Developmental models often benefit from the enhanced regenerative abilities of these early stages, but regeneration capabilities are also upheld throughout the zebrafish life course. These processes are highly attractive for regenerative medicine research.

The legacy of intense developmental and genetic analysis has given us a capacity-laden and powerful model. It is an economically attractive laboratory animal with relatively small running costs, partly due to the straightforward genetic manipulation potential and quick generational time. Data collection can be prolific through careful interbreeding of genetically altered (GAA) lines, which combine phenotypes and transgenic markers. These qualities mean that the zebrafish is more often than not an attractive prospect for biomedical endeavours, hence few academic institutes are without at least a small zebrafish facility or can collaborate easily with someone close by. Global expansion into new countries such as Brazil, South Africa, and Ghana further evidences their popularity.

Ethically we must remember that, although they shoal, zebrafish are all individuals with perceptions. Research endeavours do have the potential to cause pain, suffering, distress and lasting harm. Care and consideration of zebrafish well-being is therefore crucial to limiting adverse effects and the severity of protocols (Chapter 14).

It is paramount that we optimize husbandry processes from birth and lived experiences to end of life, through defined environmental conditions that support reproducibility.

This book illustrates why this species is such an important biomedical model for research by reviewing many avenues of investigation and detailing ideas for future research. It aims to help laboratories develop good welfare and husbandry practices, refine procedural protocols and engage with ethical obligations. Experimental design and statistical choices (Chapter 9) require critical thinking and experience, both of which are considered throughout this book.

There are many considerations in genotyping choices (Chapter 10) and procedural work to optimize data quality, colony management and ensure 3R-reduction and refinement (Chapter 8). Husbandry refinements are discussed with environmental enrichment (Chapter 2), diet (Chapter 3) and water quality (Chapter 7). Further important organizational points for facility infrastructure and management are made (Chapter 7), including biosecurity, health monitoring and environmental impact. These topics lead to an understanding of how husbandry and care can impact fish welfare and the incidence of infectious (Chapter 5) and non-infectious diseases (Chapter 4). Procedural refinements are proposed through assessments of anaesthesia and euthanasia protocols and provision of analgesia (Chapter 13), with specific consideration of severity assignments (Chapter 14). Finally, expanding areas of research benefiting from innovative technologies are considered. For instance, imaging (Chapter 11) is key to monitoring both embryologic development and adult behaviour (Chapter 12). The latter has many applications, including neurodegenerative diseases and cancer progression. Combined with the exploration of the *D. rerio* ageing process (Chapter 6), laboratory zebrafish may provide answers to current challenges for the senior Western world population. These nuances accurately reflect the true strength of this laboratory animal species: from embryo genetic manipulation to ageing, researchers have been able to develop a wide range of models over the last half-century, at a rather fast pace and reduced cost. We hope this book will help them ensure robustness for these models and optimize fish welfare.

Chapter 2

Environmental enrichment for zebrafish

Mollie E. Millington, Christian Lawrence, Lynne U. Sneddon and Claire Allen

Abstract

Environmental enrichment has a profoundly positive effect on many captive animals, including fishes. This chapter will discuss employing enrichment for zebrafish, which is one of the most popular experimental fish models globally. Topics such as why enrichment is recommended, what impact it has on zebrafish behaviour and physiology and the available enrichment options that are logistically implementable are discussed. The focus is on laboratory facilities, and our recommendations include proposals for future directions and challenges that enrichment strategies need to overcome to become widely used.

Provision of live feeds and social group housing are widely accepted as enhancing the quality of life. There is also growing evidence to support the addition of images such as gravel to the bottom of the tank, along with adequate adjustable tank-specific lighting with optional high sunlight-like intensity. Finally, variable water flow to the tank is known to affect group dynamics and potentially offers enriching experiences. All five of these options simulate natural conditions from the wild but do not generally interfere with stock management, health status or research endeavours. Use of enrichments, such as plastic plants and actual gravel, is often debated but not favoured by research facilities due to several technical issues. These include reducing the visibility of fish for health checks, increased

Claire Allen and Jean-Philippe Mocho (eds) **Zebrafish**
DOI: 10.1079/9781800629431.0002, © CAB International, 2024

biofilm development and contamination and toxicity concerns. Evidence-based enrichment options are currently flexible, leading to a lack of standardization between laboratories. Research-based requirements may dictate variability in the enrichment provision that each facility decides upon. Nevertheless, future research aimed at understanding how enrichment strategies influence the health and welfare of the animals, technicians' workload and experimental output and outcome are necessary to overcome many of the barriers to adopting enrichment in zebrafish care.

Keywords

zebrafish, environmental enrichment, welfare, husbandry, natural behaviour

Introduction to environmental enrichment

The relationships between animals and humans are complex. Humans control the lives of many species of animals, for better or worse. Further investigation is needed to determine whether interactions are positive, negative or neutral for both animals and humans (Hosey & Melfi, 2014). As we turn our attention to the zebrafish, we must consider that we have the important task of enriching their life the way they want, not the way we think they want. We must be careful not to anthropomorphize their life experience, despite our intuitions to do so. We do, however, have a moral and often legal obligation to dissect out their preferences and create an optimal environment for all animals in captivity. Research involving animal models should aim to minimize and manage pain, suffering, distress and lasting harm in all experimental studies.

Enrichment is meant to provide shelter (if deemed necessary), stimulation and present the laboratory animal with novel experiences to alleviate boredom (Burn et al., 2020; Oliveira & Galhardo, 2009). We have a moral obligation to provide enrichment for zebrafish in a laboratory setting. Moreover, as a vertebrate research model, zebrafish are legally protected in highly regulated ways, paralleling mammalian species. As such, the zebrafish demands attention and optimization of its captive experience. Consider this: if zebrafish could communicate with us, what would their opinion be of living alone in a barren tank for weeks on end? This is our idea of the worst-case scenario for a gregarious species such as zebrafish, and research is starting to reveal that it is indeed a major stressor or depressor to be held in social isolation in a barren environment (Fulcher et al., 2017; Anneser et al., 2020).

The problem is that when there are over 15 identical fish in one tank and the technician needs to feed 600 tanks twice daily, spending time getting to know the fish is difficult, if not impossible. The real issue is that the solution to providing an optimal environment with suitable enrichment for so many small animals is not always clear and may influence research output. The consequences of some forms of enrichment could be to sensitize fish, affecting them negatively, and/or creating habituated fish, where the animals no longer respond. Ideally we need to decipher that the fish are acclimatized to their environmental conditions long term and in a positive way. The challenge for the zebrafish community is to find a resolution to improving welfare via the provision of enrichment, without interfering with the objectives of scientific investigation.

How do we find the right answer as to what enrichment should be provided to laboratory zebrafish? The answer should be based on both ethics and science, with critical thinking and an understanding of the impact of enrichment on health and welfare and how this may influence data collection. The complexities of these decisions need to be acknowledged. Millions of fish worldwide probably do live in barren conditions. This is partly due to a lack of understanding of their needs. Tanks, racks and facilities vary wildly between zebrafish laboratories, so it may be unfeasible to ask facilities to have common housing designs.

Environmental enrichment implementation and design will certainly take time to unravel, but research is progressing and the National Centre for the 3Rs (NC3Rs), a UK organization dedicated to replacing, reducing and refining (the 3Rs) the use of animals in research, is interested in developing this area (see Chapter 8 for more information). At present they are taking mere 'baby steps' but momentum is building. In general, it is commonly perceived that the zebrafish community is certainly behind the times, particularly since research that involves regulated work (according to European and UK law, zebrafish are protected when they have developed to the point of independent feeding) is becoming more elaborate for this species.

There is increasing pressure from regulating bodies and the general public to improve the welfare of animals. Enrichment is one of the most commonly discussed topics by the industry, since everyone is looking for an effective, cheap, safe and low-maintenance solution to the problem. A straightforward answer could be implemented globally, but there are challenges in assessing the effectiveness of different types of enrichment. Below we discuss the empirical evidence surrounding the use of enrichment in zebrafish, the types of enrichment tested and the barriers to adopting environmental enrichment as well as future perspectives in its implementation.

What is environmental enrichment?

The concept of enrichment can be defined as increasing the complexity of an animal's environment in captivity. In the laboratory context, environmental enrichment is implemented to improve animal welfare. Studies have demonstrated in many laboratory animal species that providing enrichment promotes natural behaviours and reduces stress and stereotypes. It is also known to encourage neural development, reduce aggression, improve learning and memory and mitigate disease impact (Magnusson & Brim, 2014; Coleman & Novak, 2017; Piña et al., 2018; Jones et al., 2021; Stevens et al., 2021).

Environmental enrichment comprises many forms, including social contact, sensory stimulation, enclosure or tank size and design (including resources within the housing), nutrition, physical exertion and novel objects (Singhal et al., 2014; Stevens et al., 2021).

Gregarious species are recommended to be housed in social groups, but the composition of groups requires careful consideration. In contrast, territorial species fare better in isolation. The size and complexity of the housing structure should be considered, as should the stocking density. Furthermore, animals should have a sufficiently large space to have unrestricted movements.

Why use enrichment?

Animals kept in captivity are typically governed by local and/or national legislation, as well as industry-agreed guiding principles that ensure the animals are housed humanely. In 1965, the Brambell report (Brambell, 1965) inspired the 'Five Freedoms', which many animal welfare advocates see as a solid baseline on how to maintain animals in captivity. Brambell's list was further developed by the American Farm Animal Welfare Council for domestic animals in 1979 (Food and Agriculture Organization, 2019) to create the following five points.

1. freedom from hunger and thirst by ready access to fresh water and a diet to maintain full health and vigour
2. freedom from discomfort by providing an appropriate environment, including shelter and a comfortable resting area
3. freedom from pain, injury or disease by prevention or rapid diagnosis and treatment
4. freedom to express normal behaviour by providing sufficient space, proper facilities and company of the animal's own kind
5. freedom from fear and distress by ensuring conditions and treatment that avoid mental suffering.

While the Five Freedoms for humane living conditions are widely applied in many captive animal species, it has proved challenging to apply them all effectively for zebrafish. One challenge associated with this has been identifying what the fish find enriching. There are many qualitative and quantitative ways in which enrichment can be measured (sometimes with contradictory results), and this has delayed the establishment of zebrafish-specific industry standards for best welfare and enrichment practices. No two studies appear to use the same enrichment, tank design and line of zebrafish. Furthermore, they may all house the fish in slightly different water chemistry. As with many aspects of zebrafish husbandry, it has proved very challenging to develop a standard operating procedure (SOP) that will be suitable for every establishment, primarily because no two facilities are designed or managed the same way (Lidster et al., 2017).

Freedom 1. Freedom from hunger and thirst

This is the most fundamental requirement of all the freedoms and, as such, zebrafish require feeding every day. Freedom 1 is discussed in terms of the enrichment potential of the feeding event, but essentially it would be deemed illegal to withhold food from a laboratory animal for a species-specific time. Thirst is somewhat redundant for fish but of course water, with optimal physico-chemical parameters, is a requirement (Lee et al., 2022).

Freedom 2. Freedom from discomfort

This freedom can be applied to zebrafish in the laboratory setting. The absence of sharp edges, the maintenance of optimized water chemistry and appropriate shoaling opportunities are all relevant considerations for the comfort of the zebrafish. Experimental design should also consider the impacts of the procedure upon the experience of pain, suffering, distress and lasting harm for the fish and mitigate adverse effects where possible. Breeding apparatus could allow animals' access to a shallow area for spawning as well as allowing non-disruptive collection of eggs.

Freedom 3. Freedom from pain, injury or disease

Laboratory zebrafish require careful management in order to ensure pain, injury and disease are all kept to an absolute minimum. Many aspects of maintaining this freedom can be taken from freedom 2 considerations. When decisions are being made for procedural work, additional consideration should be given to this freedom. Robust health screening and pathogen management procedures, including the provision of quarantine for incoming fish, will all keep disease risks down.

Freedom 4. Freedom to display natural behaviour

One might argue that freedom 4 – displaying natural behaviour – is hard to achieve for laboratory zebrafish, which have been bred in captivity over many generations and undergone selection for research-specific traits (Table 2.1). A careful comparison of wild and laboratory zebrafish has been published, which aims to improve welfare in science (Lee et al., 2022).

In their native habitat, zebrafish are shoaling animals that often live in highly vegetated habitats (Arunachalam et al., 2013), and scientific evidence has demonstrated that captive zebrafish choose enrichment over a barren area in preference tests (Schroeder et al., 2014).

Natural behaviours such as shoaling, breeding and feeding in the correct areas of the tank are all displays of natural behaviour that should be witnessed, in each tank, daily. Solitary confinement prevents shoaling and breeding, so should be kept to a minimum. It is important to understand that group-housed prey fish will hide in the shoal, even when they are unwell; this only helps to mask their deteriorating health. They will only reveal this sickness at a late stage by leaving the shoal to rest on the bottom or pant at the surface of the water.

Freedom 5. Freedom from fear and distress

A trained eye can identify stress indicators, and certain behaviours indicate fear and distress (e.g. spending time at the bottom of the tank), but if these also relate to a phenotype caused by a genetic alteration then confusion could ensue. Examination with behavioural analysis equipment can reveal the subtleties in fish when they display fear and distress (Lopez-Luna et al., 2017; Deakin et al., 2019a). This freedom could be applied by providing novel objects in

Table 2.1 Key zebrafish characteristics and their implications for housing design and enrichment.

Zebrafish characteristics	Housing recommendations
Highly social species	Group housing. Avoid isolation and housing in pairs, but when unavoidable provide enrichment (e.g. plastic plant) and visual and olfactory cues of conspecifics Larval and juvenile fish should be housed separately from adults House in mixed-sex groups to prevent egg over-retention in females Recommended housing density of 5–10 adult fish per litre to reduce stress Females should be allowed to spawn at least once every 2 weeks to prevent becoming egg-bound
Shoal in small, mixed-sex schools	Tank size and stocking density should allow shoaling behaviour
Dominance hierarchies form via aggressive interactions	Avoid holding fish in groups of six or fewer
Natural environment has abundant aquatic vegetation for spawning and cover from predation	Provide plastic plants and an area or resource (e.g. plastic tubes or large pebbles) for shelter
Naturally forage on zooplankton and insects	Provide live food at least once a week
Olfaction and vision are important sensory systems	Use of gravel, image of gravel or dark substrate. Fish housed in isolation should have olfactory and visual cues of other fish to reduce stress
Diurnal	Use of dusk and dawn lighting to avoid sudden bright light: 14/10 light/dark regime
Preference for structural complexity within tank	Use of substrate (e.g. gravel, pebbles) and plastic plants
Use of whole water column and constant swimming	Tank design should allow swimming both vertically and horizontally Control water flow based on the age of fish (low to no flow for larval stages and increased rates of flow for adults)

the tank, such as a piece of PVC pipe or a small plant pot, which can be used as shelter and allow the animal to hide and thus taking control of its situation. A barren tank leaves no options for hiding when a fish experiences fear or aggression from a tank mate. Observable fear and distress often occur when zebrafish are housed in pairs, where hierarchies develop between the two fish and the dominant aggressor can potentially intimidate the subordinate continually.

Implementation of enrichment

While legal or regulatory requirements ensure that a favourable environment is maintained for the animals kept in each dedicated facility, and penalties are issued when they are not, here we focus on the reasoning behind providing enrichment and justify this with evidence-based outcomes. It has been shown that 'happy' and healthy animals enable more reliable scientific results (Poole, 1997), which may mean that a smaller sample size is needed to obtain statistically significant results (see Chapter 9 for more information). Many might not think that fish require enrichment since they will feed, grow, breed and shoal – all good indicators of their general well-being – in captivity. However, these indicators do not necessarily reflect the animal's internal mental state. Animals can grow and reproduce even when in poor welfare conditions. Without a large mammalian-like face that facilitates emotions being read or vocalizations heard, it can be hard to ascertain how content a fish is. Behavioural and morphological indicators can provide insight into the welfare of zebrafish and technology may be able to assist.

Structural elements can be placed within the captive fish tanks, including gravel, sand, plastic plants, overhead cover, pipes and plant pots (Table 2.1). Many of these act to provide camouflage and/or refuges (Jones et al., 2021). Enrichment is not in fact limited to novel objects, however, and studies with zebrafish specifically have also explored social contact (Collymore et al., 2015), sensory stimuli (such as auditory; Barcellos et al., 2018), light, occupational (e.g. swimming in increased water flow), enclosure colour and nutrition (Stevens et al., 2021).

While there has been an increased focus on the need for, and the impacts of, structural environmental enrichment on zebrafish, the actual implementation of this in the laboratory is still limited and variable in how it is applied. There are a number of reasons for this, but it is likely due to the fact that most of what has been published in the literature on the subject has not extended much beyond attempting to answer basic questions around what fish prefer if given a choice, or how they perform in behavioural assays if enrichment is provided or withheld. While important and necessary, these data are often a long way from being prescriptive. There is a great need for additional experiments to be conducted in the hope that we can discover the type of enrichment that best suits the fish and husbandry requirements. The experimental design is crucial for these investigations, and a cohesive approach across the global community could potentially glean optimal and maximum data. We know that fish prefer complex environments when given the choice (Schroeder et al., 2014; DePasquale et al., 2019; Figures 2.1, 2.2 and 2.3), which supports the move away from barren tanks in the first instance. We can, therefore, base future trials on facts as we continue to assess the effects of enrichment.

Natural behaviours

Zebrafish are a South Asian freshwater species that inhabit very dynamic and varied environments (Engeszer et al., 2007; Arunachalam et al., 2013). There are hallmarks in their natural behaviour and zebrafish are generalists – they live almost anywhere, eat almost anything and shoal. They could be considered strategists – adaptable and resilient in order to survive and thrive. This is one of the factors that make them popular in the domestic aquarium trade and as a laboratory animal model.

In all habitats, social fish species often swim together in two ways, by schooling or shoaling, with different levels of harmony. Schooling allows for a group of fish (always from the same species, when observed in the wild) to move together and perhaps avoid predation, thus utilizing their energy more efficiently. Shoaling is

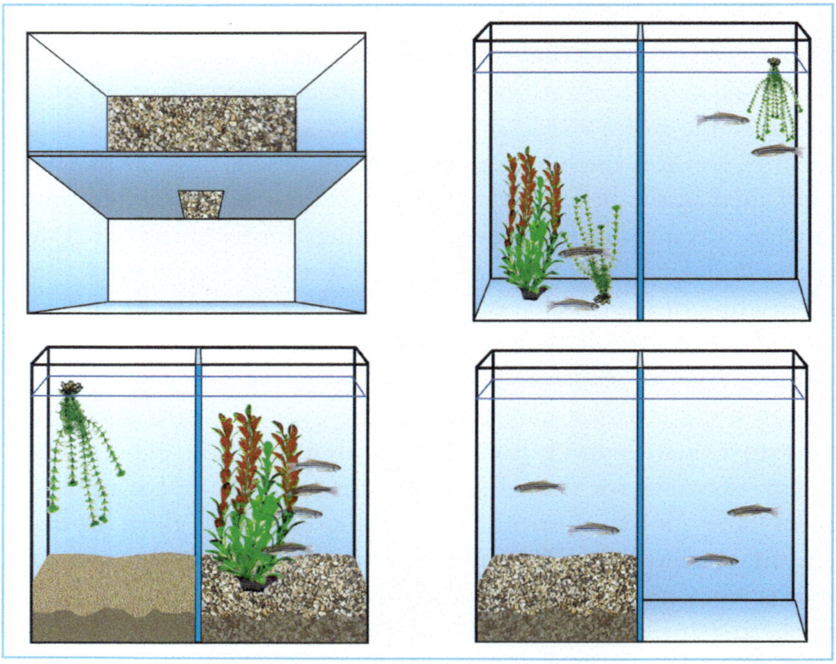

Figure 2.1 Experimental set-up tank designs for place preference testing, from Schroeder et al. (2014). Clockwise from top left: barren tank versus gravel image; floating plant versus submerged plant; floating plant with sand versus submerged plant with gravel; barren tank versus actual gravel base.

slightly different, occurring when the group is less dense and does not need to move in one direction at the same time (in the wild, different species are sometimes able to group together in this way). While schooling is well choreographed, shoaling is more informal and uncoordinated. In nature, half of all fish species are always engaged in shoaling behaviour (Shaw, 1978) and we can therefore conclude that it is important for their well-being. Shoaling is believed to improve foraging success (Pitcher et al., 1982) and can be seen at feeding times where one fish displaying feeding behaviour spreads the excitement instantaneously (personal observation). Zebrafish in captivity have been observed to perform both behaviours in a variety of tank designs and stocking densities, but schooling behaviour is harder to achieve (personal observation). It seems that, to

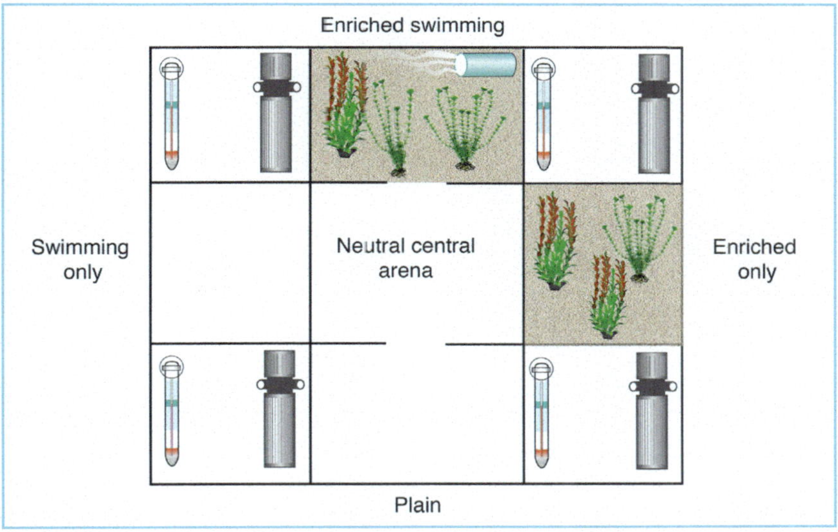

Figure 2.2 Tank design for place preference testing, based on DePasquale et al. (2019). Corners are inaccessible and contain heaters and biofilters. Clockwise from top: enriched with fast-flow rate; enriched only; barren; fast-flow rate only.

undertake schooling, zebrafish need to be in larger groups and tanks, a scenario not easily established for most laboratory zebrafish that predominantly exist in small groups.

Several comparisons between wild zebrafish, both in their natural habitat and following capture and rehoming in a laboratory setting, have helped dissect out their preferences to each other, foliage and flow. One study observed that captive AB wildtype (www.zfin.org) fish shoal less than wild-caught zebrafish, presumably due to the acclimatization of wild-type lines to the absence of prey and abundance of food (Wright et al., 2006). In the wild, group sizes are directly related to the local flow of water (Shelton et al., 2020). Higher flow rates prompt unstable, large shoals with a variety of dynamic leadership traits. Still waters enable small, closely knit groups to develop with consistent members and leaders. Interestingly, in this study the presence or absence of natural vegetation did not influence how the fish interacted with each other. An earlier study brought wild-caught zebrafish into the laboratory and observed that significantly

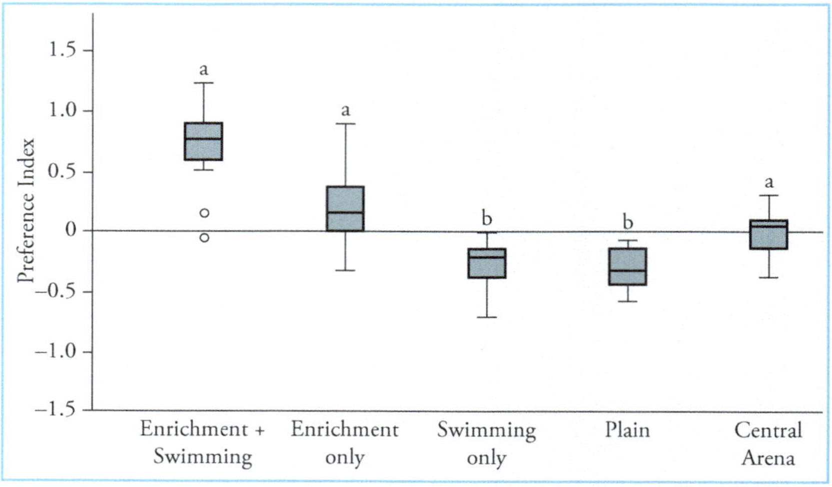

Figure 2.3 Jacob's preference for different areas of the tank, based on DePasquale et al. (2019). The graph shows significant preferences demonstrated in tested fish for an enriched, fast-flowing environment.

looser shoals occurred in barren tanks with the presence of flowing water (Suriyampola et al., 2017). The introduction of plastic plants into these experimental tanks had little impact on shoal size, and the fish were more active and aggressive. It appears that the presence of plants perhaps blocks zebrafish senses, making it harder to communicate as a shoal (Mogdans & Bleckmann, 2012). Interestingly, these studies do indicate that novel enrichment objects are not necessary for zebrafish in these scenarios.

Social context and stocking densities

Many behaviours are not clear cut, because the situation in which they are taking place may influence the way in which the fish interacts with the environment and/or its tank mates. Social aggregation is an example of this (Table 2.1). The way zebrafish associate with each other in shoals is very context specific and may be influenced by a number of factors, including flow rates, genetic relatedness, phenotype, age, tank, background, colour, group size, tank size, water depth and nutritional state (Lieggi et al., 2020). What motivates a fish to aggregate socially in one situation may not be at all applicable in another. Just because a zebrafish will choose to associate with another fish in a place preference test does not mean that it is always best to house them with other fish. This is especially the case in pairs where persistent aggression from one fish towards the other often develops over time (Harper & Lawrence, 2011; Keck et al., 2015). The ways in which these behavioural tendencies are used to develop environmental enrichment approaches must be nuanced, because the effects of these actions are hard to predict.

Typically, adult zebrafish are housed with conspecifics at varying maximum stocking densities in barren tanks (on average 5–10 fish/l; Lidster et al., 2017), although some procedures require them to be housed individually or in pairs for a short period of time. As a gregarious species, social isolation is known to be adverse (Bayne and Würbel, 2014; Collymore et al., 2015). For example, fin clipping, which is probably the most common procedure performed on laboratory zebrafish, requires fish be housed individually from one day to two weeks after a small portion of the caudal fin is removed for biopsy (see Chapter 10 for more information). Due to their small size, typical phenotype of stripes and speed of movement, it is nearly impossible to distinguish between individual fish in a group setting, so single housing is normally employed.

There are few known methods that can help make individual identification easier. One is visible elastomer (VIE) tagging, a technique that involves creating a combination of different patterns and colours from VIE injections. These forms of tagging are, however, considered invasive, are conducted under anaesthesia, can introduce infections and are liable to inflict temporary pain (Rácz et al., 2021), so they are not the most humane method. A second option is the use of computer-vision identification methods which, for example, utilize the individual's unique blue stripe patterns (Al-Jubouri et al., 2016). In specific research situations individual identification can be helpful, e.g. in behavioural studies, where many zebrafish need to be kept in the same tank for an extended period of time (see Chapter 12 for more information).

It is common for founder fish (with the first genetic alteration of its kind) to be kept single-housed for a few months after the genotyping process and as a new generation is raised and possibly screened (personal observation). Advocates of social housing would argue that this is easily enough time to establish a line. In fact, isolation of more than four weeks is considered detrimental to the well-being of the fish (personal observation), particularly for females, who may become egg-bound. This retention of eggs is due to prolonged exclusion from breeding opportunities, which eventually becomes irreversible when the oviducts become blocked by a plug of necrotic clumped eggs (Spence et al., 2008). Spawning opportunities must, therefore, be managed for solitary females with a maximum of one month's restriction (Niimi & LaHam, 1974; see Chapter 4 for more information).

One remedy that mitigates the impact of isolation is housing phenotypically different fish together to generate a temporary shoal. This group of visually distinct fish does, however, need to be observed for aggression towards each other. The phenotypically distinct fish will need to be successfully separated out at later time points, so this fish should be readily distinguishable: for instance, spots vs stripes; long vs short fin; pigmented vs albino; male vs female. There may be reluctance from scientists to blend fish into these shoals to guarantee minimal risk of losing a precious fish, but there should be genuine consideration of the fact that social isolation is detrimental to welfare and may impair the ability to mate and fertility in the long term. Collymore et al. (2015) demonstrated that single-housed zebrafish in barren tanks were depressed in both novel tanks (Figure 2.4) and light–dark tests when compared with singly housed fish with environmental enrichment. This suggests that, if single housing is absolutely necessary, environmental enrichment should be provided. Krueger et al. (2019) found that singly housed zebrafish prefer enrichment items that give the illusion of social interactions, such as mirrored paper on the side of the tank or being next to a tank of fish with no visual obstruction. This ability to see conspecifics through a barrier is commonplace in modern zebrafish tank design, with all the main suppliers offering robust and optional divisions for all sizes and dimensions of tank.

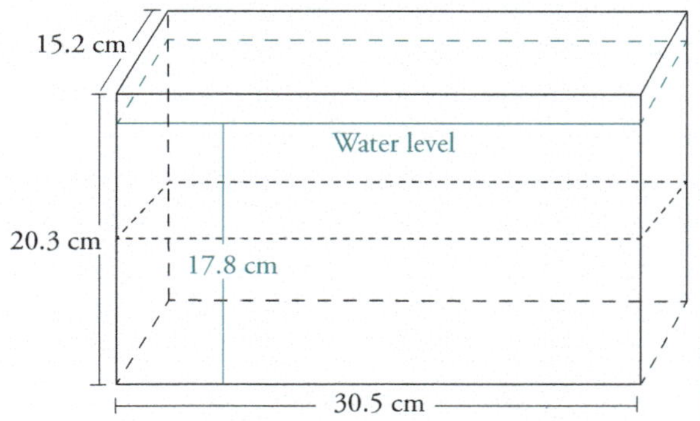

Figure 2.4 Set-up zones in a novel tank. When fish are placed into a new tank they tend to remain in the lower half. This is a stress response demonstrated by single-housed fish in barren tanks (Collymore et al., 2015).

The presence of other zebrafish decreases stress (cortisol) in some cases (Pagnussat et al., 2013) and has variable effects on anxiety (Shams et al., 2015; Tunbak et al., 2020). It is difficult to pinpoint an exact description of this in the literature but, in general, isolation is not ideal for this communal species. Interestingly, zebrafish recover much more quickly from stress when placed in a social group compared with social isolation or being held in pairs (White et al., 2017). This phenomenon is termed social buffering, and thus social context may be particularly important in regaining homeostasis after stress.

Breeding tanks

One clear example of a broadly applied environmental enrichment for laboratory zebrafish can be found in commercially available breeding equipment. Despite separation of the breeding pair, usually overnight, they are able to see each other through a clear divider and develop anticipation for breeding. Some spawning tanks have been designed to provide a depth gradient within a tank where animals can exercise their preference to breed in shallow water (Sessa et al., 2008). Further provision can be readily afforded to the breeding tank with plastic plant material that may also facilitate and enhance reproductive behaviour (Wafer et al., 2016). These simple enrichments, provision of a breeding slope and plants, can enhance productivity and they have the potential for no harm. Designs of mass-breeding tanks can include a gravel image on the bottom of the tank, good laminar flow, a breeding slope, a tray for easy embryo collection, ample space for animals and plastic plants. Care must be taken to remove both the embryos that could be trapped in the plants and the biofilm that will grow on all surfaces of the breeding tank. The former may result in stray fish growing and contaminating subsequent clutches. The latter, if left to build up, will negatively impact embryo production and overall colony health with potential pathogens.

Sensory stimuli

Sensory stimulation includes auditory (sound), visual or olfactory (odours) stimuli. For example, playing music to animals in captivity can have a calming influence (Barcellos et al., 2018). Providing new smells and visual stimuli can also promote normal exploratory behaviour.

Auditory

A general awareness of surroundings and the noise levels that are experienced in any given space is important. The potential benefits or harms should be considered for both personnel and housed animals, including zebrafish. Plant equipment can be noisy, particularly if it is ageing or failing and thus produce more noise. Zebrafish use sound to perceive their environment (Popper & Sisneros, 2022) and evolved with diverse underwater acoustics that vary in level, intensity and source. Some dangers are perceived by fish through their sounds, but the identity and nature of such sounds is generally unknown. New loud noise will initially startle zebrafish but they adjust rapidly, although recording of sounds in rural India (INDIAN Village Ambience 001| HD Sound Effects Heavy Rainy day in a green lush village in South India | Relaxing sounds of rain falling on umbrella) demonstrates the wide variety of noise in an example of their natural habitat. Aquatic acoustics alter these sounds and, in combination with additional underwater noise, presumably creates a complex sound environment. It is difficult to measure and assess this environment because measuring sound is so challenging (Popper & Sisneros, 2022). Behavioural response to sound is known to include anxiety, startle responses and freezing behaviours (Shafiei et al., 2016), but some sound may be a positive experience. Playing of classical music in zebrafish facilities seems to improve welfare (Barcellos et al., 2018; Marchetto et al., 2021). Zebrafish exposed to classical music performed better in novel tank tests (Figure 2.4). Barcellos et al. (2018) played Vivaldi for two hours twice a day, for 15 days, for three fish housed together in a 3 l tank.

If music has the potential to be a positive addition to the physical environment for a laboratory fish, how should it be applied in a real-life situation? We are far from knowing the answer to such questions. At present, if a facility manager decided to use music as an enrichment tool, they would need to make judgements on the type, volume and timing of music to be played, with each one of those factors potentially imparting positive – or negative – effects on the animals subjected to that treatment. We are all aware of different musical tastes between humans, and this applies to stakeholders utilizing the zebrafish facility. Should they have to listen to classical music all day when they prefer pop? Further research is clearly required to determine from what elements of the musical experience the fish benefited.

Consider a small fish room where a stand-alone zebrafish unit shares space with a marine display tank, air conditioner, reverse osmosis water machine and pump. The room is noisy from a human perspective and poses an occupational exposure problem. This room is perhaps too noisy for zebrafish as well. A survey was conducted to determine noise levels in the room relevant to human exposure as well as how this translates into fish tanks. A hydrophone attached to a recorder was placed in an empty tank and found an average of 150 Hz over four hours. Normally occurring activities related to animal care also produce noise in frequencies at this level, within the hearing range of zebrafish (100–4,000 Hz; Higgs et al., 2002). Since the room is cramped with a baseline noise level within zebrafish detection, care must be taken to restrict additional extraneous noise (A. McCorry, Alberta, 2022, personal communication). Until disproved, we would expect the effects of constant noise and vibration associated with the human-generated (anthropogenic) artificial environment of aquatic laboratory housing to be adverse, as discussed by Popper & Sisneros (2022). This review highlights the gaps in our knowledge. Rest assured, however, zebrafish acclimatize to their surroundings with experience and age (personal observation). Be aware though, that data acquisition could be impacted by auditory changes, levels or intensities, and so consideration and standardization is required for consistent reproducible results.

Visual

Visual stimuli, such as pictures of natural sediment materials (rather than the real thing) under the tank, have been investigated (Schroeder et al., 2014; Graham et al., 2018; Lavery et al., 2019), and studies have shown that zebrafish prefer a gravel image over a sand image (Schroeder et al., 2014; Figure 2.1). Colour preference has been explored in fishes with conflicting findings (Pryke, 2009; Oliveira et al., 2015; Roy et al., 2019). Results of studies on zebrafish have been very inconsistent, with many showing that a black substrate is preferred over lighter ones, but others conclude that the opposite is true (Bault et al., 2015; Peeters et al., 2016; Pieróg et al., 2018). It is certainly well established that they dislike white substrates (Blaser & Rosemberg, 2012; Santos et al., 2019; Serra et al., 1999), and thus white backgrounds and white-coloured resources in the tank should be avoided. Note that many automated behavioural equipment rely on a bright white background, so one ought to question the validity of behavioural data collected in this set-up. In conclusion, zebrafish prefer photos of gravel over sand in or under their tanks, and the light (white)–dark test measures time spent in the white area as a proxy for reduced anxiety, since zebrafish avoid white.

The colour of enrichment, therefore, requires serious consideration. Zebrafish are often deemed 'creatures of habit' because the precipitous contrast between what they are used to and different environments is a source of stress. For example, moving fish from their housing shelf to a benchtop for procedure may induce obvious aversive behaviour in the tank, when the benchtop's colour is in sheer contrast with that of the housing shelf: e.g. from black to white (personal observation). Thus it can be particularly useful to reduce such contrast for procedural work and when fish are kept away from their normal housing conditions. Setting pictures of gravel on benchtops seems like an easy answer. However, are we trying to mitigate contrast or simulate a natural environment? The choice of the gravel picture should address both points. We must also bear in mind that the tank bottom is not of a consistent colour along the life of zebrafish. During the first few weeks of life, biofilm builds up at the tank bottom and, depending on diet, it often varies between red and green in colour. These themes may therefore feel familiar for some zebrafish colonies.

Light management

Lighting a facility needs to serve both the health and safety requirements for the technical staff managing the animals and the researchers utilizing them. Problems arise when lighting is acceptable for the users but not for the species being managed. Science provides strong evidence that light plays a direct role in physiological development, and consequently survival, for many species (Emmer et al., 2018). Hatch rates and larval survival are better with the correct light delivery (intensity, spectrum and photoperiod), with some immediate benefits for survivability (Villamizar et al., 2014). For instance, larval stages of some fish species, in the correct light, are able to swim up from the bottom sooner, moving away from the biofilm that contains pathogens. Post hatch, light has a role in deformity rates and improves immune responses (G. Westcott, Chorleywood, 2019, personal communication; Villamizar et al., 2014). Furthermore, Wolman & Granato (2012) reported that larval zebrafish swim to bright areas when threatened. So, we should consider light as an important factor in the biology of the fish and a contributory factor to environmental enrichment. Circadian rhythms are known to be important for zebrafish well-being (Vatine et al., 2011). Sunrise and sunset can be mimicked by facilities with gradual increase and decrease in light intensity, rather than having a light cycle that comes on and goes off abruptly. There is an accumulative effect from light exposure across the full life cycle of the fish, and zebrafish see a far wider spectrum of light than humans, extending into the UV/blue. We should optimize the lighting in our tanks to enhance their experience or, at the very least, give them some high-quality daylight. This should be offered as a choice because stressed and anxious adult fish will search out dark areas, presumably to hide. A content adult will enjoy having options (see Chapter 12 for more information).

Good low-energy LED lighting is widely available and transforms the intensity of light in rooms to mimic daylight. A well-lit tank reduces the amount of movement the fish perceive outside their tank. This is thought to reduce their exposure to stressful stimulants (G. Westcott, Chorleywood, 2022, personal communication). Further investigation into the preferred exposure time to light at different intensities throughout the daylight hours is required, and care should be taken with strains such as albino that have known light sensitivities.

Nutrition

Nutritional enrichment has three main aspects: nutritional content, live versus manufactured food and frequency/schedule of feeding (Stevens et al., 2021). Many commercial feeds are available to meet the nutritional requirements of zebrafish (see Chapter 3 for more details) and so, with enrichment in mind, we will focus on the enrichment potential of feeding regimes and the provision of live foods. In general, the desire for a fish to feed with visible evidence of anticipation behaviour and minimal latency to feed is a good indicator of positive fish welfare.

Typically, zebrafish are fed ad libitum daily, but frequency varies for once, twice, to more than three times per day, with once a day being the least popular choice (Lidster et al., 2017). More frequent feeds seem to be preferential, since zebrafish fed once a day are found to be more anxious than those fed twice a day (Dametto et al., 2018). Routine within these schedules may be an important factor for the happiness of zebrafish. A study on goldfish by Gómez-Boronat et al. (2018) found that alterations in external cues, such as feeding time, altered the homeostasis of the fish and could be considered a stressor. This may also be the case for zebrafish, but the opposite could be true, and altering the schedule or timing of food delivery might provide an enriching experience.

Despite scant evidence that provision of live food improves welfare with the same positive impact as environmental enrichment, Lidster et al. (2017) reported it to be one of the most common enrichment practices implemented by zebrafish facilities. This survey found that 97% of zebrafish aquarium staff use live diets, leaving just 3% who provide only dry. The decision to remove live diets from a facility is based on evidence that live foods are a potential source of harmful pathogens (Mason et al., 2016). There are also some drawbacks to the inclusion of live feeds in the diets of fish used for experimental purposes (Chang et al., 2019b; Watts et al., 2016). In some instances, research objectives may dictate that live feeds are not an appropriate diet.

Live foods such as *Paramecium*, *Rotifera* and *Artemia* are popular diet choices. *Artemia nauplii* is perhaps the closest thing to a standard dietary item in zebrafish husbandry and is the most common type of zooplankton fed in the laboratory (Lawrence et al., 2016; Lidster et al., 2017). The two other popular live prey items, *Paramecium* (Varga, 2011) and *Rotifera* (Aoyama et al., 2015; Best et al., 2010) are of microscopic size and hence they are especially useful during the nursery rearing stage. Studies have shown improved growth, reproduction and survival with the use of live prey feeds (Fowler et al., 2019; Gonzales & Law, 2013; Monteiro et al., 2018). They also theoretically prevent boredom whereby the fish spend more time foraging, which is presumably an important expression of natural behaviour. Live feeds are a convenient, dual-purpose way to refine the environment of laboratory fish populations (Stevens et al., 2021), the proverbial 'low-hanging fruit' already included in the diet for reasons other than welfare refinement.

Physical exertion

The use of forced or voluntary exercise is considered physical enrichment. Exercise has been shown to have positive impacts upon many mammalian species. For example, a rodent running wheel within a cage allows the animal to voluntarily engage in exercise. In zebrafish, exercise using water flow improves muscle and bone development (Hasumura & Meguro, 2016; Suniaga et al., 2018) and learning ability. It also lowers anxiety levels (DePasquale & Leri, 2018; Luchiari & Chacon, 2013). Flow into tanks is necessary to maintain aeration and water quality, and zebrafish are known to experience flow speeds between 3.5 and 13.9 cm/s in nature (Arunachalam et al., 2013). Adult zebrafish prefer areas with flow as opposed to no flow, and this preference can be enhanced when enrichment, in the form of plastic plants, is available (Figure 2.2; DePasquale et al., 2019). Conversely, embryo/larval fish should not be exposed to excessive flow because they lack strength and it will reduce their body condition. Therefore it is important to consider the age of the zebrafish when choosing flow rates.

Research indicates that when the flow is increased or decreased the fish will respond slightly differently; faster flow rates encourage endurance swimming, stamina (Maierdiyali et al., 2020) and playful interactions with the currents (personal observation); slower flow rates enable inter-fish relationships to develop and foraging behaviour to occur (Shelton et al., 2020). In contrast, slower flow rates with wild-caught fish have been found to result in increased levels of aggression (Suriyampola et al., 2017), and although plasticity of zebrafish social behaviour is frequently demonstrated, wild-caught experiments will not necessarily be mirrored in captive responses. One significant difference in captive bred zebrafish is that they are known to have reduced startle responses (Robison & Rowland, 2005). The additional potential benefits to altering flow for varied experiences are, however, currently unproven but it is not impossible to imagine automated system designs that alternate flow over a set period of time. DePasquale et al. (2019) concluded that further investigation into combinations of enrichments, which include the influence of flow, are important to reveal the optimal preferences of zebrafish.

Novel objects

The use of novel objects that promote new behaviours is also termed behavioural engineering toys or occupational enrichment (Stevens et al., 2021). The purpose of these items is to stimulate the animals to engage with the novel object, prevent boredom and exert control over their environment (Burn et al., 2020; Oliveira & Galhardo, 2009). Their use could lead to play behaviour (Stevens et al., 2021).

Studies have demonstrated zebrafish have a preference for sand, substrate and plants over barren tanks (Schroeder et al., 2014; Figure 2.1) and thus there may be innate preference for natural habitat. The positive effects of improved animal welfare in laboratory zebrafish when these types of structures are added to enclosures have been documented. Wafer et al. (2016) showed that provision of plastic plant materials in tanks improved breeding and larval survival, while Lee et al. (2018) demonstrated increased survivorship in structurally enriched habitats. The provision of plastic grass in tanks also reduced cortisol levels in single- and group-housed fish, possibly indicating better adaptation to anxiety and stress (Giacomini et al., 2016). von Krogh et al. (2010) observed an increase in forebrain cell development in male zebrafish when they were housed in a tank with gravel and plants, versus fish that were kept in a barren tank. More recently, Estes et al. (2021) also found that enrichment led to relatively larger fish, with the males benefitting most.

Recently an autoclavable blue plastic shelter has become commercially available. If this structure is introduced into fry tanks in the first few days, by the time the fish have begun to exhibit shoaling behaviour they clearly shelter under the structure, often to avoid netting events. Males later appear to guard the structure, which is not classically a sought-after behaviour. When a second male approaches the tent, the guarding male accelerates out towards him seemingly chasing him away, he then returns to the inside of the structure (C. White, Portsmouth, 2023, personal communication). Some studies suggest that this type of behaviour increases the number of fertilized embryos produced (Ariyomo & Watt, 2012), and indeed embryo production and survival rates twelve hours post-fertilization have improved since using the shelters. Interestingly, the guarding male in this scenario is always much brighter in colour (C. White, Portsmouth, 2023, personal communication). The creation of such hierarchies within the tank could therefore assist breeding, but care must be taken in managing the welfare of the shoal. Territorial and dominant behaviour usually lead to fighting and aggression, but in this study there has not been an increase in wounds that could be attributed to aggressive behaviour. In fact, fewer health concerns were noted than in previous years when the fish were housed in barren tanks. Certain breeding challenges may benefit from these scenarios, so it is important to consider them within a breeding programme. This tent-like structure is certainly useful in the pre-breeding stages but perhaps needs removal once hierarchies begin to form and netting events increase. Further enrichments trialled at this facility include ornamental vegetation being attached to the lids of the tanks. The fish interact with the plant-like material, which can be easily removed when the lid is lifted. Zebrafish are observed spending more time close to/within the artificial plant material than away from it, particularly when being fed (C. White, Portsmouth, 2023, personal communication).

Not all novel object introductions into tanks that have been tested seem to have a positive effect. Wilkes et al. (2012) used black glass rods, which are relevant for toxicology studies where the use of plastics must be avoided. Several behaviours, including aggression, and the whole-body levels of the stress hormone cortisol were measured in juvenile zebrafish over a one-week period. These results indicated that following the addition of glass rod structures, as hypothesized, enrichment did not result in a measurable improvement in welfare in terms of reduced aggression. However, cortisol levels were similar to those in non-enriched fish, so these rods were also not an additional stressor to the fish. The provision of black rods may have meant that the dominance hierarchy took an extra day to resolve, and so aggression was higher for that period of time.

Other studies have shown no differences between enriched and barren housed fish. For example, unstressed fish held in an enriched home tank exhibited similar responses to fish held in a barren tank when temporarily placed in a novel tank (Marcon et al., 2018). This implies that the effects of environmental enrichment depend on overall stress experienced by the animals in holding tanks when later placed in a new environment. Alternatively, one could conclude that enrichment does not confound the results of the novel tank test. Bhat et al. (2015) found that when comparing fish obtained from the wild versus a commercial supplier, all were more aggressive when placed in novel tanks that contained environmental enrichment than when in barren tanks. Similarly, Woodward et al. (2019) found that environmental enrichment increased aggression and decreased body length of nacre mutant zebrafish (www.zfin.org); however, there was no observed effect on fertilization rates or tank space use. These authors proposed that the environmental enrichment promoted territorial behaviour when a fish from an enriched tank was placed alone in a barren novel tank with a mirror.

In the wild, fish will display aggression and territoriality. Aggression is a natural behaviour. Should we aim to remove it completely from a captive zebrafish's repertoire to improve welfare in captivity? Living within the confines of a barren tank may increase the amount of aggression, because in nature fish can swim away. Just because aggression is something that is a feature of their natural environment does not mean it should be promoted in the context of an artificial housing environment. In this case we want to reduce aggression in captivity because the features of the artificial environment make it harder for fish to control how it impacts them. The same can be said for most natural features of the environment. Not all enrichments or resources will have value in captive environments, and so some should be eliminated.

Given these contrasting results, there is frequent pushback and reluctance to provide novel objects, such as plants, within a laboratory zebrafish tank. Many stakeholders see these additional objects as a potential place to harbour harmful biofilm and pathogens that could affect the overall health of the colony and create additional cleaning burden (Chang et al., 2019a; Lidster et al., 2017). Food delivery could be a problem, since it could potentially collect on the object, perhaps out of reach of the fish, and subsequently degrade and rot. This will negatively impact local water quality parameters. Daily welfare checks are also made more difficult if fish are able to hide in the vegetation. Indeed, individuals with deteriorating health are typically more likely to hide from view, thus further impacting the potential spread of disease. Sick fish could be left hiding from sight to experience pain, suffering, distress and lasting harm. Furthermore, since zebrafish will breed daily within their home tanks, the inclusion of plants or other structures could create traps for embryos, and these fish could grow and develop. Theoretically this could create stray animals that are often a legal problem and not permitted, since each animal must be accounted for. If they are left to grow to adulthood and are indistinguishable from their genetically distinct parents, they will contaminate the shoal. The risk of this scenario occurring is fairly high because many research facilities have hundreds, if not thousands, of tanks with high stocking densities, and human error can impact daily checks particularly in unenriched environments. The difference is that in a barren tank daily embryo production is managed by laminar flow and filial cannibalism. The latter process helps maintain fish body condition.

One final consideration is that if the zebrafish are provided with plastic aquarium plants from the local pet store or Internet store, these are not autoclavable (to prevent being a source of pathogens) and some may easily fall apart and leach chemicals. This could, over time, disrupt the life support system. In some cases the fish may even ingest small fragments of the poor quality artificial plant with unknown consequences.

Research zebrafish have been bred for many generations to live in a barren environment. Having never lived in the wild, do these captive fish relate to their natural habitat instinctively? Even if they do, the real problem is that data acquisition is a priority in the laboratory setting. Scientists might argue that alternative housing conditions are another variable that could impact their research results, particularly if the science utilizes behavioural evidence. Consequently, no sudden changes in environmental enrichment in facilities undertaking longitudinal studies should proceed without careful consideration. However, where enrichment does improve welfare, scientists must concede that using healthier animals means their data will be more robust, valid, and reproducible. When considering the request for an absence of enrichment, facility managers should consider the procedure's severity and whether it is ethically justifiable to omit enrichment from the home tank. Fish left to experience solitary confinement need special consideration (see Chapter 12 for more information) and should be offered enrichment such as novel objects: for instance, a mirror or image of zebrafish on the outside of the tank, carefully selected foliage and a gravel image on the floor. The daily checks, perhaps at feeding time, need to ensure that the fish is clearly observed interacting with the tank and eats normally. Stressed fish display repetitive movements, stay low in the tank and are distinctly behaviourally different from non-stressed fish (personal observation).

In conclusion, if the evidence in favour of adding novel objects (presumably plant-like structures) into the tank is greater than the reasons to withhold this enrichment, then clearly further research into the design of plants needs to be conducted to resolve these issues.

Measuring impact with biological assays

One of the biggest challenges in enrichment research in zebrafish is how to measure the influence (whether positive or negative) of the enrichment on the animal itself. Impact can be tested with invasive measures such as blood sampling (Zang et al., 2015), monitoring neural circuits (Sumbre & de Polavieja, 2014), non-invasive place preference (Collymore et al., 2015; DePasquale et al., 2019; Schroeder et al., 2014), opercular beat rate (Reilly et al., 2008), novel tank exploratory behaviour (Collymore et al., 2015; Figure 2.4), fecundity (Wafer et al., 2016) and water sampling for cortisol (White et al., 2017; Figure 2.5). Other challenges associated

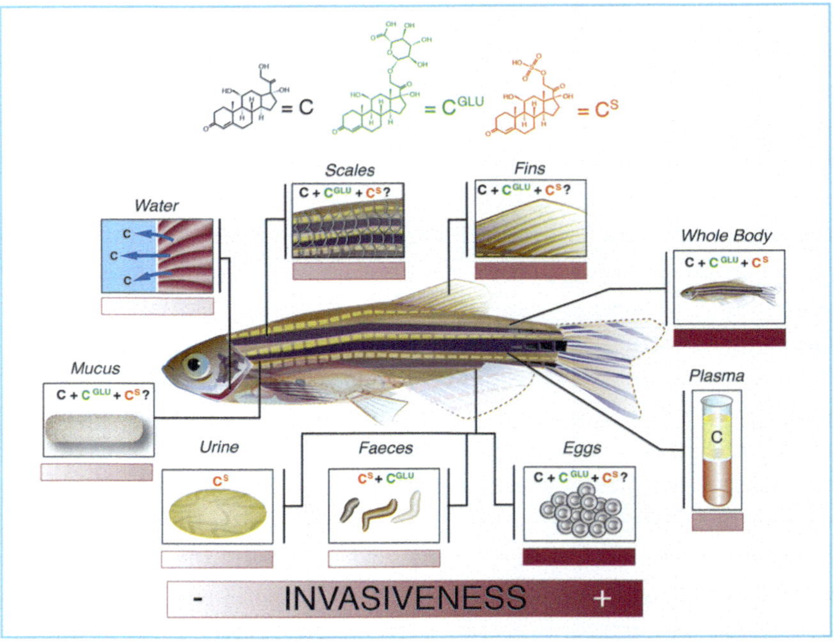

Figure 2.5 Based on Sadoul and Geffroy (2019). Potential sampling methods used to measure cortisol levels in fish (C, cortisol; C^{GLU}, cortisol glucuronidated; C^{S}, cortisol sulphated). Some methods are more invasive than others, as indicated by the colour scale.

with conducting this type of welfare research include a lack of resources (financial, staff availability, space limitations, key knowledge and skills in experimental design) given that funding agencies rarely prioritize welfare refinement. Overall, we need a common approach to measuring welfare indicators in response to the presence of enrichment to ensure that we are providing the right type, the right amount and that the welfare of our fish is not compromised in any way.

The broad plasticity of fish behaviour makes the application of environmental enrichment particularly challenging. Naturally there is a desire to identify specific behavioural actions as being absolute; especially if the goal is to use these as endpoints to evaluate welfare state. Some behaviours, such as rapid gulping of air at the water surface accompanied by opercular fanning, fall more within this category. If fish are observed to be engaging in these behaviours, especially in concert with some other feature of the environment (i.e. absence of water flow), there is a straight path to the proper management response. Other behaviours, such as freezing or rapid, evasive swimming throughout the water column in a tank, also meet these same criteria. All are stereotypical responses to negative stimuli, and the necessary management steps required to modify the environment and solve the problem are clear and easy to make. It is not environmental enrichment, but it is a similar dynamic where behaviours of the animals are used as a basis for management decisions.

Over time it will likely be that a number of other approaches will reach the level of these examples. What the industry does have, in ever-increasing availability, is the basic, first-step information from which to build these types of prescriptive guidelines for enrichment in zebrafish facilities. Importantly, the tools that can be used to evaluate fish behaviour are improving all the time (Deakin et al., 2019a; Thomson et al., 2019). This will help move the field from basic level understanding to the more nuanced view that is required to make the enrichment for zebrafish a more straight-line decision, where the action overwhelmingly leads to improved well-being of the animals.

Summary of the five most evidenced environmental enrichment measures

Live feed

According to global surveys, the most commonly employed form of environmental enrichment is the provision of life feeds (Lidster et al., 2017). The application of paramecium and rotifers to larval rearing tanks appears to improve their welfare and survival. *Artemia* can be used as food enrichment for adults since it stimulates natural foraging behaviours. Some users are concerned about the potential introduction of pathogens from live feed, and therefore welfare improvements need to be balanced against biosecurity and health.

Companionship

The shoal offers security by enhanced sensing of the situation as a whole and moving as a team. Group housing must be facilitated. Laboratory zebrafish in conventional tanks regularly shoal, but they do not usually school. Schooling is a more synchronized behaviour with displays of turning, twisting and sweeping together. Even if laboratory zebrafish are rarely seen schooling, shoaling still offers social interaction and a level of perceived safety.

Gravel

One enrichment type that stands out and which can be provided easily with relatively little effort is the image of gravel substrate under a housing tank bottom. It has been shown in different studies, and across a number of different scenarios, that zebrafish prefer to be in compartments where images of a gravel substrate have been provided (Graham et al., 2018; Schroeder et al., 2014). The gravel image offers them the opportunity to camouflage, which is an instinctive behaviour. This modification of the housing environment, which takes place outside of the immediate enclosure and therefore does not result in any immediate biological or chemical changes, is still visually perceived and preferred by the animals. This is an example of sensory enrichment (Pritchett-Corning, 2020) that could be readily applied in a fish facility at relatively low cost and with little to no disturbance to the animals and workforce. Given these factors, it is fairly likely that this form of tank modification will be one of the first broadly implemented enhancements in the zebrafish community.

Flow rates

The potential enrichment benefits of water flow into tanks are widely acknowledged by the zebrafish community (Lee et al., 2022; Lieggi et al., 2020). Water exchange is required to maintain good water quality in static tanks, and optimized, conditioned water is usually delivered to each tank continuously. The flow that is generated from the inlet (usually at the front of the tank) to the outlet (usually at the back of the tank), and the currents that it creates within the tank, should be considered an existing enrichment. Laminar flow is often induced in tank design for swift removal of waste products from the outlet; it remains unclear, however, whether this is the preference for zebrafish.

Light and seasonality

Light is an important factor in the culture of freshwater species, such as zebrafish, that come from hot, humid, tropical climates. Lee et al. (2022) draw our attention to the seasonality and sheer intensity of light in South-East Asia at noon, which does not compare to the artificial lighting provided in the laboratory facility setting. In small, barren tanks, such as those widely adopted by research facilities, we need to be very careful with intensity levels because the fish have no shade or places to retreat from this light. All light provision requires strain-specific welfare assessments, and we can hypothesize that some strains will prefer low-intensity, 'cloudy' days while others will thrive in full sunshine, for at least part of the day. We recommend wider recognition that the provision of optimal lighting, with a broad spectrum, and carefully orchestrated photoperiods (e.g. dawn and dusk) is an environmental enrichment for zebrafish. Further studies are required into the potential welfare benefits of the three contributing factors (intensity, spectrum and photoperiods) during a variable annual cycle. For example, recent research has already demonstrated improved growth and reproductive performance with longer photoperiods (Abdollahpour et al., 2020).

Conclusions

It is widely understood that better husbandry and welfare conditions in captivity produce healthier animals, including livestock and many laboratory animal species. Their life experiences from birth are, however, extremely far removed from the natural habitats and lives of their wild counterparts. Optimized care creates a setting where animal products and research data are more likely to be standardized and reproducible in their biological readouts and, therefore, of higher value in human endeavours. So, we can surmise that healthier zebrafish give more robust results, and a drive to better and deeper understand their husbandry requirements will help to develop refinements in experimental procedures. We simply cannot develop an objective to recapitulate their natural habitat that mimics the life experiences of wild zebrafish. Captive bred zebrafish cannot exert their instinct to escape unless we offer them hiding places and routes into areas of perceived safety. The presence of stressors, as an experience, is not recommended. Because research colonies are, in the most part, confined to relatively small home tanks, we can predict that health and well-being would deteriorate if we introduced negative stimulation in the form of a threat. Indeed, Buenhombre et al. (2021) reviewed the evidence that aquarium-related stressors and unpredictable chronic stress result in reduced quality of life and increased anxiety.

Due to the robust nature of this species and their adaptability, measurable outcomes such as fecundity and survival may not indicate optimal well-being. These two measures are not the sole indicators of a content, healthy fish, and welfare analysis tools such as the Fish Behaviour Index (FBI) should be employed to investigate actual response following invasive and adverse procedures (Deakin et al., 2019a). The desire to survive and reproduce is primal but, as research shows, there are so many additional provisions that we can introduce into their lives to create a better and more content existence. In some studies there is no effect of enrichment on the data and so enrichment is not a confounding factor. Nevertheless, experimental design needs to withstand certain variables and studies should always include control groups, which offer practical enrichment to reduce stress and/or anxiety.

Proponents for providing enrichment for zebrafish say that complex environments yield better study results (Hebb, 1947; Baybe & Wurbel, 2014). Behavioural studies that measure place preference have demonstrated that fish choose to be in social settings and/or where there is gravel, sand or plants (Schroeder et al., 2014).

This behaviour presumably offers them protection from threats that they instinctively want to avoid. Zebrafish do indeed come from a complex environment but, as a prey species, much of their natural behaviour could be attributed to predator avoidance (Gerlai, 2013). Acclimatization is a big factor so that they feel secure in familiar surroundings: for instance, the long-term absence of prey threat may be perceived by the fish, making them less reactive when handled. Because startle reflexes are reduced in captive-bred zebrafish, we can surmise that fish might respond less. They are not experiencing the presence of predators that stimulate flee and hiding responses, so perhaps social housing and gravel images are sufficient enrichment to offer a sense of protection. This goes back to the Five Freedoms, where fish are held in a comfortable environment that supports the performance of natural behaviour and, as such, are allowed to thrive in a situation more similar to the wild habitat. Lee et al. (2022) tackle the need to understand their natural environment in our research endeavours in order to inform enrichment choices. Consideration should be given to these factors, and experimental design should minimize 'contingent suffering', a term given to captive animals experiencing reduced welfare as a result of poor health and abnormal behaviour (Richmond, 2002).

Some elements of enrichment provide what we could describe as pleasurable experiences, where the fish are considered to be more content because they have been given enhanced life experiences: the interaction with a shoal; the chase for live foods; swimming against a flow of freshwater; and optimized lighting with naturally recapitulated sunrises and sunsets. All of these measures, together with images of gravel, are our five recommendations for environmental enrichment. All are possible within a research facility without the introduction of objects into the tank environment. It is assumed that solitary confinement is avoided and/or kept to a minimum. Together they offer a broad spectrum of provision and a good starting point until such time as further evidence suggests specific combinations, the optimal delivery of each enrichment and currently unknown additional options.

The next steps are to build upon the available evidence to continue to establish that no harm is being done. Research has shown that zebrafish choose enrichment as described previously. In the long term, if specific enrichment types are prescribed on a global scale, each item or condition needs to be tested with each genetically altered and wild-type strain, of which there are hundreds, and supported with data analysis as mentioned above. A cohesive and coordinated approach needs to be agreed by zebrafish laboratories across the world to test enrichment on different strains, and this could inform a standardized, evidence-based enrichment protocol. Automated behavioural analysis would help remove observer bias and ensure that the protocol is consistent. The fractal analysis proposed by Deakin et al. (2019b) is a good starting point. Thomson et al. (2019) have also developed a software-monitoring tool, the chromatic fish analyser, to determine changes in zebrafish welfare via video analysis. If this technology is not available, the DETAILS framework set forth by Jones et al. (2021) will help those publishing their findings to provide the appropriate level of detail to help others implement successful enrichment within their own facilities. Finally, manufacturers may be able to develop in-tank products that are autoclavable, minimize water flow disruptions, prohibit debris collection and prevent fish from hiding. Perhaps a tall order? Once we have achieved these goals there will no longer be a question of whether enrichment should be provided for fish species: we will know, and it will be implemented.

For now, our proposed enrichment approaches can be applied, mostly because they are already in place for other reasons and/or because they involve changes that are relatively easy to make. A key feature of each option is a very low likelihood to do harm. If the practice is already part of the environment it is not as likely to be detrimental. If the enrichment can be added (or subtracted) to the animals' environments with minimal disturbance, it is also far less likely to create problems. This does not mean that changing or modifying the environment even in small ways does not have the potential to impact reproducibility of the science being done. It does mean though that general reluctance to make changes, which is a major barrier to the adoption of some practices, is a problem in some important instances. Ultimately, by improving the welfare of zebrafish through environmental enrichment, we hope that this will enhance the quality of the research which should be the main goal in scientific studies.

Acknowledgement

We would like to thank the Biological Resource Facility (BRF) Aquatics team for all their hard work managing the zebrafish facility at the Francis Crick Institute, UK, particularly throughout the COVID pandemic of 2020–2021. In addition, thank you to Dr. Joe Brock at the Francis Crick Institute for producing the table and figures in this chapter and the Zebrafish Husbandry Association (ZHA) global community for their advice and support.

We would also like to thank Aleah McCorry (Canada), Gyles Westcott (UK) and Charlotte White (UK) for their contributions to specific topics within the chapter.

References

Abdollahpour, H., Falahatkar, B., & Lawrence, C. (2020). The effect of photoperiod on growth and spawning performance of zebrafish, Danio rerio. *Aquaculture Reports, 17*. https://doi.org/10.1016/j.aqrep.2020.100295

Al-Jubouri, Q., Al-Nuaimy, W., Al-Taee, M., Al-Saaidah, B., Al-Libawy, H., & Young, I. (2016). Identification and extraction of new feature for zebrafish. *Discrimination*. 9th International Conference *on* Developments *in eSystems* Engineering 19, *23*.

Anneser, L., Alcantara, I. C., Gemmer, A., Mirkes, K., Ryu, S., & Schuman, E. M. (2020). The neuropeptide Pth2 dynamically senses others via mechanosensation. *Nature, 588*(7839), 653–657. https://doi.org/10.1038/s41586-020-2988-z

Aoyama, Y., Moriya, N., Tanaka, S., Taniguchi, T., Hosokawa, H., & Maegawa, S. (2015). A novel method for rearing zebrafish by using freshwater rotifers (*Brachionus calyciflorus*). *Zebrafish, 12*(4), 288–295. https://doi.org/10.1089/zeb.2014.1032

Ariyomo, T. O., & Watt, P. J. (2012). The effect of variation in boldness and aggressiveness on the reproductive success of zebrafish.

Animal Behaviour, 83(1), 41–46. https://doi.org/10.1016/j.anbehav.2011.10.004

Arunachalam, M., Raja, M., Vijayakumar, C., Malaiammal, P., & Mayden, R. L. (2013). Natural history of zebrafish (*Danio rerio*) in India. *Zebrafish, 10*(1), 1–14. https://doi.org/10.1089/zeb.2012.0803

Barcellos, H. H. A., Koakoski, G., Chaulet, F., Kirsten, K. S., Kreutz, L. C., Kalueff, A. V., & Barcellos, L. J. G. (2018). The effects of auditory enrichment on zebrafish behavior and physiology. *PeerJ, 6*, e5162. https://doi.org/10.7717/peerj.5162

Bault, Z. A., Peterson, S. M., & Freeman, J. L. (2015). Directional and color preference in adult zebrafish: Implications in behavioral and learning assays in neurotoxicology studies. *Journal of Applied Toxicology, 35*(12), 1502–1510. https://doi.org/10.1002/jat.3169

Bayne, K., & Würbel, H. (2014). The impact of environmental enrichment on the outcome variability and scientific validity of laboratory animal studies. *Revue Scientifique et Technique, 33*(1), 273–280. https://doi.org/10.20506/rst.33.1.2282

Best, J., Adatto, I., Cockington, J., James, A., & Lawrence, C. (2010). A novel method for rearing first-feeding larval zebrafish: Polyculture with Type L saltwater rotifers (*Brachionus plicatilis*). *Zebrafish, 7*(3), 289–295. https://doi.org/10.1089/zeb.2010.0667

Bhat, A., Greulich, M. M., & Martins, E. P. (2015). Behavioral plasticity in response to environmental manipulation among zebrafish (*Danio rerio*) populations. *PLOS ONE, 10*(4), e0125097. https://doi.org/10.1371/journal.pone.0125097

Blaser, R. E., & Rosemberg, D. B. (2012). Measures of anxiety in zebrafish (*Danio rerio*): Dissociation of black/white preference and novel tank test. *PLOS ONE, 7*(5), e36931. https://doi.org/10.1371/journal.pone.0036931

Brambell, F. W. R., and Technical Committee to Enquire into the Welfare of Animals kept under Intensive Livestock Husbandry Systems. (1965). *Report of the technical committee. Animals kept under intensive livestock husbandry systems.* HM Stationery Office.

Buenhombre, J., Daza-Cardona, E. A., Sousa, P. & Gouveia, A. J. (2021). Different influences of anxiety models, environmental enrichment, standard conditions and intraspecies variation (sex, personality and strain) on stress and quality of life in adult and juvenile zebrafish: A systematic review. *Neuroscience and Biobehavioral Reviews, 131*, 765–791.

Burn, C. C., Raffle, J., & Bizley, J. K. (2020). Does "playtime" reduce stimulus-seeking and other boredom-like behaviour in laboratory ferrets? *Animal Welfare, 29*(1), 19–26. https://doi. org/10.7120/09627286.29.1.019

Chang, C. T., Lewis, J., & Whipps, C. M. (2019a). Source or sink: Examining the role of biofilms in transmission of *Mycobacterium spp.* in Laboratory Zebrafish. In *Zebrafish, 16*(2), 197–206. https://doi.org/10.1089/zeb.2018.1689

Chang, C. T., Benedict, S., & Whipps, C. M. (2019b). Transmission of Mycobacterium chelonae and Mycobacterium marinum in laboratory zebrafish through live feeds. *Journal of Fish Diseases, 42*(10), 1425–1431. https://doi.org/10.1111/jfd.13071

Coleman, K., & Novak, M. A. (2017). Environmental Enrichment in the 21st Century. *ILAR Journal, 58*(2), 295–307. https://doi.org/10.1093/ilar/ilx008

Collymore, C., Tolwani, R. J., & Rasmussen, S. (2015). The behavioral effects of single housing and environmental enrichment on adult zebrafish (*Danio rerio*). *Journal of the American Association for Laboratory Animal Science, 54*(3), 280–285.

Dametto, F. S., Fior, D., Idalencio, R., Rosa, J. G. S., Fagundes, M., Marqueze, A., Barreto, R. E., Piato, A., & Barcellos, L. J. G. (2018). Feeding regimen modulates zebrafish behavior. *PeerJ, 6*, e5343. https://doi.org/10.7717/peerj.5343

Deakin, A. G., Buckley, J., AlZu'bi, H. S., Cossins, A. R., Spencer, J. W., Al'Nuaimy, W., Young, I. S., Thomson, J. S., & Sneddon, L. U. (2019a). Automated monitoring of behaviour in zebrafish after invasive procedures. *Scientific Reports, 9*(1), 9042. https://doi.org/10.1038/s41598-019-45464-w

Deakin, A. G., Spencer, J. W., Cossins, A. R., Young, I. S., & Sneddon, L. U. (2019b). Welfare challenges influence the complexity of movement: Fractal analysis of behaviour in zebrafish. *Fishes, 4*(1), 1–16. https://doi.org/10.3390/fishes4010008

DePasquale, C., & Leri, J. (2018). The influence of exercise on anxiety-like behavior in zebrafish (*Danio rerio*). *Behavioural Processes, 157*, 638–644. https://doi.org/10.1016/j.beproc.2018.04.006

DePasquale, C., Fettrow, S., Sturgill, J., & Braithwaite, V. A. (2019). The impact of flow and physical enrichment on preferences in zebrafish. *Applied Animal Behaviour Science, 215*, 77–81. https://doi.org/10.1016/j.applanim.2019.03.015

Emmer, K. M., Russart, K. L. G., Walker, W. H., Nelson, R. J., & DeVries, A. C. (2018). Effects of light at night on laboratory animals and research outcomes. *Behavioral Neuroscience*, *132*(4), 302–314. https://doi.org/10.1037/bne0000252

Engeszer, R. E., Patterson, L. B., Rao, A. A., & Parichy, D. M. (2007). Zebrafish in the wild: A review of natural history and new notes from the field. *Zebrafish*, *4*(1), 21–40. https://doi.org/10.1089/zeb.2006.9997

Estes, J. M., Altemara, M. L., Crim, M. J., Fletcher, C. A., & Whitaker, J. W. (2021). Behavioral and Reproductive Effects of Environmental Enrichment and Pseudoloma neurophilia infection on Adult zebrafish (*Danio rerio*). *Journal of the American Association for Laboratory Animal Science*, *60*(3), 249–258. https://doi.org/10.30802/AALAS-JAALAS-20-000113

Farm Animal Welfare Committee. (1993). Report on priorities for animal welfare. *Research and Development*. Fine Arts Work Center in Provincetown.

Food and Agriculture Organization. (2019). MSU extension website. https://www.canr.msu.edu/news/an_animal_welfare_history_lesson_on_the_five_freedoms Retrieved April 5, 2022. The Zebrafish Information Network (Zfin). http://www.zfin.org Retrieved April 5, 2022.

Fowler, L. A., Williams, M. B., Dennis-Cornelius, L. N., Farmer, S., Barry, R. J., Powell, M. L., & Watts, S. A. (2019). Influence of commercial and laboratory diets on growth, body composition, and reproduction in the zebrafish *Danio rerio*. *Zebrafish*, *16*(6), 508–521. https://doi.org/10.1089/zeb.2019.1742

Fulcher, N., Tran, S., Shams, S., Chatterjee, D., & Gerlai, R. (2017). Neurochemical and behavioral responses to unpredictable chronic mild stress following developmental isolation: The zebrafish as a model for major depression. *Zebrafish*, *14*(1), 23–34. https://doi.org/10.1089/zeb.2016.1295

Gerlai, R. (2013). Antipredatory behavior of zebrafish: Adaptive function and a tool for translational research. *Evolutionary Psychology*, *11*(3), 591–605. https://doi.org/10.1177/147470491301100308

Giacomini, A. C., Abreu, M. S., Zanandrea, R., Saibt, N., Friedrich, M. T., Koakoski, G., Gusso, D., Piato, A. L., & Barcellos, L. J. G. (2016). Environmental and pharmacological manipulations blunt the stress response of zebrafish in a similar manner. *Scientific Reports, 6*, 28986. https://doi.org/10.1038/srep28986

Gómez-Boronat, M., Sáiz, N., Delgado, M. J., de Pedro, N., & Isorna, E. (2018). Time-lag in feeding schedule acts as a stressor that alters circadian oscillators in goldfish. *Frontiers in Physiology, 9*, 1749. https://doi.org/10.3389/fphys.2018.01749

Gonzales, J. M., & Law, S. H. W. (2013). Feed and feeding regime affect growth rate and gonadosomatic index of adult zebrafish (*Danio rerio*). *Zebrafish, 10*(4), 532–540. https://doi.org/10.1089/zeb.2013.0891

Graham, C., von Keyserlingk, M. A. G., & Franks, B. (2018). Zebrafish welfare: Natural history, social motivation and behaviour. *Applied Animal Behaviour Science, 200*, 13–22. https://doi.org/10.1016/j.applanim.2017.11.005

Harper, C., & Lawrence, C. (2011). Husbandry. In M. A. Suckow (Ed.), *The laboratory zebrafish* (pp. 47–59). CRC Press.

Hasumura, T., & Meguro, S. (2016). Exercise quantity-dependent muscle hypertrophy in adult zebrafish (*Danio rerio*). *Journal of Comparative Physiology. Part B, 186*(5), 603–614. https://doi.org/10.1007/s00360-016-0977-1

Hebb, D. O. (1947). The effects of early experience on problem-solving at maturity. *American Psychologist, 2*, 306–307.

Higgs, D. M., Souza, M. J., Wilkins, H. R., Presson. J. C. & Popper, A. N. (2002). Age- and size-related changes in the inner ear and hearing ability of the adult zebrafish (*Danio rerio*). *Journal of the Association for Research in Otolaryngology, 3*(2), 174–184.

Hosey, G., & Melfi, V. (2014). Human-animal interactions, relationships and bonds: A review and analysis of the literature. *International Journal of Comparative Psychology, 27*(1), 117–142. https://doi.org/10.46867/ijcp.2014.27.01.01

Jones, N. A. R., Webster, M. M., & Salvanes, A. G. V. (2021). Physical enrichment research for captive fish: Time to focus on the DETAILS. *Journal of Fish Biology, 99*(3), 704–725. https://doi.org/10.1111/jfb.14773

Keck, V. A., Edgerton, D. S., Hajizadeh, S., Swift, L. L., Dupont, W. D., Lawrence, C., & Boyd, K. L. (2015). Effects of habitat complexity on pair-housed zebrafish. *Journal of the American Association for Laboratory Animal Science, 54*(4), 378–383.

Krueger, L. D., Thurston, S. E., Kirk, J., Elsaeidi, F., Freeman, Z. T., Goldman, D., Lofgren, J. L., & Keller, J. M. (2019). Enrichment preferences of singly housed zebrafish (*Danio rerio*). *Journal of the American Association for Laboratory Animal Science, 59*(2), 148–155. https://doi.org/10.30802/AALAS-JAALAS-19-000078

Lavery, M., Braithwaite, V., Miller, N., & Mason, G. (2019). Identifying enriched housing conditions for zebrafish (*Danio rerio*) that vary along a scale of preference. In R. C. Newberry & B. O. Braastad (Eds.). *Proceedings of the 53rd Congress of the ISAE: Animals Lives Worth Living* p. 388. Wageningen Academic Publishers.

Lawrence, C., Eisen, J. S., & Varga, Z. M. (2016). Husbandry and health program survey synopsis. *Zebrafish, 13*, Suppl. 1, S5–S7. https://doi.org/10.1089/zeb.2016.1309

Lee, C. J., Paull, G. C., & Tyler, C. R. (2019). Effects of environmental enrichment on survivorship, growth, sex ratio and behaviour in laboratory maintained zebrafish *Danio rerio*. *Journal of Fish Biology, 94*(1), 86–95. https://doi.org/10.1111/jfb.13865

Lee, C. J., Paull, G. C., & Tyler, C. R. (2022). Improving zebrafish laboratory welfare and scientific research through understanding their natural history. *Biological Reviews of the Cambridge Philosophical Society*, 97(3), 1038–1056. https://doi.org/10.1111/brv.12831

Lidster, K., Readman, G. D., Prescott, M. J., & Owen, S. F. (2017). International survey on the use and welfare of zebrafish *Danio rerio* in research. *Journal of Fish Biology, 90*(5), 1891–1905. https://doi.org/10.1111/jfb.13278

Lieggi, C., Kalueff, A. V., Lawrence, C., & Collymore, C. (2020). The influence of behavioral, social, and environmental factors on reproducibility and replicability in aquatic animal models. *ILAR Journal, 60*(2), 270–288. https://doi.org/10.1093/ilar/ilz019

Lopez-Luna, J., Al-Jubouri, Q., Al-Nuaimy, W., & Sneddon, L. U. (2017). Reduction in activity by noxious chemical stimulation is ameliorated by immersion in analgesic drugs in zebrafish. *Journal of Experimental Biology, 220*(8), 1451–1458. https://doi.org/10.1242/jeb.146969

Luchiari, A. C., & Chacon, D. M. M. (2013). Physical exercise improves learning in zebrafish, *Danio rerio*. *Behavioural Processes, 100*, 44–47. https://doi.org/10.1016/j.beproc.2013.07.020

Magnusson, K. R., & Brim, B. L. (2014). The aging brain. *Reference module in biomedical research.*

Maierdiyali, A., Wang, L., Luo, Y., & Li, Z. (2020). Effect of tank size on zebrafish behavior and physiology. *Animals (Basel), 10*(12), 2353. https://doi.org/10.3390/ani10122353

Marchetto, L., Barcellos, L. J. G., Koakoski, G., Soares, S. M., Pompermaier, A., Maffi, V. C., Costa, R., da Silva, C. G., Zorzi, N. R., Demin, K. A., Kalueff, A. V., & de Alcantara Barcellos, H. H. (2021). Auditory environmental enrichment prevents anxiety-like behavior, but not cortisol responses, evoked by 24-h social isolation in zebrafish. *Behavioural Brain Research, 404*, 113169. https://doi.org/10.1016/j.bbr.2021.113169

Marcon, M., Mocelin, R., Benvenutti, R., Costa, T., Herrmann, A. P., De Oliveira, D. L., Koakoski, G., Barcellos, L. J. G., & Piato, A. (2018). Environmental enrichment modulates the response to chronic stress in zebrafish. *Journal of Experimental Biology, 221*(4), jeb176735. https://doi.org/10.1242/jeb.176735

Mason, T., Snell, K., Mittge, E., Melancon, E., Montgomery, R., McFadden, M., Camoriano, J., Kent, M. L., Whipps, C. M., & Peirce, J. (2016). Strategies to mitigate a Mycobacterium marinum outbreak in a zebrafish research facility. *Zebrafish, 13*, Suppl. 1, S77–S87. https://doi.org/10.1089/zeb.2015.1218

Mogdans, J., & Bleckmann, H. (2012). Coping with flow: Behavior, neurophysiology and modeling of the fish lateral line system. *Biological Cybernetics, 106*(11–12), 627–642. https://doi.org/10.1007/s00422-012-0525-3

Monteiro, J. F., Martins, S., Farias, M., Costa, T., & Certal, A. C. (2018). The impact of two different cold-extruded feeds and feeding regimens on zebrafish survival, growth and reproductive performance. *Journal of Developmental Biology, 6*(3), 15. https://doi.org/10.3390/jdb6030015

Niimi, A. J., & LaHam, Q. N. (1974). Influence of breeding time interval on egg number, mortality, and hatching of zebrafish Brachydanio rerio. *Canadian Journal of Zoology, 52*(4), 515–517. https://doi.org/10.1139/z74-063

Oliveira, J., Silveira, M., Chacon, D., & Luchiari, A. (2015). The zebrafish world of colors and shapes: Preference and discrimination. *Zebrafish, 12*(2), 166–173. https://doi.org/10.1089/zeb.2014.1019

Oliveira, R., & Galhardo, L. (2009). Psychological stress and welfare in fish. *Annual Review of Biomedical Data Science, 11*, 1–20.

Pagnussat, N., Piato, A. L., Schaefer, I. C., Blank, M., Tamborski, A. R., Guerim, L. D., Bonan, C. D., Vianna, M. R., & Lara, D. R. (2013). One for all and all for one: The importance of shoaling on behavioral and stress responses in zebrafish. *Zebrafish, 10*(3), 338–342. https://doi.org/10.1089/zeb.2013.0867

Peeters, B. W. M. M., Moeskops, M., & Veenvliet, A. R. J. (2016). Color preference in *Danio rerio*: Effects of age and anxiolytic treatments. *Zebrafish, 13*(4), 330–334. https://doi.org/10.1089/zeb.2015.1150

Pieróg, M., Guz, L., Doboszewska, U., Poleszak, E., & Wlaź, P. (2018). Effects of alprazolam treatment on anxiety-like behavior induced by color stimulation in adult zebrafish. *Progress in Neuro-Psychopharmacology and Biological Psychiatry, 82*, 297–306. https://doi.org/10.1016/j.pnpbp.2017.08.025

Piña, J. A., Namba, M. D., Leyrer-Jackson, J. M., Cabrera-Brown, G., & Gipson, C. D. (2018). Social influences on nicotine-related behaviors. *International Review of Neurobiology, 140*, 1–32. https://doi.org/10.1016/bs.irn.2018.07.001

Pitcher, T. J., Magurran, A. E., & Winfield, I. J. (1982). Fish in larger shoals find food faster. *Behavioral Ecology and Sociobiology, 10*(2), 149–151. https://doi.org/10.1007/BF00300175

Poole, T. (1997). Happy animals make good science. *Laboratory Animals, 31*(2), 116–124. https://doi.org/10.1258/002367797780600198

Popper, A. N., & Sisneros, J. A. (2022). The Sound world of zebrafish: A critical review of hearing assessment. *Zebrafish*, 19(2), 37–48. https://doi.org/10.1089/zeb.2021.0063

Pritchett-Corning, K. R. (2020). Environmental complexity and research outcomes. *ILAR Journal, 60*(2), 239–251. https://doi.org/10.1093/ilar/ilaa007

Pryke, S. R. (2009). Is red an innate or learned signal of aggression and intimidation? *Animal Behaviour, 78*(2), 393–398. https://doi.org/10.1016/j.anbehav.2009.05.013

Rácz, A., Allan, B., Dwyer, T., Thambithurai, D., Crespel, A., & Killen, S. S. (2021). Identification of individual zebrafish (*Danio rerio*): A refined protocol for VIE tagging whilst considering animal welfare and the principles of the 3 Rs. *Animals (Basel), 11*(3), 616. https://doi.org/10.3390/ani11030616

Reilly, S. C., Quinn, J. P., Cossins, A. R., & Sneddon, L. U. (2008). Behavioural analysis of a nociceptive event in fish: Comparisons between three species demonstrate specific responses. *Applied Animal Behaviour Science, 114*(1–2), 248–259. https://doi.org/10.1016/j.applanim.2008.01.016

Richmond, J. (2002). Refinement, reduction, and replacement of animal use for regulatory testing: Future improvements and implementation within the regulatory framework. *ILAR Journal*, *43*(1), Suppl., S63–S68. https://doi.org/10.1093/ilar.43.suppl_1.s63

Robison, B. D., & Rowland, W. (2005). A potential model system for studying the genetics of domestication: Behavioral variation among wild and domesticated strains of zebra danio (*Danio rerio*). *Canadian Journal of Fisheries and Aquatic Sciences*, *62*(9), 2046–2054. https://doi.org/10.1139/f05-118

Roy, T., Suriyampola, P. S., Flores, J., López, M., Hickey, C., Bhat, A., & Martins, E. P. (2019). Color preferences affect learning in zebrafish, *Danio rerio*. *Scientific Reports*, *9*(1), 14531. https://doi.org/10.1038/s41598-019-51145-5

Sabet, S. S., Wesdorp, K., van Dooren, D., & Slabbekoorn, H. (2016). Sound affects behavior of captive zebrafish: Always consider the potential for acoustic effects on your laboratory tests. *Proceedings of Meetings on Acoustics. Proceedings of the Meetings on Acoustics Acoustical Society of America*, *27*(1), 10. https://doi.org/10.1121/2.0000248

Sadoul, B., & Geffroy, B. (2019). Measuring cortisol, the major stress hormone in fishes. *Journal of Fish Biology*, *94*(4), 540–555. https://doi.org/10.1111/jfb.13904

Santos, B. R. D., Santos, R. C. D., Dias, C. A. G. M., Maximino, C., & Gouveia, A., Jr. (2019). White environment can be used as an aversive stimulus in zebrafish inhibitory avoidance learning. *Zebrafish*, *16*(5), 443–450. https://doi.org/10.1089/zeb.2019.1743

Schroeder, P., Jones, S., Young, I. S., & Sneddon, L. U. (2014). What do zebrafish want? Impact of social grouping, dominance and gender on preference for enrichment. *Laboratory Animals*, *48*(4), 328–337. https://doi.org/10.1177/0023677214538239

Serra, E. L., Medalha, C. C., & Mattioli, R. (1999). Natural preference of zebrafish (*Danio rerio*) for a dark environment. *Brazilian Journal of Medical and Biological Research*, *32*(12), 1551–1553. https://doi.org/10.1590/s0100-879x1999001200016

Sessa, A. K., White, R., Houvras, Y., Burke, C., Pugach, E., Baker, B., Gilbert, R., Thomas Look, A. T., & Zon, L. I. (2008). The effect of a depth gradient on the mating behavior, oviposition site preference, and embryo production in the zebrafish, *Danio rerio*. *Zebrafish*, *5*(4), 335–339. https://doi.org/10.1089/zeb.2008.0535

Shafiei Sabet, S., Van Dooren, D. & Slabbekoorn, H. (2016). Son et lumiere: Sound and light effects on spatial distribution and swimming behavior in captive zebrafish. *Environment and Pollution*, *212*, 480–488.

Shams, S., Chatterjee, D., & Gerlai, R. (2015). Chronic social isolation affects thigmotaxis and whole-brain serotonin levels in adult zebrafish. *Behavioural Brain Research, 292*, 283–287. https://doi.org/10.1016/j.bbr.2015.05.061

Shaw, E. (1978). Schooling fishes. *American Scientist, 66*(2), 166–175.

Shelton, D. S., Shelton, S. G., Daniel, D. K., Raja, M., Bhat, A., Tanguay, R. L., Higgs, D. M., & Martins, E. P. (2020). Collective behavior in wild zebrafish. *Zebrafish, 17*(4), 243–252. https://doi.org/10.1089/zeb.2019.1851

Singhal, G., Jaehne, E. J., Corrigan, F., & Baune, B. T. (2014). Cellular and molecular mechanisms of immunomodulation in the brain through environmental enrichment. *Frontiers in Cellular Neuroscience, 8*, 97. https://doi.org/10.3389/fncel.2014.00097

Spence, R., Gerlach, G., Lawrence, C., & Smith, C. (2008). The behaviour and ecology of the zebrafish, *Danio rerio*. *Biological Reviews of the Cambridge Philosophical Society, 83*(1), 13–34. https://doi.org/10.1111/j.1469-185X.2007.00030.x

Stevens, C. H., Reed, B. T., & Hawkins, P. (2021). Enrichment for laboratory zebrafish-A review of the evidence and the challenges. *Animals (Basel), 11*(3), 698. https://doi.org/10.3390/ani11030698

Sumbre, G., & de Polavieja, G. G. (2014). The world according to zebrafish: How neural circuits generate behavior. *Frontiers in Neural Circuits, 8*, 91. https://doi.org/10.3389/fncir.2014.00091

Suniaga, S., Rolvien, T., Vom Scheidt, A., Fiedler, I. A. K., Bale, H. A., Huysseune, A., Witten, P. E., Amling, M., & Busse, B. (2018). Increased mechanical loading through controlled swimming exercise induces bone formation and mineralization in adult zebrafish. Scientific Reports, 8(1), 3646. https://doi.org/10.1038/s41598-018-21776-1.

Suriyampola, P. S., Sykes, D. J., Khemka, A., Shelton, D. S., Bhat, A., & Martins, E. P. (2017). Water flow impacts group behavior in zebrafish (*Danio rerio*). *Behavioral Ecology, 28*(1), 94–100. https://doi.org/10.1093/beheco/arw138

Thomson, J. S., Al-Temeemy, A. A., Isted, H., Spencer, J. W., & Sneddon, L. U. (2019). Assessment of behaviour in groups of zebrafish (*Danio rerio*) using an intelligent software monitoring tool, the chromatic fish analyser. *Journal of Neuroscience Methods, 328*, 108433. https://doi.org/10.1016/j.jneumeth.2019.108433

Tunbak, H., Vazquez-Prada, M. C., Ryan, T. M., Kampff, A. R., & Dreosti, E. (2020). Whole-brain mapping of socially isolated zebrafish reveals that lonely fish are not loners. *eLife*, *9*, e55863. https://doi.org/10.7554/eLife.55863

Varga, Z. M. (2011). Aquaculture and husbandry at the zebrafish international resource center. *Methods in Cell Biology*, *104*, 453–478. https://doi.org/10.1016/B978-0-12-374814-0.00024-0

Vatine, G., Vallone, D., Gothilf, Y., & Foulkes, N. S. (2011). It's time to swim! Zebrafish and the circadian clock. *FEBS Letters*, *585*(10), 1485–1494. https://doi.org/10.1016/j.febslet.2011.04.007

Villamizar, N., Vera, L. M., Foulkes, N. S., & Sánchez-Vázquez, F. J. (2014). Effect of lighting conditions on zebrafish growth and development. *Zebrafish*, *11*(2), 173–181. https://doi.org/10.1089/zeb.2013.0926

von Krogh, K., Sørensen, C., Nilsson, G. E., & Øverli, Ø. (2010). Forebrain cell proliferation, behavior, and physiology of zebrafish, Danio rerio, kept in enriched or barren environments. *Physiology and Behavior*, *101*(1), 32–39. https://doi.org/10.1016/j.physbeh.2010.04.003

Wafer, L. N., Jensen, V. B., Whitney, J. C., Gomez, T. H., Flores, R., & Goodwin, B. S. (2016). Effects of environmental enrichment on the fertility and fecundity of zebrafish (*Danio rerio*). *Journal of the American Association for Laboratory Animal Science*, *55*(3), 291–294.

Watts, S. A., Lawrence, C., Powell, M., & D'Abramo, L. R. (2016). The vital relationship between nutrition and health in zebrafish. *Zebrafish*, *13*, Suppl. 1, S72–S76. https://doi.org/10.1089/zeb.2016.1299

White, L. J., Thomson, J. S., Pounder, K. C., Coleman, R. C., & Sneddon, L. U. (2017). The impact of social context on behaviour and the recovery from welfare challenges in zebrafish, *Danio rerio. Animal Behaviour, 132, 189–199.*

Wilkes, L., Owen, S. F., Readman, G. D., Sloman, K. A., & Wilson, R. W. (2012). Does structural enrichment for toxicology studies improve zebrafish welfare? *Applied Animal Behaviour Science*, *139*(1–2), 143–150. https://doi.org/10.1016/j.applanim.2012.03.011

Wolman, M., & Granato, M. (2012). Behavioral genetics in larval zebrafish: Learning from the young. *Developmental Neurobiology*, *72*(3), 366–372. https://doi.org/10.1002/dneu.20872

Woodward, M. A., Winder, L. A., & Watt, P. J. (2019). Enrichment increases aggression in zebrafish. *Fishes*, *4*(1), 22. https://doi.org/10.3390/fishes4010022

Wright, D., Nakamichi, R., Krause, J., & Butlin, R. K. (2006). QTL analysis of behavioral and morphological differentiation between wild and laboratory zebrafish (*Danio rerio*). *Behavior Genetics, 36*(2), 271–284. https://doi.org/10.1007/s10519-005-9029-4

Zang, L., Shimada, Y., Nishimura, Y., Tanaka, T., & Nishimura, N. (2015). Repeated blood collection for blood tests in adult zebrafish. *Journal of Visualized Experiments: JoVE, 102*(102), e53272. https://doi.org/10.3791/53272

Chapter 3

Dietary requirements for zebrafish

Joana Monteiro, Sandra Martins, Rita Almeida, Carolina Cabrera and Ana Catarina Certal

Abstract

Zebrafish is the second most used research model for a vast array of human conditions and diseases. Their natural ecology, anatomy and physiology allow them to adapt well to a wide spectrum of nutrients sources. Therefore, ever since zebrafish were brought into captivity, fish facilities have successfully maintained their colonies using the feeds and feeding routines that best adapted to the reality of each. As it became clear that nutrition affects fish physiology and behaviour and could potentially affect experimental outcomes in uncontrolled ways, the community's concern with the animals' diet increased. The available literature is not consensual in what the best feeds are, how much food fish should eat or how many meals per day should be given. Nutritional details apart, it is widely accepted that diet standardization would improve the reproducibility and comparison of results obtained by different laboratories, which ultimately translates into more accurate scientific results and faster translation of biomedical findings into human applications. Nevertheless, the implementation of reference diets still faces important challenges. This chapter includes a non-extensive literature review of the feeding ecology, anatomy and physiology of zebrafish, their nutritional requirements throughout life and how the quality and quantity of different nutrients may affect them. Available feeding options will also be addressed, as well as the extent to which these meet the known

Claire Allen and Jean-Philippe Mocho (eds) **Zebrafish**
DOI: 10.1079/9781800629431.0003, © CAB International, 2024

dietary requirements of zebrafish. Finally, diet standardization is discussed.

Keywords

zebrafish, feeds, nutrition, diet, standardization, live-feeds

Zebrafish feeding in the wild

Mimicking the wild zebrafish diet is likely the best approach to ensure adequate feeding for fish in captivity, although information on zebrafish nutritional requirements in its natural habitat is incomplete. For instance, there is no detailed information on how feeding habits change throughout the life cycle, such as preferred prey size, nutritional value of different feeds and circadian feeding patterns, including frequency and amount of daily biomass intake.

Despite this, several aspects of wild zebrafish ecology are known, including information on their diet that is relevant for optimization of feeds and feeding schemes. For instance, it is known that zebrafish are found mainly in shallow, low-flowing streams and standing water pools in South Asia and that they are omnivorous, feeding primarily on allochthonous items (Arunachalam et al., 2013; Engeszer et al., 2007; Parichy, 2015; Spence et al., 2007, 2008). During larval stages they feed mostly on zooplankton (Harper & Lawrence, 2016; Westerfield, 2000), while juveniles and adults also eat insects and occasionally phytoplankton, filamentous algae, vascular plant material, crustacean eggs and inorganic elements (Arunachalam et al., 2013; Engeszer et al., 2007; Spence et al., 2007). The fact that some of these items are found at the water surface and others in the water column or substratum confirmed zebrafish's natural ability to feed at various vertical locations and demonstrated that wild zebrafish can easily adjust their feeding habits to the available nutrient sources. As so, it is not surprising that laboratory zebrafish adapt well to a variety of feeding protocols. For example, Carvalho et al. (2006), Fowler et al. (2019), Lawrence et al. (2015) and Monteiro et al. (2018) each addressed different feeds and feeding protocols, yet all found high survival, growth and breeding rates under at least one of the dietary regimens tested. However, despite the benefit and comfort of this easy adaptation for laboratory fish facilities, it has hampered the important and collective goal of producing standardized diets (Watts et al., 2016; Watts and D'Abramo, 2021).

Relation between anatomy, physiology and feeding

Feeding apparatus

The relation among the jaws, teeth and feeding is essential for animal survival (Hammer et al., 2016) and consequently needs to be considered during the development or selection of a feed. For example, the oral cavity and craniofacial structure both play roles related to food intake (Abbate et al., 2006). In this regard, zebrafish count on some evolutionary morphological novelties, typical of cyprinids, that grant them increased capacity for ecological adaptation as detailed by Hernandez and Staab (2015). Briefly, they start feeding exogenously around 5 days post fertilization (dpf), at which point the process relies uniquely on head elevation and movement of the lower jaw. They have a toothless, terminal mouth and use the teeth present on the last lower pharyngeal jaws to chew their prey against a chewing pad attached to the base of the skull (Aigler et al., 2014). Interestingly, even though replacement teeth are present throughout life, these develop faster and in higher numbers in larvae and early juveniles as compared with their adult counterparts, possibly accompanying dietary changes as the fish develops (Van der Heyden et al., 2001). Following metamorphosis they develop a small, sesamoid bone, the kinethmoid, that enhances the retractability of the upper jaw. The morphology of this bone, along with the terminal to up-turned mouth, provide decreased but faster protrusion of the premaxillary, which favours pelagic-surface feeding on elusive prey, also using ram feeding coupled to suction feeding. In addition they are capable of bottom feeding thanks to the palatal organ, a thick muscle on the roof of the buccal chamber that helps separate debris from food items (Hernandez & Staab, 2015).

Because the morphology of the zebrafish's feeding apparatus may be affected by developmental defects as a consequence of inbreeding depression, it is worth considering the level of genetic variability of a line as a factor that could affect the animals' feeding ability. Many facilities keep track of the pedigrees of the fish lines they maintain, which provides hints about the level of genetic heterogeneity of in-house fish colonies. It is the sharing of pedigrees when transferring a line to another facility that needs more attention because in most cases the importer/exporter doesn't request/volunteer such details.

If, on one hand, the morphology of the feeding apparatus impacts directly on feeding habits, on the other, dietary choices can also play a role in the development of such structures. Several studies in mice and cichlid fish have shown that the type of diet directly influences jaw morphology (Hammer et al., 2016). Considering the great diversity of feeds used across animal facilities worldwide, it would be interesting to investigate whether diet types affect the morphology of the zebrafish feeding apparatus over time and across generations, because this could impact the feeding ability of fish transferred between facilities and from the wild into the domestic environment. In fact, some facilities have reported that fish recently derived from wild stocks have somewhat different feeding habits, such as feeding more frequently from the bottom rather than from the water column (L. Carvalho, Lisbon, Portugal, 2019, personal communication; M. Sampaio, Lisbon, Portugal, 2020, personal communication). Although these observations were not further investigated, it is worth considering domestication as a factor that could possibly affect the animals' feeding ability.

Sensory organs

Sensory organs are used during foraging and hunting (Aigler et al., 2014). For example, chemosensory taste bud cells are used to sort food from non-edible items (Hansen et al., 2002) and vision is used for prey detection and capture (Bianco et al., 2011). Zebrafish can rely on both chemosensory and visual cues from as early as 5 dpf thanks to the conjunction of two factors. First, the neural circuitry processing olfactory and visual stimuli and transducing them into behaviour is developed. Second, the swimming bladder becomes inflated, allowing for free movement in the water column in response to sensory cues (Lindsay & Vogt, 2004; Strähle et al., 2012). The fact that zebrafish modulate their behaviour in response to feeds indicates that the feeds themselves have the potential to modify the animals' behaviour through visual, somatosensory and olfactory stimuli (Huang et al., 2015; Kopp et al., 2016; Novak et al., 2005). In this regard, and despite man-made feeds typically including attractive cues such as smell and colour, live prey are naturally more attractive and stimulating.

Digestion and nutrient absorption

Because zebrafish are stomachless, once a prey is captured it is passed from the pharynx to the oesophagus and directly into the intestine. There, feeds are digested with the help of enzymes synthesized by the pancreas and liver. The intestine is divided into anterior, middle and posterior portions. The anterior gut, also known as the intestinal bulb, is larger and has a major role in nutrient absorption (Wallace et al., 2005). The whole gut has a specific microbiota that starts developing following opening of the intestinal tract (~5 dpf; Quinlivan & Farber, 2017) and undergoes stage-specific changes during development (Stephens et al., 2016), probably accompanying changes in both diet and nutrient absorption needs. Both wild-type and laboratory-reared zebrafish seem to share a common core microbiota, composed of bacteria from the same taxa (Roeselers et al., 2011). However, there is room for some variation that adds richness and complexity to the microbiota as, for example, the presence of distinct fungal species (Siriyappagouder et al., 2018). Microbiota variations could be related to differences in water quality, exposure to infectious microorganisms and viruses and also to diet. In fact there is a close and reciprocal relationship between gut microbiota and fish diet. On the one hand intestinal bacterial populations affect dietary lipid uptake and metabolism

(Quinlivan & Farber, 2017) and on the other, hand diet composition, especially fat content, has a major impact on the gut bacterial community (Sheng et al., 2018; Wong et al., 2015). Importantly, several aspects of zebrafish biology, for example, immunology, nutrition and gastrointestinal development, are sensitive to the composition of the local microbial community (López Nadal et al., 2020; Roeselers et al., 2011). Therefore, when performing research in such fields, standardized dietary protocols would be greatly beneficial, by abolishing differences in the microbiota composition of animals across aquaculture facilities that could result in phenotypic variation and poor experimental reproducibility.

Basic nutritional requirements

Most laboratory lines of zebrafish are the product of many generations in captivity, and it is not known the extent to which they differ from wild fish (Crim & Lawrence, 2021), including their dietary requirements. In fact, zebrafish nutritional needs are not fully described for either wild or captive animals.

For the first 4–5 dpf, zebrafish feed on their protein-rich and lipid-rich yolk, the precise composition of which changes as the fish develop (Fraher et al., 2016). At 5 dpf the digestive tract is fully functional, the yolk depleted and fish start feeding exogenously. In most research facilities the zebrafish diet comprises one or two species of live prey and one or more manufactured feeds, starting with small-diameter options and working up to larger feeds in a stepwise process as the animals grow. Many facilities offer exclusively manufactured feeds for most of the life course of the fish, based on the fact that most include a combination of ingredients selected to meet the expected nutritional requirements of the species and life stage (Figure 3.1; see Chapter 4 for more information; Watts, 2012).

An adequate diet is fundamental for reproduction. In the adult zebrafish, increased resource availability improves oocyte maturation, breeding rates and sperm fertilization. A nourishing diet also increases the size and lipid content of the yolk in the oocytes, which benefits the offspring by increasing early embryo survival

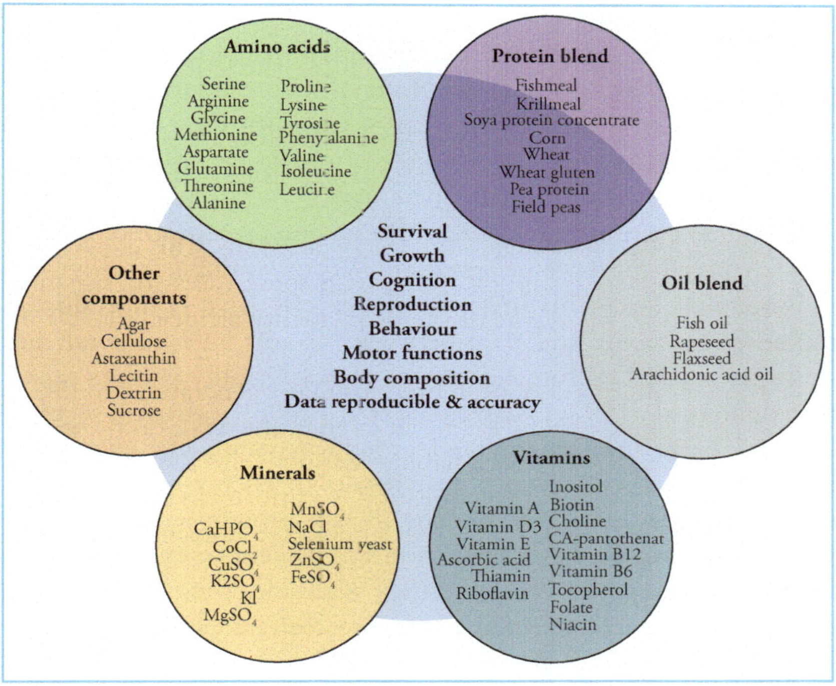

Figure 3.1 Diagrammatic depiction of nutrients and their sources in zebrafish feeds.

and growth and improving their energy balance during adulthood (Newman et al., 2016). Overfeeding does not seem to be a problem for zebrafish reproduction, because they do not appear to develop obesity-related infertility and parental diet does not seem to influence weight gain in the offspring (Newman et al., 2016). It should be noted that feeding frequencies can affect growth and breeding differently. Lister et al. (2017) reported that zebrafish are usually fed ad libitum and that twice to three or more times per day is the preferred protocol (see Chapter 7 for more information on feed distribution; Lidster et al., 2017).

Proteins

Proteins are the basic requirement for body growth and help ensure an efficient usage of feeds and a healthy body composition. As so, there may be a temptation to overestimate the protein needs of a species. However, protein content in feeds for aquatic species should be regulated to avoid the toxic effects of its breakdown in the fish water. Briefly, the breakdown of proteins from feed leftovers and fish faeces releases ammonia into the water. In the absence of adequate water renewal or efficient biological filters, ammonia accumulates and eventually may cause disease and even death of the fish (see Chapter 7 for more information; Wilson, 2012).

Protein requirements in terms of quantity have been evaluated for other cyprinids and vary between 30 and 53% of total diet (Ulloa et al., 2011). Information on protein quality is more sparse (Carvalho et al., 1997; Ostaszewska et al., 2008). Zebrafish regulate food intake to meet energy and protein demands but give priority to protein needs, even if that results in energy overconsumption (Fernandes et al., 2016). The dietary protein requirement of fish cannot thus be dissociated from the protein-to-energy balance of the diet (Gonzales & Law, 2013; Kaushik et al., 2011; Watts et al., 2012), since an optimal dietary protein/energy ratio will maximize the allocation of protein for growth and minimize its use as an energy source for metabolic purposes (Fernandes et al., 2016). Dietary protein requirements for zebrafish juvenile growth is 37.6% for maximum weight gain and 44.8% for maximum protein retention, with a protein/energy ratio of 22.5 g/MJ that ensures a proper balance between crude protein and energy (Fernandes et al., 2016). In the transition from juvenile to adulthood, the quantity of dietary protein sufficient to meet this specific stage and growth requirement is about 32% (O'Brine et al., 2015). This age-related decrease is due to a significant reduction in the growth rate (two to three times lower) as zebrafish get closer to the adult stage and maximum size (Fernandes et al., 2016; Lawrence et al., 2012a).

In zebrafish the protein source has a significant impact on length, body weight and body composition (lean and fat mass). Comparing wheat protein-, hydrolysate protein- and soy protein-based diets, wheat gluten protein and hydrolysate protein can be associated with a higher percentage of body fat; on the other hand, casein-based diets can lead to a lower percentage of body fat. Feeding fish on diets with wheat gluten and casein can result in smaller and lighter animals whereas a soy protein-only diet performs similarly to a mixed-protein diet for these endpoints (Smith et al., 2013). Future studies on protein zebrafish requirements specifically for each developmental phase are needed and are of great importance.

Lipids

Lipids are complex chemicals with a multi-functional nature (Powell et al., 2015). Essential fatty acids are among the most relevant lipids because most animals cannot synthesize them and therefore have to obtain them from their diet (Cunnane, 2003). Each fish species has its own specific essential dietary fatty acid requirements, such as 18- to 22-carbon polyunsaturated fatty acids (PUFA) and long-chain PUFA (LC-PUFA) of the ω6 and ω3 families. Their greatest benefits may be realized only when both families are present in the correct proportions (Powell et al., 2015), but such requirements may vary according to environmental factors such as temperature (Meinelt et al., 1999, 2000). Thus, it is easy to understand why consensus regarding the recommended dietary levels for these nutrients is hard to achieve (Powell et al., 2015). For zebrafish, the specific ratio of ω3 and ω6 PUFA and LC-PUFA continues to be debated (Powell et al., 2015), but a total dietary lipid content of approximately 12% is known to promote growth and survival (Smith et al., 2013).

Essential fatty acids are also important in reproduction. For instance, their content in broodstock diet affects the fatty acid composition of eggs, especially the yolk sac, in teleosts. A lower ω3/ω6 ratio improves reproductive performance in zebrafish (Meinelt et al., 1999), while insufficient dietary ω6 (PUFA and LC-PUFA) leads to decreased growth and quality of eggs (Abi-Ayad, 1997; Santiago & Reyes, 1993). Increasing ω6 fatty acids in the diet results in larger, leaner female zebrafish with reduced adiposity, but individual variation in response to higher levels of ω6 fatty acids can markedly vary fat-free mass (Powell, et al., 2015). Arachidonic acid (ARA) also plays an important role in reproductive development and performance. In teleost fish, this lipid has been reported to improve egg quality due to its role in regulating oocyte meiosis and steroidogenesis (Bell et al., 1997; Henken et al., 1987; Pati & Habibi, 2002;

Sorbera et al., 2001). It remains uncertain, however, what the optimum level of ARA should be in ovaries or fertilized eggs (Gonzales & Law, 2013), particularly in zebrafish (Jaya-Ram et al., 2008; Kaushik et al., 2011; Meinelt et al., 1999). Polyunsaturated fatty acids are very important in zebrafish reproduction: fish can have selective accumulation of these nutrients in gonadal tissue so that during periods of starvation gonadal development is not compromised (Gonzales & Law, 2013; Jaya-Ram et al., 2008).

Changes in combined PUFA and LC-PUFA ratios alone are sufficient to cause significant changes in other physiologic outcomes (Powell et al., 2015). For example, very high intake of ω6 PUFA and LC-PUFA promotes inflammation, resulting in chronic inflammatory diseases associated with metabolic syndromes (Fritsche, 2008; Simopoulos, 2008). In zebrafish, gene expression levels for pro-inflammatory response proteins decrease when combined PUFA and LC-PUFA dietary ω6 fatty acids are proportionally high relative to combined PUFA and LC-PUFA ω3 fatty acids (Powell et al., 2015). Excessive dietary cholesterol also has a negative impact on physiological functions in zebrafish, such as high weight gain, high triglycerides and plaque formation in vascular tissues (Stoletov et al., 2009).

Carbohydrates

In fish, gluconeogenic and glycolytic regulation (glucose metabolism) appears to be influenced by variation in carbohydrate supply, in contrast to mammals where this role is played by the liver (Palmer & Ryman, 1972). Diets containing digestible carbohydrates typically promote faster growth and provide metabolic intermediates for the synthesis of other biologically important compounds (Grisdale-Helland & Helland, 1997). Furthermore, the mechanisms of carbohydrate digestion, absorption and metabolism are different among different fish species, and carbohydrate requirements have not yet been determined in zebrafish. Nevertheless it is possible to speculate about their carbohydrate needs based on evidence that omnivorous fish, such as zebrafish, can use higher amounts of dietary carbohydrates and exhibit a greater degree of protein sparing (Deng et al., 2000; Jantrarotai et al., 1998; Krogdahl, 2005). The same was observed for herbivorous fish (Deng et al., 2000; Jantrarotai et al., 1998; Krogdahl, 2005), while carnivorous fish have a relatively lower ability to use carbohydrates (Palmer & Ryman, 1972).

Minerals and vitamins

A good nutritional diet is not complete without an adequate supply of minerals and vitamins (Watts et al., 2012). Although the available literature on zebrafish needs for these nutrients is still limited, a growing body of literature does provide some relevant information (Lall & Kaushik, 2021). For example, selenium plays a relevant role in reproduction, the immune system and locomotion (Penglase et al., 2014), boron is required for gamete development (Rowe & Eckhert, 1999) and zinc induces cell differentiation in the gills (Zheng et al., 2010). Several papers provide examples of how an excess or insufficient amount of certain minerals can affect zebrafish, including iron, zinc and copper (Zhao et al., 2014), manganese and iron (Herman et al., 2021; Marins et al., 2019). Such evidence provides the basis for speculation about the acceptable concentration limits for each mineral. A similar rationale holds true for identifying the concentration ranges of vitamins. For example, vitamin C at 1,000 mg/kg of feed increases growth rates, body weight and fecundity (Mehrad et al., 2011) and vitamin E (alpha-tocopherol) plays a role in offspring survival, development (Head et al., 2020: Lebold et al., 2011), fecundity, fertilization rate and hatching rate (Utomo et al., 2005). Nutrients of the 1-C cycle, mostly folate, methionine and choline, or the co-factor vitamins B6 and B12, are essential for several vital metabolic processes and regulate energy metabolism, methylation and transamination reactions. Continuous 1-C parental deficiency in zebrafish reduces fecundity and massively affects embryonic transcriptomes related to immune response, blood coagulation and redox regulation (Skjærven et al., 2016).

Available feeds for laboratory zebrafish

Because there are several feeding options available and all have advantages and disadvantages, selecting one may be challenging. To overcome that, it is useful to define the key features, starting from the criteria that a perfect feed would fulfil. An ideal feed would be attractive (colour, motion, taste, smell); stable in the water for long enough to retain its full nutritional value while fish are eating; and again stable as a means of minimizing the detrimental effect of its breakdown on water quality. It would be highly digestible to maximize nutrient uptake, minimize waste and contribute to good water quality; be nutritionally complete, with known and stable composition across batches; and it would not pose any biosecurity risk. Other requisites are commercial accessibility, simple storage with long shelf life, fast and easy preparation and provision to large numbers of fish. Finally, feeds should promote fish well-being, typically assessed by survival, growth and breeding performance. Perfect feeds do not exist, but over the years new and improved products have been introduced on the commercial circuit.

There are many types of fish food available, including manufactured (processed), dried, fresh, frozen and live feeds. The most popular options for zebrafish are live feeds and manufactured feeds, although some facilities include other types (Frederickson et al., 2021). Of note, many of the feeds available are not specific for this species, but rather general formulae for ornamental fish and aquaculture, to which zebrafish happen to respond well. There is a vast list of available literature that compares different diets. A non-extensive list of examples includes Farias and Certal (2016), Fowler et al. (2019), Frederickson et al. (2021), Gonzales & Law (2013), Goolish et al. (2011), Hensley & Leung (2010), Kolb et al. (2018), Lawrence et al. (2012a, 2015), Monteiro et al. (2018), Nowosad et al. (2017), Reed & Jennings (2011) and Siccardi et al. (2009). These and other studies used different protocols and feeds, yet the various approaches resulted in good welfare indicators including survival, growth and reproduction. The positive consequence of such evidence is that managers are able to choose the option that best suits their facility, usually considering practical aspects including price, local availability, storage, time and ease of preparation. On the negative side, nutrition influences at least some aspects of fish anatomy, behaviour (Dametto et al., 2018) and physiology (López Nadal et al., 2020; Watts et al., 2016), none of which is usually addressed when evaluating the quality and efficiency of feeds. Consequently, the unregulated and heterogeneous feed choices across fish facilities are likely biasing experimental work in distinct and uncontrolled ways that compromise experimental reproducibility and scientific outcomes (Crim & Lawrence, 2021; Watts & D'Abramo, 2021).

Live feeds

Based on evidence that live feeds, especially rotifers and *Artemia*, greatly benefit larval and juvenile survival and growth and even breeding performance (Farias & Certal, 2016; Fowler et al., 2019; Goolish et al., 2011; Lawrence et al., 2012a, 2016; Martins et al., 2016; Monteiro et al., 2018; Reed & Jennings, 2011), many facilities adopt these as part of their husbandry routine. It is anticipated that such an age-benefit relation is caused by specific larval nutritional demands that are difficult to fulfil using human-made alternatives. At the same time, live feeds offer benefits that no other food type can replace in terms of attractiveness and stimulation of natural predatory behaviour. In fact, live feeds are frequently included as part of the environmental enrichment program towards improved welfare (see Chapter 2 for more information; Martins et al., 2016; Stevens et al., 2021). Notwithstanding, the necessity to include live feeds in the zebrafish diet is not consensual and several laboratories opt to step away from the labour (Lawrence et al., 2012a) and biosecurity risk (Chang et al., 2019) posed by live feeds, and instead adopt described protocols for efficiently raising zebrafish using only manufactured feeds (Carvalho et al., 2006; Hensley & Leung, 2010; Kaushik et al., 2011).

Rotifers

Rotifers are commonly used as first feed due to their small size and low swimming speed, their compatibility with the small mouth of early larvae and because they are conducive to high efficiency for larval survival and normal growth. In addition, they can replace *Artemia* for juvenile zebrafish (Lawrence et al., 2015; Monteiro et al., 2018), reducing costs and simplifying facility operation by including a single live feed for raising fish. Protocols for establishment and maintenance of rotifer populations and how to feed them to zebrafish larvae are well described (Aoyama et al., 2015; Lawrence et al., 2012b, 2016). Saltwater *Brachionus plicatilis* was the first rotifer species included in zebrafish feeding programmes and its application continues to be widespread. It has an optimal growth temperature of 18–25ºC and size ranging from 130 to 340 nm. It is frequently referred to as L-type (L, large), to distinguish it from the S-type (S, small) *Brachionus rotundiformis* that grows best at 28–35ºC and that carries an advantage because larval feed relies on smaller size (100–210 nm) (Lavens & Sorgeloos, 1996). An increasing number of facilities are opting instead for the freshwater rotifer *Brachionus calyciflorus*, which prefers a salinity level closer to that used in many zebrafish holding systems. As such, neither species is subjected to major salinity changes when placed in contact with one another, both benefiting in regard to survival and growth (Aoyama et al., 2015).

Maintaininging stable populations of rotifers is time consuming and often challenging, because density tends to oscillate and troubleshooting can be difficult (Harper & Lawrence, 2016; Wilson, 2012). The solution to this issue may lie with an alternative rotifer species and a novel preservation technique presented by Nakayama et al. (2018). They used the freshwater bdelloid rotifer, a species known to undergo cryptobiosis (an ametabolic state of life) in response to desiccation, to create dried rotifers sheets on filter paper. These organisms can be revived up to 12 months later by immersion in water, leading to 14 dpf zebrafish survival curves similar to those for freshwater rotifers (*B. calyciflorus*). Although more research is needed, if these dried rotifers sheets become commercially available at prices affordable to most facilities and results continue to prove positive, they may well become a relevant alternative to the classic species and protocols.

Artemia

Artemia sp. is captured from the wild, which carries some downsides that have led to a growing number of facilities turning to *Artemia*-free nutritional programs (Carvalho et al., 2006; Lawrence et al., 2015). Specifically, if *Artemia* is exposed to any contaminants or pathogens in the environment its nutritional quality and availability fluctuate according to available nutritional resources and environmental conditions; and this incurs high costs as a consequence of the limited level of wild production. Nevertheless, *Artemia* is still widely offered to juvenile and adult zebrafish as a nutritional complement (Fowler et al., 2019), as environmental enrichment (Stevens et al., 2021) or both.

Artemia can be obtained from commercial suppliers in the form of encapsulated cysts that require chemical decapsulation prior to hatching, or as decapsulated dry cysts ready to hatch. It is also possible to purchase magnetized cysts that can be separated from the nauplii during harvest by simply using a magnetized tube, thus preventing fish from eating the empty cysts. Among *Artemia* species, *A. franciscana* and *A. salina* are the more popular options but there are inter-specific variations among all species in terms of size, hatching rate and efficiency, tolerance to temperature and salinity, and nutritional value that should be considered before selecting a particular one. Regardless of the choice, *Artemia* should be offered to fish soon after hatching while the nauplii are in the nutritionally rich, instar I stage (Lavens & Sorgeloos, 1996). Detailed protocols for *Artemia* decapsulation and hatching are available (Martins et al., 2016). Alternatively, non-hatching decapsulated *Artemia* cysts can also be used as an alternative to the nauplii, with associated reduction in both time and costs associated with hatching (Tye et al., 2015), although in this case their value as environmental enrichment for stimulation of prey-capture natural behaviour is lost.

Artemia and rotifers frequently lack some essential nutrients. One way to compensate for that is to enrich them with the missing nutrients such as fatty acids, vitamins, minerals and probiotics to improve rearing success. For example, *Artemia* is frequently deficient in PUFA and, to overcome this, many aquaculture facilities enrich it with PUFA using marine oil emulsions. The rotifer nutritional profile depends on the microalgae on which they feed, the reason why it is recommended to feed rotifers with multiple microalgal species over monospecific diets. In addition, rotifers can be enriched by nutrient encapsulation in the algae that they usually eat (Samat et al., 2020).

Other available live feeds

Paramecia are maintained in concentrated populations and usually fed with yeast. They can be used as a less expensive alternative to rotifers but are nutritionally poorer and as laborious to maintain in stable populations (Flores et al., 2019; Wilson, 2012). Other organisms available include *Daphnia* spp. (Herman et al., 2021) and various nematodes (microworms), but these are not commonly used due to the paucity of available information on their biosecurity and nutritional adequacy for zebrafish.

Manufactured feeds

Manufactured, or processed, fish feeds are very popular among the zebrafish husbandry community because in general they have long shelf life, high digestibility and stability and are ready to use, greatly reducing the labour involved. They can be formulated in a species- and developmental stage-specific manner, although this is underexplored in the case of zebrafish. Zebrafish-specific formulae have started to appear in the commercial circuit in the last decade but are sparse, in part because zebrafish respond well to more general formulations designed for ornamental fish or commercial aquaculture; in part due also to incomplete information on their nutritional requirements for all life stages and to lack of interest from suppliers in pursuing that information for such a small niche market as that of the zebrafish within the global aquaculture industry. As if that was not enough, the criteria for approval by competent authorities such as the US Food and Drug Administration and the European Food Safety Authority vary and so, for a product to be distributed globally, companies may have to find alternative formulations that comply with different regulatory authorities while preserving similar efficacy and nutritional composition.

Available processed feeds are mainly composed of protein, fibres, fats (oils), vitamins and minerals, to which are added binding agents that provide water stability to the feed, preservatives that help extend the shelf life and attractants to enhance palatability. All ingredients are ground up and delivered as pellets, flakes, powders or slow-dissolving gels that come in different sizes, to fit the oral size of larvae, juveniles and adults. Feeds have different colours depending on the ingredients used, and as a means of stimulating visual attractiveness, although colour doesn't seem to be a determining factor for zebrafish, based on the variety of colours of feeds (brown to orange to green) that have proved efficient for this species (Frederickson et al., 2021; Monteiro et al., 2018). The shape of feed particles is relevant because the lower their surface/volume ratio the lower the oxidation rate and thus the encapsulated nutrients remain stable for longer. Besides, both shape and weight determine how long particles float and how fast they sink in the water column. Particles that float for some time before slowly sinking are more easily captured by zebrafish and any other species with similar feeding apparatus. Finally, the production method is also relevant. Feeds produced by cold extrusion are typically more stable in water, so fish have more time to ingest nutrients under optimal conditions before these start to degrade. In fact, some cold-extruded feeds have been associated with improved growth and breeding in zebrafish (Farias & Certal, 2016; Frederickson et al., 2021; Lawrence, 2015; Monteiro et al., 2018).

Table 3.1 shows the publicly available (online and/or packages) composition of some of the most common processed feeds used in zebrafish research facilities. Some were specifically designed for zebrafish, others are general formulae for small tropical cyprinids. Brands are omitted.

Current challenges

There is a widespread lack of detailed nutritional information about the feeds provided to zebrafish. Commercial suppliers traditionally

Table 3.1 Known composition of seven commercially available processed feeds frequently used in zebrafish research facilities. Percentage (%) minims for crude protein, crude fat and phosphorus and maxima for moisture, crude fibre and crude ash. Information available from individual suppliers.

Feeds (brands omitted)	Crude protein % (min)	Crude fat % (min)	Crude fibre % (max)	Moisture % (max)	Crude ash % (max)	Phosphorus % (min)
A	55	7	6.4	5.5	10.5	–
B	59	14	0.2	–	14	1.3
C	45	6	2.5	6	–	–
D	40	4.5	0.5	7	4.5	0.45
E	46	7	2	8	–	1.4
F	55	15	1.5	12	12	1.5
G	63	14	1.8	–	12	–

keep the detailed nutritional composition of their feeds private (Table 3.1), and live feeds are naturally susceptible to oscillations in their nutritional content. In addition, live feeds are frequently cultured in aquatic facilities with no analysis of their nutritional composition having been undertaken. Poor nutritional details are not a big issue for the majority of fish facilities, where feed choices are validated based on simple welfare indicators (survival, growth and breeding). However, available information is often insufficient for assessing correlations between the quality and quantity of specific nutrients and the physiological aspects under investigation. Consequently it is also not possible to predict how the use of various feeds across facilities may affect experimental reproducibility. Some laboratories produce their own purified feeds with tightly controlled composition, but again this only solves the issue locally. Feed standardization seems to be the answer, as described later in this chapter.

Another challenge is to achieve accuracy and consistency in feed dosages. Theoretical daily feed portions have been calculated (Lawrence et al., 2012a), but turning theory into practice has proved quite difficult. The two main types of feeding device are syringes/guns and squeeze bottles, none of which are very ergonomic or precise, with an error range that depends on the operator. Feeding robots eliminate the human error factor, but the robots themselves need frequent verification and are expensive. A promising new option is gel diets available in units of uniform weight, that can easily be administered by hand in a consistent manner (for example, one gel unit per every *x* number of fish).

Due to the difficulty in controlling the precise quantity of feeds provided per tank and per fish, there is a common tendency to overfeed animals. This is a mistake that arises from the generalized misconception that underfeeding, but not overfeeding, is a sign of poor husbandry practice. The truth is that both scenarios may prejudice the animals directly (Arslan-Ergul et al., 2016; Stankiewicz et al., 2017). Increased debris accumulation due to overfeeding also induces faster water quality degradation and increases the frequency of cleaning or replacement of recirculating system components, with incremental costs for materials and human resources. During the 2020 lockdown due to the SARS-CoV-2 pandemic, many fish facilities were forced to reduce feeding frequency and/or portions due to shortage of supply, shortage of staff and as a preventive measure to maximize the autonomy of recirculating systems operating with good water quality. Although there are no studies published about how these measures affected the fish, several facility managers have shared empirical observations that there was no deterioration in fish health and, in at least one case, breeding even improved (Danio Zoom Friday meetings, online, 2021, personal communication). These observations suggest that at least some facilities are overfeeding their fish, either due to administrative deviation from theoretical goals or overestimation of nutritional needs. Considering the welfare, financial and time management benefits of maintaining precise feeding routines, some facilities are now revisiting their protocols.

Feed biosecurity: risks and prevention

Live feeds pose a much higher biosecurity risk for fish than formulated feeds because they are much more exposed to environmental contaminants, toxins and pathogens. Therefore, it is important that facility managers and researchers are aware of their potential hazards as well as preventive and mitigation strategies before introducing live feeds into an aquaculture system. In this section, the most relevant pathogens that can use zebrafish live feeds as transmission vectors will be discussed.

All three organisms most typically used as live feeds for zebrafish – *Artemia*, rotifers and *Paramecium* – are frequently used as bioindicators in ecotoxicology assays, due to their tendency to accumulate compounds and microorganisms from the water (Cruzeiro et al., 2017; Houneida et al., 2012; Janssen et al., 1994; Kokalj et al., 2018). This clearly illustrates their capacity to serve as transmission vectors. In many cases, *Artemia*, rotifers and *Paramecium* feed on, or accumulate, microorganisms that do not represent a risk by themselves but may constitute a real threat for the fish that eat them (Chang et al., 2019; Lavens & Sorgeloos, 1996; Peterson et al., 2013). *Daphnia* is also highly susceptible to contamination (Cruzeiro et al., 2017; Ebert, 1995; Herman et al., 2021), which is one of the reasons why its usage as live feed is not more frequent.

Artemia cysts are caught from the wild and are, therefore, potentially exposed to many biological and chemical contaminants present in the natural environment. *Artemia* can act as a reservoir or carrier for bacterial pathogens such as *Erwinia*, *Micrococcus*, *Staphylococcus*, *Bacillus* and *Víbrio* (Kumaran & Citarasu, 2016). Of note, the genus *Vibrio* includes major disease-causing bacteria that are highly virulent in aquaculture species, including zoonotic bacteria responsible for acute intestinal infections such as cholera (Austin, 2010). This is a major reason for maintaining well-defined biosecurity programmes for live feed production (Vanderberghe et al., 2003). Rotifers and *Paramecium* are usually cultured in house, and therefore it should be easier to control their sanitization. However, it is possible to find several bacterial species in rotifer cultures, mainly belonging to the genera *Pseudomonas*, *Vibrio*, *Moraxella* and *Flavobacterium* (Verdonck et al., 1994). It is important to make conscious choices when selecting the microalgae that will serve as feed for rotifers, because they themselves are a potential vector of bacterial pathogens (Avila-Villa et al., 2011).

Artemia, rotifers and *Paramecium* are also susceptible to the incorporation and transmission of different bacteria of the genus *Pseudomonas*, including the zoonotic *P. aeruginosa* (Benie et al., 2017; Gortz, 1982; Llamas & van der Sar, 2014), as well as several *Mycobacterium* spp. including the zoonotic *M. marinum* and *M. chelonae* (Chang, 2019). Of note, the chronic disease caused by *Mycobacterium* spp. is the second most common microbial infection in zebrafish research colonies (Peterson et al., 2013). Also of concern is *Edwardsiella ictaluri* (Hawke et al., 2013), a severe systemic pathogen that can be ingested by both *Artemia* (Situmorang et al., 2014) and ciliated protozoans (King & Shotts, 1988), including *Paramecium*. Finally, health programmes should include screens for the presence of *Pseudoloma neurophilia*, a potentially lethal parasite that infects the central nervous system and muscle and that is the most common pathogen found in laboratory zebrafish (Sanders et al., 2012). Although there are no reports of *P. neurophilia* contamination via live feeds, rotifers and *Artemia* can be the host for other parasites of the same phylum such as *Microsporidia* (Ebert, 1995; Gorbunov & Koscva, 2001; Rode et al., 2013; see Chapter 5 for more information).

Most zebrafish pathogens are difficult to eradicate from the populations, chiefly because there are very few treatments that are efficient for fish and that do not harm the nitrogen-cycle bacteria of the biological filters typically used in zebrafish recirculating housing systems. With this in mind, biosecurity and health programmes based on prevention, monitoring and early detection of pathogens are the best way to prevent animal loss or poor welfare in fish facilities. Several authors and institutions have published their own programmes (Borges et al., 2016; Kent et al., 2009; Martins et al., 2016; Murray, 2016) that can be used as guidelines. Nevertheless, sanitation and health programmes should be custom designed to account for the particularities of each facility. Measures used to prevent feed-mediated biological contamination of fish include prioritization of pathogen-free certified feeds, ensuring appropriate storage conditions (temperature, humidity and light), respecting expiry dates and routine cleaning and disinfection of feed containers. *Artemia* chemical decapsulation also decreases the risks of *Artemia*-mediated pathogen transmission (Martins et al., 2016), although this will deal only with pathogens present in the outer surface of cysts. Training facility staff on biosecurity and wearing adequate personal protection equipment (hair caps, gowns, shoe covers, face masks and gloves) is very important for preventing transmission of zoonotic pathogens between feed and staff. Finally, sending frequent feed samples for microbiological screening, including the growth media for live feeds, will help in the early detection of microorganisms, in many cases before fish manifest clinical signs.

Live feeds as vehicles for molecule delivery to zebrafish

Live feeds are very efficient vehicles for the delivery of drugs, nutrients or pathogens to fish, thanks to their ability to incorporate a product of interest either orally or by phagocytosis, pinocytosis or endocytosis (Aguilar-Aguila et al., 1994; Gomez-Gil et al., 1998; Hontoria et al., 1994) – a process called bioencapsulation (Rolando, 2001). After transformation into live capsules, these are fed to the final receptor organism, in which the products are released in the gastrointestinal tract (Flores et al., 2019). This approach has allowed the controlled infection of zebrafish in different life stages and subsequent disease investigation. For example, *Artemia* and rotifers have been bioencapsulated with different *Víbrio* species (Gómez-Gil et al., 1998; Martínez-Díaz et al., 2003), and *Paramecium* was transformed into an infection vector for the study of host–pathogen interactions and hallmarks of pathogenicity (Flores et al., 2019). On the other extreme, live feeds can be bioencapsulated with medicines (Rodriguez et al., 2011; Roiha et al., 2010) and even probiotics for increasing protection against pathogens such as *Vibrio* spp. (Giarma et al., 2017; Zoumpourtikoudi et al., 2018) and *Edwardsiella* spp. (Loh et al., 2018).

Finally, a non-viral technology for delivering nanomaterials and macromolecules into live cells, called cell-penetrating peptides (CPPs), has been successfully applied to rotifers: these incorporated two CPPs carrying plasmid DNAs, red fluorescent proteins and semiconductor quantum dots (Liu et al., 2013). This approach holds great potential for the development of new methods to deliver exogenous genes, proteins and nanoparticles to fish.

Standardization of diets

Standards for zebrafish housing, water quality (physical and chemical parameters), light and enrichment are well recognized and widely accepted due to a general recognition of what is best for the animals, based on welfare indicators that include survival, growth, breeding performance and healthy external appearance and swimming behaviour. Standards for diets have, to date, relied on the same welfare indicators and, because zebrafish show convincing responses to a large variety of feeding regimens, many feeds have been accepted without looking at their specific nutritional composition. The problem is that both the quantity and source of specific nutrients affect a myriad of physiological phenomena that, to date, have been left out of the equation. For example, a fat- or sugar-rich diet can alter mitochondrial activity (Chao et al., 2018) and induce activation of inflammatory and lipogenic gene expression, lipoprotein oxidation, hypertriglyceridaemia, hypercholesterolaemia, insulin resistance by modulation of B cell plasticity, obesity and hepatic steatosis (Chen et al., 2018; Fang et al., 2014; Gut et al., 2017; Schlegel & Gut, 2015). High-calorie diets can lead to metabolically induced premature ageing and associated perturbation of sleep, circadian functions and adult neurogenesis (Stankiewicz et al., 2017), whereas diet restriction seems to affect cell senescence mechanisms differently in young and old zebrafish (Arslan-Ergul et al., 2016). Diet also affects the immune system (López Nadal et al., 2020; Martin and Król, 2017; Ulloa et al., 2011) and behaviours such as shoaling preferences and foraging success (Krause et al., 1999). Consequently, the current heterogeneity of feeding protocols across facilities is a source of experimental biases that may jeopardize the scientific value of zebrafish-based research.

The solution relies on the standardization of feeds and feeding routines, similar to that carried out for many decades for other model organisms. The most effective and simple approach would be to generate and use worldwide a single standard reference diet for each main developmental stage (larvae, juveniles and adults). The next step would then be the development of additional standard reference diets to feed animals according to their research purpose. In fact, nutritional programmes have already been used as a tool during the establishment of zebrafish human disease models (Heckler & Kroll, 2017; Landgraf et al., 2017; Martin & Król, 2017; Schlegel & Gut, 2015), but these lack standardization.

We are far from a position of having all the data required to produce perfect feeds, but the available information is sufficient to develop good zebrafish reference diets. The next step is the development of strategies that guarantee commercial suppliers the continuity of their business. This will encourage the sharing of ingredients lists and quantities and nurture collaborations towards the production of standard reference diets and their distribution worldwide. In parallel, institutions have an important role in motivating researchers and fish facility staff to embrace change from deep-rooted protocols, and in providing the necessary human resources, infrastructure and equipment to make those changes possible. In fact, feed standardization has been identified as a priority for several years and discussed extensively in meetings and publications (Watts et al., 2012; Watts & D'Abramo, 2021). What is now likely missing is a push from competent international authorities to start the engine that will drive all previously identified actions and ultimately make zebrafish feed standardization a reality.

Concluding remarks

In this chapter several aspects of zebrafish anatomy, physiology and behaviour were pinpointed as affecting, and being conditioned by, the diet at each life stage. This high adaptability of zebrafish to varying water conditions and feeding schemes has been one of the keys for this species' success as a research model over the past few decades, allowing its culture by many types of facility. However, the same plasticity has also been responsible for delaying the development of standard reference feeding protocols that are fundamental for research reproducibility and accuracy. Over the years, fish facilities have been publishing data on the implementation of optimized protocols that are now being established in many other facilities. We can only assume that standardization will follow.

References

Abbate, F., Germanà, G. P., De Carlos, F., Montalbano, G., Laurà, R., Levanti, M. B., & Germanà, A. (2006). The oral cavity of the adult zebrafish (Danio rerio). *Anatomia, Histologia, Embryologia, 35*(5), 299–304. https://doi.org/10.1111/j.1439-0264.2006.00682.x

Abi-Ayad, A., Mélard, C., & Kestemont, P. (1997). Effect of n-3 fatty acids in Eurasian perch broodstock diet on egg fatty acid composition and larva stress resistance. *Aquaculture International, 5*(2), 161–168.

Aguilar-Aguila, A., Mansir, A. T., & Manríquez, A. R. (1994). Using brine shrimp as a drug carrier for therapeutic applications in aquaculture. *Aquacultural Engineering, 13*(4), 301–309. https://doi.org/10.1016/0144-8609(94)90017-5

Aigler, S. R., Jandzik, D., Hatta, K., Uesugi, K., & Stock, D. W. (2014). Selection and constraint underlie irreversibility of tooth loss in cypriniform fishes. *Proceedings of the National Academy of Sciences of the United States of America, 111*(21), 7707–7712. https://doi.org/10.1073/pnas.1321171111

Aoyama, Y., Moriya, N., Tanaka, S., Taniguchi, T., Hosokawa, H., & Maegawa, S. (2015). A novel method for rearing zebrafish by using freshwater rotifers (*Brachionus calyciflorus*). *Zebrafish, 12*(4), 288–295. https://doi.org/10.1089/zeb.2014.1032

Arslan-Ergul, A., Erbaba, B., Karoglu, E. T., Halim, D. O., & Adams, M. M. (2016). Short-term dietary restriction in old zebrafish changes cell senescence mechanisms. *Neuroscience*, 334, 64–75. https://doi.org/10.1016/j.neuroscience.2016.07.033

Arunachalam, M., Raja, M., Vijayakumar, C., Malaiammal, P., & Mayden, R. L. (2013). Natural history of zebrafish (Danio rerio) in India. *Zebrafish, 10*(1), 1–14. https://doi.org/10.1089/zeb.2012.0803

Austin, B. (2010). Vibrios as causal agents of zoonoses. *Veterinary Microbiology, 140*(3–4), 310–317. https://doi.org/10.1016/j.vetmic.2009.03.015

Avila-Villa, L. A., Martínez-Porchas, M., Gollas-Galván, T., López-Elías, J. A., Mercado, L., Murguia-López, Á., Mendoza-Cano, F., & Hernández-López, J. (2011). Evaluation of different microalgae species and artemia (artemia franciscana) as possible vectors of necrotizing hepatopancreatitis bacteria. *Aquaculture, 318*(3–4), 273–276. https://doi.org/10.1016/j.aquaculture.2011.05.043

Bell, J. G., Farndale, B. M., Bruce, M. P., Navas, J. M., & Carillo, M. (1997). Effects of broodstock dietary lipid on fatty acid compositions of eggs from sea bass (Dicentrarchus labrax). *Aquaculture*, *149*(1–2), 107–119. https://doi.org/10.1016/S0044-8486(96)01436-6

Benie, C. K. D., Dadié, A., Guessennd, N., N'gbesso-Kouadio, N. A., N'zebo Désiré Kouame, N.D., N'golo, D.C., Aka, S., Dako, E., Dje, K.M., & Dosso, M. (2017). Characterization of Virulence Potential of Pseudomonas Aeruginosa Isolated from Bovine Meat, Fresh Fish, and Smoked Fish. *European journal of microbiology and immunology* *7*(1), 55–64.

Bianco, I. H., Kampff, A. R., & Engert, F. (2011). Prey capture behavior evoked by simple visual stimuli in larval zebrafish. *Frontiers in Systems Neuroscience*, 5, 101. https://doi.org/10.3389/fnsys.2011.00101

Borges, A. C., Pereira, N., Franco, M., Vale, L., Pereira, M., Cunha, M. V., Amaro, A., Albuquerque, T., & Rebelo, M. (2016). Implementation of a zebrafish health program in a research facility: A 4-year retrospective study. *Zebrafish*, *13*, Suppl. 1, S115–S126. https://doi.org/10.1089/zeb.2015.1230

Carvalho, A. P., Araújo, L., & Santos, M. M. (2006). Rearing zebrafish (Danio rerio) larvae without live food: Evaluation of a commercial, a practical and a purified starter diet on larval performance. *Aquaculture Research*, *37*(11), 1107–1111. https://doi.org/10.1111/j.1365-2109.2006.01534.x

Carvalho, A. P., Escaffre, A.-M., Oliva Teles, A. O., & Bergot, P. (1997). First feeding of common carp larvae on diets with high levels of protein hydrolysates. *Aquaculture International*, *5*(4), 361–367. https://doi.org/10.1023/A:1018368208323

Chang, C. T., Benedict, S., & Whipps, C. M. (2019). Transmission of Mycobacterium chelonae and Mycobacterium marinum in laboratory zebrafish through live feeds. *Journal of Fish Diseases*, *42*(10), 1425–1431. https://doi.org/10.1111/jfd.13071

Chao, Y. J., Wu, W. H., Balazova, M., Wu, T. Y., Lin, J., Liu, Y. W., & Hsu, Y.-H. H. (2018). Chlorella diet alters mitochondrial cardiolipin contents differentially in organs of Danio rerio analyzed by a lipidomics approach. *PLOS ONE*, *13*(3), e0193042. https://doi.org/10.1371/journal.pone.0193042

Chen, B., Zheng, Y. M., & Zhang, J. P. (2018). Comparative study of different diets-induced NAFLD models of zebrafish. *Frontiers in Endocrinology*, 9, 366. https://doi.org/10.3389/fendo.2018.00366

Crim, M. J., & Lawrence, C. (2021). A fish is not a mouse: Understanding differences in background genetics is critical for reproducibility. *Lab Animal, 50*(1), 19–25. https://doi.org/10.1038/s41684-020-00683-x

Cruzeiro, C., Amaral, S., Rocha, E., & Rocha, M. J. (2017). Determination of 54 pesticides in waters of the Iberian Douro River estuary and risk assessment of environmentally relevant mixtures using theoretical approaches and artemia salina and Daphnia magna bioassays. *Ecotoxicology and Environmental Safety*, 145, 126–134. https://doi.org/10.1016/j.ecoenv.2017.07.010

Cunnane, S. C. (2003). Problems with essential fatty acids: Time for a new paradigm? *Progress in Lipid Research, 42*(6), 544–568. https://doi.org/10.1016/s0163-7827(03)00038-9

Dametto, F. S., Fior, D., Idalencio, R., Rosa, J. G. S., Fagundes, M., Marqueze, A., Barreto, R. E., Piato, A., & Barcellos, L. J. G. (2018). Feeding regimen modulates zebrafish behavior. *PeerJ, 6*, e5343. https://doi.org/10.7717/peerj.5343

Defoirdt, T., Boon, N., Sorgeloos, P., Verstraete, W., & Bossier, P. (2007). Alternatives to antibiotics to control bacterial infections: Luminescent vibriosis in aquaculture as an example. *Trends in Biotechnology, 25*(10), 472–479. https://doi.org/10.1016/j.tibtech.2007.08.001

Deng, D. F., Refstie, S., Hemre, G.-I., Crocker, C. E., Chen, H. Y., Cech, Jr., J. J., & Hung, S. S. O. (2000). A new technique of feeding, repeated sampling of blood and continuous collection of urine in white sturgeon. *Fish Physiology and Biochemistry, 22*(3), 191–197. https://doi.org/10.1023/A:1007803307617

Ebert, D. (1995). The ecological interactions between a microsporidian parasite and its host Daphnia magna. *Journal of Animal Ecology, 64*(3), 361–369. https://doi.org/10.2307/5897

Engeszer, R. E., Patterson, L. B., Rao, A. A., & Parichy, D. M. (2007). Zebrafish in the wild: A review of natural history and new notes from the field. *Zebrafish, 4*(1), 21–40. https://doi.org/10.1089/zeb.2006.9997

Fang, L., Liu, C., & Miller, Y. I. (2014). Zebrafish models of dyslipidemia: Relevance to atherosclerosis and angiogenesis. *Translational Research: The Journal of Laboratory and Clinical Medicine, 163*(2), 99–108. https://doi.org/10.1016/j.trsl.2013.09.004

Farias, M., & Certal, A. C. (2016). Different feeds and feeding regimens have an impact on zebrafish larval rearing and breeding performance. *Symbiosis Open access Journal of Aquatic Research, 1*, 1–8.

Fernandes, H., Peres, H., & Carvalho, A. P. (2016). Dietary protein requirement during juvenile growth of zebrafish (Danio rerio). *Zebrafish, 13*(6), 548–555. https://doi.org/10.1089/zeb.2016.1303

Flores, E., Thompson, L., Sirisaengtaksin, N., Nguyen, A. T., Ballard, A., & Krachler, A. M. (2019). Using the protozoan Paramecium caudatum as a vehicle for food-borne infections in zebrafish larvae. *Journal of Visualized Experiments: JoVE, 143*(143), e58949. https://doi.org/10.3791/58949

Fowler, L. A., Williams, M. B., Dennis-Cornelius, L. N., Farmer, S., Barry, R. J., Powell, M. L., & Watts, S. A. (2019). Influence of commercial and laboratory diets on growth, body composition, and reproduction in the zebrafish Danio rerio. *Zebrafish, 16*(6), 508–521. https://doi.org/10.1089/zeb.2019.1742

Fraher, D., Sanigorski, A., Mellett, N. A., Meikle, P. J., Sinclair, A. J., & Gibert, Y. (2016). Zebrafish embryonic lipidomic analysis reveals that the yolk cell is metabolically active in processing lipid. *Cell Reports, 14*(6), 1317–1329. https://doi.org/10.1016/j.celrep.2016.01.016

Frederickson, S. C., Steinmiller, M. D., Blaylock, T. R., Wisnieski, M. E., Malley, J. D., Pandolfo, L. M., & Castranova, D. (2021). Comparison of juvenile feed protocols on growth and spawning in zebrafish. *Journal of the American Association for Laboratory Animal Science, 60*(3), 298–305. https://doi.org/10.30802/AALAS-JAALAS-20-000105

Fritsche, K. L. (2008). Too much linoleic acid promotes inflammation-doesn't it? *Prostaglandins, Leukotrienes, and Essential Fatty Acids, 79*(3–5), 173–175. https://doi.org/10.1016/j.plefa.2008.09.019

Giarma, E., Amanetidou, E., Toufexi, A., & Touraki, M. (2017). Defense systems in developing artemia franciscana nauplii and their modulation by probiotic bacteria offer protection against a Vibrio anguillarum challenge. *Fish and Shellfish Immunology*, 66, 163–172. https://doi.org/10.1016/j.fsi.2017.05.008

Gomez-Gil, B., Herrera-Vega, M. A., Abreu-Grobois, F. A., & Roque, A. (1998). Bioencapsulation of two different vibrio species in nauplii of the brine shrimp (artemia franciscana). *Applied and Environmental Microbiology, 64*(6), 2318–2322. https://doi.org/10.1128/AEM.64.6.2318-2322.1998

Gonzales, J. M., & Law, S. H. W. (2013). Feed and feeding regime affect growth rate and gonadosomatic index of adult zebrafish (Danio rerio). *Zebrafish, 10*(4), 532–540. https://doi.org/10.1089/zeb.2013.0891

Goolish, E. M., Okutake, K., & Lesure, S. (2011). Growth and survivorship of larval zebrafish Danio rerio on processed diets. *North American Journal of Aquaculture*, *61*(3), 189–198.

Gorbunov, A. K., & Kosova, A. A. (2001). Parasites in rotifers from the Volga Delta. *Hydrobiologia*, *446–447*(1), 51–55.

Gortz, H.D. (1982). Infections of Paramecium bursaria with bacteria and yeasts. *Journal of Cell Science*, 58(1), 445–453.

Grisdale-Helland, B., & Helland, S. J. (1997). Replacement of protein by fat and carbohydrate in diets for Atlantic salmon (Salmo salar) at the end of the freshwater stage. *Aquaculture*, *152*(1–4), 167–180. https://doi.org/10.1016/S0044-8486(97)00003-3

Gut, P., Reischauer, S., Stainier, D. Y. R., & Arnaout, R. (2017). Little fish, big data: Zebrafish as a model for cardiovascular and metabolic disease. *Physiological Reviews*, *97*(3), 889–938. https://doi.org/10.1152/physrev.00038.2016

Hammer, C. L., Atukorala, A. D. S., & Franz-Odendaal, T. A. (2016). What shapes the oral jaws? Accommodation of complex dentition correlates with premaxillary but not mandibular shape. *Mechanisms of Development*, 141, 100–108. https://doi.org/10.1016/j.mod.2016.04.001

Hansen, A., Reutter, K., & Zeiske, E. (2002). Taste bud development in the zebrafish, Danio rerio. *Developmental Dynamics*, *223*(4), 483–496. https://doi.org/10.1002/dvdy.10074

Harper, C., & Lawrence, C. (2016). *The laboratory zebrafish* (1st ed). CRC Press.

Hawke, J. P., Kent, M., Rogge, M., Baumgartner, W., Wiles, J., Shelley, J., Savolainen, L. C., Wagner, R., Murray, K., & Peterson, T. S. (2013). Edwardsiellosis caused by Edwardsiella ictaluri in laboratory populations of zebrafish Danio rerio. *Journal of Aquatic Animal Health*, *25*(3), 171–183. https://doi.org/10.1080/08997659.2013.782226

Head, B., La Du, J., Tanguay, R. L., Kioussi, C., & Traber, M. G. (2020). Vitamin E is necessary for zebrafish nervous system development. *Scientific Reports*, *10*(1), 15028. https://doi.org/10.1038/s41598-020-71760-x

Heckler, K., & Kroll, J. (2017). Zebrafish as a model for the study of microvascular complications of diabetes and their mechanisms. *International Journal of Molecular Sciences*, *18*(9). https://doi.org/10.3390/ijms18092002

Henken, A. M., Boon, J. B., Cattel, B. C., & Lobée, H. W. J. (1987). Differences in growth rate and feed utilization between male and female African catfish, Clarias gariepinus (Burchell 1822). *Aquaculture, 63*(1–4), 221–232. https://doi.org/10.1016/0044-8486(87)90074-3

Hensley, M. R., & Leung, Y. F. (2010). A convenient dry feed for raising zebrafish larvae. *Zebrafish, 7*(2), 219–231. https://doi.org/10.1089/zeb.2010.0652

Herman, P., Fehér, M., Molnár, Á., Harangi, S., Sajtos, Z., Stündl, L., Fábián, I., & Baranyai, E. (2021). Iron and Manganese Retention of Juvenile zebrafish (Danio rerio) Exposed to Contaminated Dietary Zooplankton (Daphnia pulex)-a Model Experiment. *Biological Trace Element Research, 199*(2), 732–743. https://doi.org/10.1007/s12011-020-02190-z

Hernandez, L. P., & Staab, K. L. (2015). Bottom feeding and beyond: How the premaxillary protrusion of cypriniforms allowed for a novel kind of suction feeding. *Integrative and Comparative Biology, 55*(1), 74–84. https://doi.org/10.1093/icb/icv038

Hontoria, F., Crowe, J. H., Crowe, L. M., & Amat, F. (1994). Potential use of liposomes in larviculture as a delivery system through *artemia* nauplii. *Aquaculture, 127*(2–3), 255–264. https://doi.org/10.1016/0044-8486(94)90431-6

Houneida, B., Berrebah, H., Berredjem, M., & Djebar, M. R. (2012). Effect of novel phosphoramidate on growth and respiratory metabolism of *Paramecium* aurelia. *Journal of Natural Science, Biology, and Medicine, 3*(1), 48–51. https://doi.org/10.4103/0976-9668.95949

Huang, Y., Wang, X. L., Zhang, J. W., & Wu, K. S. (2015). Impact of endocrine-disrupting chemicals on reproductive function in zebrafish (Danio rerio). *Reproduction in Domestic Animals, 50*(1), 1–6. https://doi.org/10.1111/rda.12468

Janssen, C. R., Ferrando, M. D., & Persoone, G. (1994). Ecotoxicological studies with the freshwater rotifer Brachionus calyciflorus. IV. Rotifer behavior as a sensitive and rapid sublethal test criterion. *Ecotoxicology and Environmental Safety, 28*(3), 244–255. https://doi.org/10.1006/eesa.1994.1050

Jantrarotai, W., Sitasit, P., Jantrarotai, P., Viputhanumas, T., & Srabua, P. (1998). Protein and Energy Levels for Maximum Growth, Diet Utilization, Yield of Edible Flesh and Protein Sparing of Hybrid Clarias Catfish (Clarias macrocephalus×Clarias gariepinus). *Journal of the World Aquaculture Society, 29*(3), 281–289. https://doi.org/10.1111/j.1749-7345.1998.tb00648.x

Jaya-Ram, A., Kuah, M.-K., Lim, P.-S., Kolkovski, S., & Shu-Chien, A. C. (2008). Influence of dietary HUFA levels on reproductive performance, tissue fatty acid profile and desaturase and elongase mRNAs expression in female zebrafish Danio rerio. *Aquaculture*, *277*(3–4), 275–281. https://doi.org/10.1016/j.aquaculture.2008.02.027

Karga, J., & Mandal, S. C. (2017). Effect of different feeds on the growth, survival and reproductive performance of zebrafish, Danio rerio (Hamilton, 1822). *Aquaculture Nutrition*, *23*(2), 406–413. https://doi.org/10.1111/anu.12407

Kaushik, S., Georga, I., & Koumoundouros, G. (2011). Growth and body composition of zebrafish (Danio rerio) larvae fed a compound feed from first feeding onward: Toward implications on nutrient requirements. *Zebrafish*, *8*(2), 87–95. https://doi.org/10.1089/zeb.2011.0696

Kent, M. L., Feist, S. W., Harper, C., Hoogstraten-Miller, S., Law, J. M., Sánchez-Morgado, J. M., Tanguay, R. L., Sanders, G. E., Spitsbergen, J. M., & Whipps, C. M. (2009). Recommendations for control of pathogens and infectious diseases in fish research facilities. *Comparative Biochemistry and Physiology. Toxicology and Pharmacology*, *149*(2), 240–248. https://doi.org/10.1016/j.cbpc.2008.08.001

Kim, J. D., & Kaushik, S. J. (1992). Contribution of digestible energy from carbohydrates and estimation of protein/energy requirements for growth of rainbow trout (Oncorhynchus mykiss). *Aquaculture*, *106*(2), 161–169. https://doi.org/10.1016/0044-8486(92)90200-5

King, C. H., & Shotts, E. B. (1988). Enhancement of Edwardsiella tarda and Aeromonas salmonicida through ingestion by the ciliated protozoan Tetrahymena pyriformis. *FEMS Microbiology Letters*, *51*(2–3), 95–99. https://doi.org/10.1111/j.1574-6968.1988.tb02976.x

Kokalj, A. J., Kunej, U., & Skalar, T. (2018). Screening study of four environmentally relevant microplastic pollutants: Uptake and effects on Daphnia magna and artemia franciscana. *Chemosphere*, 208, 522–529. https://doi.org/10.1016/j.chemosphere.2018.05.172

Kolb, A., Hildebrandt, F., & Lawrence, C. (2018). Effects of diet and social housing on reproductive success in adult zebrafish, Danio rerio. *Zebrafish*, *15*(5), 445–453. https://doi.org/10.1089/zeb.2018.1599

Kopp, R., Billecke, N., Legradi, J., den Broeder, M. D., Parekh, S. H., & Legler, J. (2016). Bringing obesity to light: Rev-erbα, a central player in light-induced adipogenesis in the zebrafish? *International Journal of Obesity*, *40*(5), 824–832. https://doi.org/10.1038/ijo.2015.240

Krause, J., Hartmann, N., & Pritchard, V. L. (1999). The influence of nutritional state on shoal choice in zebrafish, Danio rerio. *Animal Behaviour*, *57*(4), 771–775. https://doi.org/10.1006/anbe.1998.1010

Krogdahl, A., Hemre, G.-I., & Mommsen, T. P. (2005). Carbohydrates in fish nutrition: Digestion and absorption in postlarval stages. *Aquaculture Nutrition*, *11*(2), 103–122. https://doi.org/10.1111/j.1365-2095.2004.00327.x

Kumaran, T., & Citarasu, T. (2016). Isolation and characterization of Vibrio species from shrimp and artemia culture and evaluation of the potential virulence factor. *Intellectual Property Rights*, *4*, 1. http://dx.doi.org/10.4172/2375-4516.1000153.

Lall, S. P., & Kaushik, S. J. (2021). Nutrition and metabolism of minerals in fish. *Animals: An Open Access Journal from MDPI*, *11*(9), 2711. https://doi.org/10.3390/ani11092711

Landgraf, K., Schuster, S., Meusel, A., Garten, A., Riemer, T., Schleinitz, D., Kiess, W., & Körner, A. (2017). Short-term overfeeding of zebrafish with normal or high-fat diet as a model for the development of metabolically healthy versus unhealthy obesity. *BMC Physiology*, *17*(1), 4. https://doi.org/10.1186/s12899-017-0031-x

Lavens, P., & Sorgeloos, P. (1996). Manual on the production and use of live food for aquaculture. *FAO Fisheries Technical Paper No. 361*. http://www.fao.org/docrep/003/W3732E/W3732E00.HTM Retrieved October 2021.

Lawrence, C., Best, J., Cockington, J., Henry, E. C., Hurley, S., James, A., Lapointe, C., Maloney, K., & Sanders, E. (2016). The complete and updated "rotifer polyculture method" for rearing first feeding zebrafish. *Journal of Visualized Experiments: JoVE*, *17*(107), e53629. https://doi.org/10.3791/53629

Lawrence, C., Best, J., James, A., & Maloney, K. (2012). The effects of feeding frequency on growth and reproduction in zebrafish (Danio rerio). *Aquaculture*, 368–369, 103–108. https://doi.org/10.1016/j.aquaculture.2012.09.022

Lawrence, C., James, A., & Mobley, S. (2015). Successful Replacement of artemia salina nauplii with Marine Rotifers (Brachionus plicatilis) in the Diet of Pre Adult zebrafish (Danio rerio). *Zebrafish*, *12*(5), 366–371. https://doi.org/10.1089/zeb.2015.1118

Lawrence, C., Sanders, E., & Henry, E. (2012). Methods for culturing saltwater rotifers (Brachionus plicatilis) for rearing larval zebrafish. *Zebrafish*, *9*(3), 140–146. https://doi.org/10.1089/zeb.2012.0771

Lebold, K. M., Jump, D. B., Miller, G. W., Wright, C. L., Labut, E. M., Barton, C. L., Tanguay, R. L., & Traber, M. G. (2011). Vitamin E deficiency decreases long-chain PUFA in zebrafish (Danio rerio).

Journal of Nutrition, 141(12), 2113–2118. https://doi.org/10.3945/jn.111.144279

Lindsay, S. M., & Vogt, R. G. (2004). Behavioral responses of newly hatched zebrafish (Danio rerio) to amino acid chemostimulants. *Chemical Senses, 29*(2), 93–100. https://doi.org/10.1093/chemse/bjh009

Liu, B. R., Liou, J. S., Chen, Y. J., Huang, Y. W., & Lee, H. J. (2013). Delivery of nucleic acids, proteins, and nanoparticles by arginine-rich cell-penetrating peptides in rotifers. *Marine Biotechnology, 15*(5), 584–595. https://doi.org/10.1007/s10126-013-9509-0

Llamas, M. A., & van der Sar, A. M. (2014). Assessing Pseudomonas virulence with nonmammalian host: Zebrafish. *Methods in Molecular Biology (Clifton, NJ), 1149*, 709–721. https://doi.org/10.1007/978-1-4939-0473-0_55

Loh, J. Y., Kay, G. L., & Ting, A. S. Y. (2018). Bioencapsulation and Colonization Characteristics of Lactococcus lactis lactis CF4MRS in artemia franciscana: A Biological Approach for the Control of edwardsiellosis in Larviculture. *Marine Biotechnology (New York), 20*(3), 353–362. https://doi.org/10.1007/s10126-018-9813-9

López Nadal, A., Ikeda-Ohtsubo, W., Sipkema, D., Peggs, D., McGurk, C., Forlenza, M., Wiegertjes, G. F., & Brugman, S. (2020). Feed, microbiota, and gut immunity: Using the zebrafish model to understand fish health. *Frontiers in Immunology, 11*, 114. https://doi.org/10.3389/fimmu.2020.00114

Marins, K., Lazzarotto, L. M. V., Boschetti, G., Bertoncello, K. T., Sachett, A., Schindler, M. S. Z., Chitolina, R., Regginato, A., Zanatta, A. P., Siebel, A. M., Magro, J. D., & Zanatta, L. (2019). Iron and manganese present in underground water promote biochemical, genotoxic, and behavioral alterations in zebrafish (Danio rerio). *Environmental Science and Pollution Research International, 26*(23), 23555–23570. https://doi.org/10.1007/s11356-019-05621-0

Martin, S. A. M., & Król, E. (2017). Nutrigenomics and immune function in fish: New insights from omics technologies. *Developmental and Comparative Immunology*, 75, 86–98. https://doi.org/10.1016/j.dci.2017.02.024

Martínez-Díaz, S. F., Álvarez-González, C. A., Legorreta, M. M., Vázquez-Juárez, R., & Barrios-González, J. (2003). Elimination of the associated microbial community and bioencapsulation of bacteria in the rotifer Brachionus plicatilis. *Aquaculture International, 11*(1–2), 95–108.

Martins, S., Monteiro, J. F., Vito, M., Weintraub, D., Almeida, J., & Certal, A. C. (2016). Toward an integrated zebrafish health management program supporting cancer and neuroscience research. *Zebrafish*, *13*, Suppl. 1, S47–S55. https://doi.org/10.1089/zeb.2015.1198

Mehrad, B., Jafaryan, H., & Animal, M. T. A. B. (2011). Impact of different dietary vitamin C contents on growth, survival, fecundity and egg diameter in the zebrafish, Danio rerio (Pisces, Cyprinidae). *Animal Biology and Animal Husbandry*, *3*(1), 18–25.

Meinelt, B. T., Schulz, C., Wirth, M., Kürzinger, H., & Steinberg, C. (1999). Dietary fatty acid composition influences the fertilization rate of zebrafish (Danio rerio Hamilton-Buchanan). *Journal of Applied Ichthyology*, *15*(1), 19–23. https://doi.org/10.1046/j.1439-0426.1999.00121.x.

Meinelt, T., Schulz, C., Wirth, M., Kürzinger, H., & Steinberg, C. (2000). Correlation of diets high in n-6 polyunsaturated fatty acids with high growth rate in zebrafish (Danio rerio). *Comparative Medicine*, *50*(1), 43–45.

Monteiro, J. F., Martins, S., Farias, M., Costa, T., & Certal, A. C. (2018). The impact of two different cold-extruded feeds and feeding regimens on zebrafish survival, growth and reproductive performance. *Journal of Developmental Biology*, *6*(3). https://doi.org/10.3390/jdb6030015

Murray, K. N., Varga, Z. M., & Kent, M. L. (2016). Biosecurity and health monitoring at the zebrafish international resource center. *Zebrafish*, *13*, Suppl. 1, S30–S38. https://doi.org/10.1089/zeb.2015.1206

Nakayama, H., Katayama, K., Onabe, Y., Sato, A., Nishimura, N., & Shimada, Y. (2018). Dried rotifer sheet: A novel live feed for rearing first-feeding larvae. *Zebrafish*, *15*(3), 291–294. https://doi.org/10.1089/zeb.2017.1553

Newman, T., Jhinku, N., Meier, M., & Horsfield, J. (2016). Dietary intake influences adult fertility and offspring fitness in Zebrafish. In P. Schausberger (Ed.), *PLoS ONE*, *11*(11), e0166394. https://doi.org/10.1371/journal.pone.0166394

Novak, C. M., Jiang, X., Wang, C., Teske, J. A., Kotz, C. M., & Levine, J. A. (2005). Caloric restriction and physical activity in zebrafish (Danio rerio). *Neuroscience Letters*, *383*(1–2), 99–104. https://doi.org/10.1016/j.neulet.2005.03.048

Nowosad, J., Kucharczyk, D., & Targońska, K. (2017). Enrichment of zebrafish Danio rerio (Hamilton, 1822) diet with polyunsaturated fatty acids improves fecundity and larvae quality. *Zebrafish*, *14*(4), 364–370. https://doi.org/10.1089/zeb.2017.1416

O'Brine, T. M., Vrtělová, J., Snellgrove, D. L., Davies, S. J., & Sloman, K. A. (2015). Growth, oxygen consumption, and behavioral responses of Danio rerio to variation in dietary protein and lipid levels. *Zebrafish, 12*(4), 296–304. https://doi.org/10.1089/zeb.2014.1008

Oliveri, A. N., Knuth, M., Glazer, L., Bailey, J., Kullman, S. W., & Levin, E. D. (2020). Zebrafish show long-term behavioral impairments resulting from developmental vitamin D deficiency. *Physiology and Behavior, 224*, 113016. https://doi.org/10.1016/j.physbeh.2020.113016

Ostaszewska, T., Dabrowski, K., Hliwa, P., Gomółka, P., & Kwasek, K. (2008). Nutritional regulation of intestine morphology in larval cyprinid fish, silver bream (Vimba vimba). *Aquaculture Research, 39*(12), 1268–1278. https://doi.org/10.1111/j.1365-2109.2008.01989.x

Palmer, T. N., & Ryman, B. E. (1972). Studies on oral glucose intolerance in fish. *Journal of Fish Biology, 4*(2), 311–319. https://doi.org/10.1111/j.1095-8649.1972.tb05680.x

Parichy, D. M. (2015). Advancing biology through a deeper understanding of zebrafish ecology and evolution. *eLife*, 4, 4.e05635. https://doi.org/10.7554/eLife.05635

Pati, D., & Habibi, H. R. (2002). Involvement of protein kinase C and arachidonic acid pathways in the gonadotropin-releasing hormone regulation of oocyte meiosis and follicular steroidogenesis in the goldfish ovary. *Biology of Reproduction, 66*(3), 813–822. https://doi.org/10.1095/biolreprod66.3.813

Penglase, S., Hamre, K., Rasinger, J. D., & Ellingsen, S. (2014). Selenium status affects selenoprotein expression, reproduction, and F_1 generation locomotor activity in zebrafish (Danio rerio). *British Journal of Nutrition* 14, *111*(11), 1918–1931.

Peterson, T. S., Ferguson, J. A., Watral, V. G., Mutoji, K. N., Ennis, D. G., & Kent, M. L. (2013). Paramecium caudatum enhances transmission and infectivity of Mycobacterium marinum and M. chelonae in zebrafish Danio rerio. *Diseases of Aquatic Organisms, 106*(3), 229–239. https://doi.org/10.3354/dao02649

Portha, B., Fournier, A., Kioon, M. D. A., Mezger, V., & Movassat, J. (2014). Early environmental factors, alteration of epigenetic marks and metabolic disease susceptibility. *Biochimie*, 97, 1–15. https://doi.org/10.1016/j.biochi.2013.10.003

Powell, M. L., Pegues, M. A., Szalai, A. J., Ghanta, V. K., D'Abramo, L. R., & Watts, S. A. (2015). Effects of the dietary ω3:ω6 fatty acid ratio on body fat and inflammation in zebrafish (Danio rerio). *Comparative Medicine*, *65*(4), 289–294.

Quinlivan, V. H., & Farber, S. A. (2017). Lipid uptake, metabolism, and transport in the larval zebrafish. *Frontiers in Endocrinology*, 8, 319. https://doi.org/10.3389/fendo.2017.00319

Reed, B., & Jennings, M. (2011). *Guidance on the housing and care of zebrafish, Danio rerio* (1st ed). Royal Society for the Prevention of Cruelty to Animals.

Rode, N. O., Lievens, E. J. P., Segard, A., Flaven, E., Jabbour-Zahab, R., & Lenormand, T. (2013). Cryptic microsporidian parasites differentially affect invasive and native artemia spp. *International Journal for Parasitology*, *43*(10), 795–803. https://doi.org/10.1016/j.ijpara.2013.04.009

Rodriguez, L., Livengood, E. J., Miles, R. D., & Chapman, F. A. (2011). Uptake of metronidazole in artemia at different developmental life stages. *Journal of Aquatic Animal Health*, *23*(2), 100–102. https://doi.org/10.1080/08997659.2011.587745

Roeselers, G., Mittge, E. K., Stephens, W. Z., Parichy, D. M., Cavanaugh, C. M., Guillemin, K., & Rawls, J. F. (2011). Evidence for a core gut microbiota in the zebrafish. *ISME Journal*, *5*(10), 1595–1608. https://doi.org/10.1038/ismej.2011.38

Roiha, I. S., Otterlei, E., Litlabø, A., & Samuelsen, O. B. (2010). Uptake and elimination of florfenicol in Atlantic cod (Gadus morhua) larvae delivered orally through bioencapsulation in the brine shrimp artemia franciscana. *Aquaculture*, *310*(1–2), 27–31. https://doi.org/10.1016/j.aquaculture.2010.10.014

Rolando, G. F. (2001). Artemia bioencapsulation I. Effect of particle sizes on the filtering behavior of artemia franciscana. *Journal of Crustacean Biology*, *21*(2), 435–442https://doi.org/10.1163/20021975-99990144.

Rowe, R. I., & Eckhert, C. D. (1999). Boron is required for zebrafish embryogenesis. *Journal of Experimental Biology*, *202*(12), 1649–1654. https://doi.org/10.1242/jeb.202.12.1649

Samat, N. A., Yusoff, F. M., Rasdi, N. W., & Karim, M. (2020). Enhancement of live food nutritional status with essential nutrients for improving aquatic animal health: A review. *Animals: An Open Access Journal from MDPI*, *10*(12), 2457. https://doi.org/10.3390/ani10122457

Sanders, J. L., Watral, V., & Kent, M. L. (2012). Microsporidiosis in zebrafish research facilities. *ILAR Journal*, *53*(2), 106–113. https://doi.org/10.1093/ilar.53.2.106

Santiago, C. B., & Reyes, O. S. (1993). Effects of dietary lipid source on reproductive performance and tissue lipid levels of Nile tilapia Oreochromis niloticus (Linnaeus) broodstock. *Journal of Applied Ichthyology*, *9*(1), 33–40. https://doi.org/10.1111/j.1439-0426.1993.tb00385.x

Schlegel, A., & Gut, P. (2015). Metabolic insights from zebrafish genetics, physiology, and chemical biology. *Cellular and Molecular Life Sciences*, *72*(12), 2249–2260. https://doi.org/10.1007/s00018-014-1816-8

Sheng, Y., Ren, H., Limbu, S. M., Sun, Y., Qiao, F., Zhai, W., Du, Z. Y., & Zhang, M. (2018). The presence or absence of intestinal microbiota affects lipid deposition and related genes expression in zebrafish (Danio rerio). *Frontiers in Microbiology*, 9, 1124. https://doi.org/10.3389/fmicb.2018.01124

Siccardi, A. J., Garris, H. W., Jones, W. T., Moseley, D. B., D'Abramo, L. R., & Watts, S. A. (2009). Growth and survival of zebrafish (Danio rerio) fed different commercial and laboratory diets. *Zebrafish*, *6*(3), 275–280. https://doi.org/10.1089/zeb.2008.0553

Simopoulos, A. P. (2008). The importance of the omega-6/omega-3 fatty acid ratio in cardiovascular disease and other chronic diseases. *Experimental Biology and Medicine*, *233*(6), 674–688. https://doi.org/10.3181/0711-MR-311

Siriyappagouder, P., Kiron, V., Lokesh, J., Rajeish, M., Kopp, M., & Fernandes, J. (2018). The intestinal mycobiota in wild zebrafish comprises mainly Dothideomycetes while Saccharomycetes predominate in their laboratory-reared counterparts. *Frontiers in Microbiology*, 9, 387. https://doi.org/10.3389/fmicb.2018.00387

Situmorang, M. L., Dierckens, K., Mlingi, F. T., Van Delsen, B., & Bossier, P. (2014). Development of a bacterial challenge test for gnotobiotic Nile tilapia Oreochromis niloticus larvae. *Diseases of Aquatic Organisms*, *109*(1), 23–33. https://doi.org/10.3354/dao02721

Skjærven, K. H., Jakt, L. M., Dahl, J. A., Espe, M., Aanes, H., Hamre, K., & Fernandes, J. M. O. (2016). Parental vitamin deficiency affects the embryonic gene expression of immune-, lipid transport- and apolipoprotein genes. *Scientific Reports*, *6*(1), 34535. https://doi.org/10.1038/srep34535

Smith, D. L., Barry, R. J., Powell, M. L., Nagy, T. R., D'Abramo, L. R., & Watts, S. A. (2013). Dietary protein source influence on body size and composition in growing zebrafish. *Zebrafish*, *10*(3), 439–446. https://doi.org/10.1089/zeb.2012.0864

Sorbera, L. A., Asturiano, J. F., Carrillo, M., & Zanuy, S. (2001). Effects of polyunsaturated fatty acids and prostaglandins on oocyte maturation in a marine teleost, the European sea bass (Dicentrarchus labrax). *Biology of Reproduction, 64*(1), 382–389. https://doi.org/10.1095/biolreprod64.1.382

Spence, R., Fatema, M. K., Ellis, S., Ahmed, Z. F., Wahab, M. A., Ahmed, Z. F., & Smith, C. (2007). Diet, growth and recruitment of wild zebrafish in Bangladesh. *Journal of Fish Biology, 71*(1), 304–309. https://doi.org/10.1111/j.1095-8649.2007.01492.x

Spence, R., Gerlach, G., Lawrence, C., & Smith, C. (2008). The behaviour and ecology of the zebrafish, Danio rerio. *Biological Reviews of the Cambridge Philosophical Society, 83*(1), 13–34. https://doi.org/10.1111/j.1469-185X.2007.00030.x

Stankiewicz, A. J., McGowan, E. M., Yu, L., & Zhdanova, I. V. (2017). Impaired sleep, circadian rhythms and neurogenesis in diet-induced premature aging. *International Journal of Molecular Sciences, 18*(11), 2243. https://doi.org/10.3390/ijms18112243

Stevens, C. H., Reed, B. T., & Hawkins, P. (2021). Enrichment for laboratory zebrafish-A review of the evidence and the challenges. *Animals: An Open Access Journal from MDPI, 11*(3), 11(3):698. https://doi.org/10.3390/ani11030698

Stephens, W. Z., Burns, A. R., Stagaman, K., Wong, S., Rawls, J. F., Guillemin, K., & Bohannan, B. J. (2016). The composition of the zebrafish intestinal microbial community varies across development. *ISME Journal, 10*(3), 644–654. https://doi.org/10.1038/ismej.2015.140

Stoletov, K., Fang, L., Choi, S. H., Hartvigsen, K., Hansen, L. F., Hall, C., Pattison, J., Juliano, J., Miller, E. R., Almazan, F., Crosier, P., Witztum, J. L., Klemke, R. L., & Miller, Y. I. (2009). Vascular lipid accumulation, lipoprotein oxidation, and macrophage lipid uptake in hypercholesterolemic zebrafish. *Circulation Research, 104*(8), 952–960. https://doi.org/10.1161/CIRCRESAHA.108.189803

Strähle, U., Scholz, S., Geisler, R., Greiner, P., Hollert, H., Rastegar, S., Schumacher, A., Selderslaghs, I., Weiss, C., Witters, H., & Braunbeck, T. (2012). Zebrafish embryos as an alternative to animal experiments – A commentary on the definition of the onset of protected life stages in animal welfare regulations. *Reproductive Toxicology, 33*(2), 128–132. https://doi.org/10.1016/j.reprotox.2011.06.121

Tye, M., Rider, D., Duffy, E. A., Seubert, A., Lothert, B., & Schimmenti, L. A. (2015). Nonhatching decapsulated artemia cysts as a replacement

to artemia nauplii in juvenile and adult zebrafish culture. *Zebrafish, 12*(6), 457–461. https://doi.org/10.1089/zeb.2014.1031
Ulloa, P. E., Iturra, P., Neira, R., & Araneda, C. (2011). Zebrafish as a model organism for nutrition and growth: Towards comparative studies of nutritional genomics applied to aquacultured fishes. *Reviews in Fish Biology and Fisheries, 21*(4), 649–666. https://doi.org/10.1007/s11160-011-9203-0
Utomo, N. B. P., Junior, M. Z., Yusuf, T. L., Mokoginta, I., & Bintang, M. (2005). The effect of different levels of vitamin E on the reproductive performance of zebrafish (Danio rerio). *Jurnal Akuakultur Indonesia, 4*(2), 125–129. https://doi.org/10.19027/jai.4.125-129
Van der Heyden, C., Wautier, K., & Huysseune, A. (2001). Tooth succession in the zebrafish (Danio rerio). *Archives of Oral Biology, 46*(11), 1051–1058. https://doi.org/10.1016/s0003-9969(01)00064-4
Vanderberghe, J., Thompson, F. L., Gomez-Gil, B., & Swings, J. (2003). Phenotypic diversity amongst Vibrio isolates from marine aquatic systems. *Aquaculture, 219*(1–4), 9–20.
Verdonck, L., Swings, J., Kersters, K., Dehasque, M., Sorgeloos, P., & Léger, P. (1994). Variability of the microbial environment of rotifer Brahionis plicatilis and artemia production systems. *Journal of the World Aquaculture Society, 25*(1), 55–59. https://doi.org/10.1111/j.1749-7345.1994.tb00804.x
Wallace, K. N., Akhter, S., Smith, E. M., Lorent, K., & Pack, M. (2005). Intestinal growth and differentiation in zebrafish. *Mechanisms of Development, 122*(2), 157–173. https://doi.org/10.1016/j.mod.2004.10.009
Waterland, R. A. (2014). Epigenetic mechanisms affecting regulation of energy balance: Many questions, few answers. *Annual Review of Nutrition, 34*(1), 337–355. https://doi.org/10.1146/annurev-nutr-071813-105315
Watts, S. A., & D'Abramo, L. R. (2021). Standardized reference diets for zebrafish: Addressing nutritional control in experimental methodology. *Annual Review of Nutrition, 41*(41), 511–527. https://doi.org/10.1146/annurev-nutr-120420-034809
Watts, S. A., Lawrence, C., Powell, M., & D'Abramo, L. R. (2016). The vital relationship between nutrition and health in zebrafish. *Zebrafish, 13*, Suppl. 1, S72–S76. https://doi.org/10.1089/zeb.2016.1299

Watts, S. A., Powell, M., & D'Abramo, L. R. (2012). Fundamental approaches to the study of zebrafish nutrition. *ILAR Journal, 53*(2), 144–160. https://doi.org/10.1093/ilar.53.2.144

Westerfield, M. (2000). *The zebrafish book. A guide for the laboratory use of zebrafish (Danio rerio)* (4th ed). University of Oregon Press.

Wilson, C. (2012). Aspects of larval rearing. *ILAR Journal, 53*(2), 89–94.

Wong, S., Stephens, W. Z., Burns, A. R., Stagaman, K., David, L. A., Bohannan, B. J. M., Guillemin, K., & Rawls, J. F. (2015). Ontogenetic differences in dietary fat influence microbiota assembly in the zebrafish gut. *mBio,* A. Maldonado, M. G. Dominguez Bello et al. (Eds.), *6*(5), e00687–e00615. https://doi.org/10.1128/mBio.00687-15

Xu, J., & Sinclair, K. D. (2015). One-carbon metabolism and epigenetic regulation of embryo development. *Reproduction, Fertility, and Development, 27*(4), 667–676. https://doi.org/10.1071/RD14377

Zhao, L., Xia, Z., & Wang, F. (2014). Zebrafish in the sea of mineral (iron, zinc, and copper) metabolism. *Frontiers in Pharmacology, 5*, 33. https://doi.org/10.3389/fphar.2014.00033

Zheng, D., Kille, P., Feeney, G. P., Cunningham, P., Handy, R. D., & Hogstrand, C. (2010). Dynamic transcriptomic profiles of zebrafish gills in response to zinc supplementation. *BMC Genomics, 11*(1), 553. https://doi.org/10.1186/1471-2164-11-553

Zoumpourtikoudi, V., Pyrgelis, N., Chatzigrigoriou, M., Tasakis, R. N., & Touraki, M. (2018). Interactions among yeast and probiotic bacteria enhance probiotic properties and metabolism offering augmented protection to artemia franciscana against Vibrio anguillarum. *Microbial Pathogenesis,* 125, 497–506. https://doi.org/10.1016/j.micpath.2018.10.022

Chapter 4

Non-infectious diseases of zebrafish

Katrina N. Murray and Michael L. Kent

Abstract

Some non-infectious diseases of zebrafish result from poor water quality parameters while others are idiopathic. As for other disease states, non-infectious diseases should be limited to promote optimal animal welfare and health and robust, replicable research. The basis for preventing, detecting and correcting non-infectious disease is an understanding of predisposing environmental and genetic factors, clinical signs of disease and histological changes.

Keywords

zebrafish, non-infectious, disease, water quality, idiopathic, histological

Introduction

Non-infectious diseases of laboratory zebrafish can result in a wide range of presentations, from acute morbidities and mortalities to apparently subclinical disease with no discernible changes in appearance and behaviour. However, even apparently normal fish can exhibit micro-anatomic organ changes that result in clinical disease

Claire Allen and Jean-Philippe Mocho (eds) **Zebrafish**
DOI: 10.1079/9781800629431.0004,

during elevated stress, or that could have long-term effects on fecundity and growth. Importantly, these changes in otherwise normal-appearing fish used in research may cause unintended variation in experimental endpoints and thus may compromise the validity of data obtained using these animals (Kent et al., 2012).

A number of non-infectious diseases have been attributed to sub-optimal water quality while others are idiopathic, with unknown aetiology. For diseases of water quality, diagnosis can be made by associating clinical signs and histological changes with specific water parameters. Water quality parameters can fluctuate with seasonal changes at the water source, time of day, demand on the water system, procedural set-ups that occur off the main water system in static water, and frequency of use of particular areas on a system. Also, it is very important to recognize that, whereas zebrafish are tolerant of a wide variety of water quality parameters, rapid change in a parameter (e.g. conductivity, temperature, pH) can cause severe disease in what would otherwise be an acceptable range (see Chapter 7 for more information). In these situations, a diagnosis may be made on clinical and histological signs and by ruling out other factors. Knowledge of water quality conditions and husbandry history is important, because many water quality-associated diseases do not present with specific clinical or histological changes. Other idiopathic diseases described in this chapter are diagnosed by evaluation of organ anatomy in histological sections.

Even when infectious agents are identified, non-infectious disease should still be considered. Non-infectious background diseases may increase susceptibility to infectious agents, including opportunistic pathogens, and optimization of parameters that impact the prevalence of non-infectious diseases can be an important component of controlling infections in a facility (see Chapter 5 for more information). We have observed a high prevalence of certain idiopathic lesions, in particular wild-type lines. An appreciation for these types of changes is important when choosing particular zebrafish lines for studies. Furthermore, health-monitoring strategies should be designed to detect non-infectious diseases because resulting tissue changes may have repercussions for interpretation of research endpoints in control and experimental animals, reproducibility of results and the ability to compare datasets between zebrafish lines and facilities (see Chapters 7, 9 and 14 for more information; Kent et al., 2012; Mocho et al., 2022).

In this chapter we describe the common non-infectious diseases of laboratory zebrafish. Pathogenesis, clinical signs, diagnosis and treatment are presented. Water quality-related diseases include ammonia, nitrite, chlorine and chloramine toxicity, heavy metal toxicosis, gas supersaturation and nephrocalcinosis. For idiopathic diseases we describe egg-associated inflammation, hepatic megalocytosis, spinal deformities, operculum malformations, bile duct proliferation and common neoplasms.

Ammonia toxicity

Ammonia (NH_3) occurs in a fish environment as a result of fish metabolism and from the breakdown of organic material, including uneaten food. Total ammonia nitrogen (TAN) includes both ionized (NH_4+) and unionized (NH_3) forms. The unionized form is most toxic to fish and is favoured at higher pH and water temperature. Ammonia negatively affects cellular metabolism and structures (Rosenberg, 2012) and increases tissue oxygen consumption (Schwedler et al., 1985). Affected fish may hyperventilate at the water surface, be recumbent, anorexic or display erratic swimming (Daoust & Ferguson, 1984). Ammonia toxicity can also result in immunosuppression, decreased growth and mortality (Colt, 1979; Walters & Plumb, 1980). Gill epithelial hypertrophy and hyperplasia may occur (Figure 4.1), although this is a nonspecific effect of poor water quality and not specific to ammonia exposure (Braun et al., 2009).

Diagnosis of ammonia toxicity is made by measuring unionized ammonia in fish water. NH_3 can be calculated from TAN using temperature and pH tables (see Chapter 7 for more information). Elevated ammonia is typically associated with other water quality changes (e.g. increased bacterial counts). Although it is unclear whether a certain amount of ammonia would be acceptable as an isolated water quality change, the current recommendation is that

Table 4.1 Non-infectious diseases of laboratory zebrafish.

Disease	Clinical signs	Diagnosis	Treatment/management
Ammonia toxicity	Hyperventilation, recumbency, anorexia, erratic swimming, immunosuppression, decreased growth, mortality	Measurement of NH_3 in the water	Increase water exchange, ammonia adsorbent or neutralizer, improve biological filtration
Nitrite toxicity	Hyperventilation, lethargy, brown blood and gills, decreased growth, mortality	NO_2^- >1 mg/l in water and clinical signs	Increase water exchange, add chloride, improve biological filtration
Chlorine and chloramine toxicity	Respiratory distress and mortality	Detection of any chlorine in the water	Carbon filtration, chlorine neutralizer, ±ammonia binder
Supersaturation	Bubbles in tissues, exophthalmia, disequilibrium and hyperventilation on bottom of tank	Clinical signs and history, saturometer reading >100%	Identify source, aeration, allow water to equilibrate with atmospheric pressure before fish exposure, vacuum and passive degassers
Nephrocalcinosis	Usually subclinical	Basophilic debris in renal tubules and ducts in histological sections	Decrease stocking density, increase aeration and water exchange
Metal toxicity	Gill pathology, respiratory distress, decreased embryo hatching and survival	Measurement of toxic levels of metal in water, feed or tissue	Reverse osmosis, carbon and ion exchange filters, metal chelation, increase calcium and water hardness
Egg-associated inflammation	Subclinical to distended abdomen, external ulcers and decreased fecundity are possible	Inflammation associated with degenerating egg material in histological sections	Regular spawning may help

Hepatic megalocytosis	Subclinical	Enlarged hepatocyte nucleus and cytoplasm in histological sections	Carbon filter if toxin suspected
Spinal deformities	Kyphosis, lordosis, scoliosis and platyspondyly	Gross and histological spinal deformity	Investigate water parameters, nutrition, trauma, toxins, genetics, infectious agents
Operculum malformation	Unilateral or bilateral, usually eversion of opercular flap, exposed gills	Gross and histological malformation of operculum	Investigate genetic and environmental parameters
Bile duct hyperplasia	Subclinical	Multiple tortuous ducts, often with fibrous connective tissue in histological sections	No treatment; prevalence may be higher in some fish lines
Neoplasia	Usually non-specific (weight loss, lethargy, swelling, exophthalmia, colour change)	Histopathology, impression smears of focal masses	Consider toxin and infectious causes if multiple cases (especially intestinal carcinomas and lymphosarcomas)

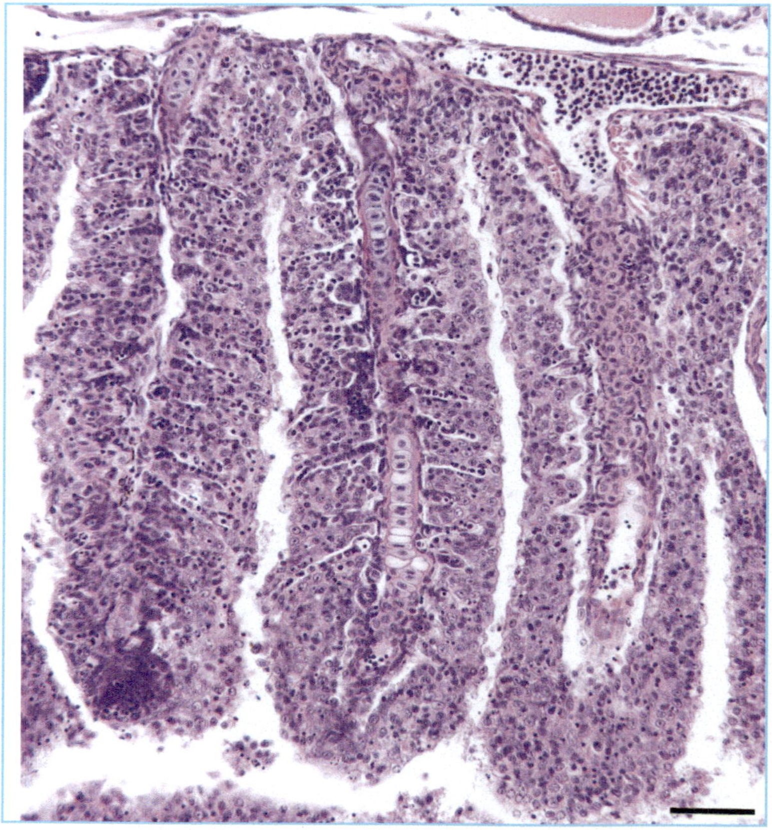

Figure 4.1 Gill epithelial hyperplasia and hypertrophy resulting in fusion of secondary lamellae. Scale bar, 50 μm. Sagittal section stained with haematoxylin and eosin.

measured ammonia should be zero. Immediate methods to reduce ammonia include increasing water changes and using an ammonia adsorbent or neutralizing agent. Lowering pH and raising salinity will reduce the amount of unionized ammonia, favouring the ionized form (Wedemeyer, 1996). However, changing these parameters can also be detrimental and should be done carefully. Improving biological filtration to speed the conversion of ammonia to nitrite and nitrate should be considered. Additional factors that may be evaluated to minimize ammonia accumulation include stocking densities, food quantity and preparation and areas of no or low water flow.

Nitrite toxicity

Nitrite (NO_2–) toxicity can occur alone or with ammonia toxicity, as ammonia is converted to nitrite in the nitrogen cycle (NH_3 + O_2 —> NO_2–). Nitrites convert haemoglobin to methaemoglobin by oxidizing ferrous iron ($Fe+^2$) to the ferric form ($Fe+^3$). Methaemoglobin does not bind oxygen and gives the blood and gills a brown colour. Clinical signs of methaemoglobinemia, or brown blood disease, include lethargy, being by the water inlet, hyperventilation (despite water being well oxygenated), decreased growth and mortality (Kroupova et al., 2005; Lewis & Morris, 1986). Microanatomically, hypertrophy of gill chloride cells may be observed (Gaino et al., 1984). Gill epithelial hypertrophy and hyperplasia may also occur as a non-specific response to poor water quality (Wedemeyer & Yasutake, 1978).

Nitrite toxicosis is diagnosed by clinical signs and measuring nitrites >0.1 mg/l (Wedemeyer, 1996). Water changes should be increased to reduce circulating nitrites. Chloride may be added to decrease the amount of nitrite absorbed by the fish. A dose of 3 mg of chloride for every 1 mg of nitrite was used to treat methaemoglobinemia in channel catfish (Bowser et al., 1983). As with ammonia toxicity, long-term control and prevention may require optimization of biological filtration, stocking densities and feeding protocols (see Chapter 7 for more information).

Chlorine and chloramine toxicity

Zebrafish may be exposed to chlorine from residues of bleach used to surface disinfect embryos and sanitize equipment, or from the municipal water source if carbon filters are not working properly (see Chapter 7 for more information). In water, chlorine will occur as hypochlorous acid (HOCl) and hypochlorite ion (OCl–), both of which are strong oxidizers. Chloramines, which are more toxic than chlorine, are formed when chlorine reacts with nitrogenous compounds in the water (Wedemeyer, 1996). Chlorine and chloramines result in severe, acute gill necrosis with subsequent respiratory distress and mortality.

Measuring any amount of chlorine in the water is diagnostic for chlorine toxicity. Treatment should involve identifying the source of chlorine exposure. Carbon filters should be active and properly installed. If bleach is used to surface-sanitize embryos or equipment, sulphur compounds can be used to neutralize residual chlorine. Sodium thiosulfate is effective when used in a ratio of 7.4 ppm to 1 ppm chlorine (Jensen, 1989). The neutralization reaction will release ammonia from chloramines and therefore concomitant ammonia treatment may be needed. ChlorAmX® binds chlorine and chloramines along with ammonia.

Supersaturation or gas bubble disease

Water is supersaturated when the total pressure of dissolved gases is greater than atmospheric pressure. In a zebrafish facility this can happen due to various causes, including (1) when air is drawn into a pressurized system, as with a leak on the suction side of a pump, (2) when cold water is pumped in and heated under pressure and (3) importantly for zebrafish recirculating systems, when air is trapped in pressurized plumbing supplying incoming water to the aquarium. When fish are exposed to supersaturated water, tissues equilibrate with the dissolved gases in water and bubbles emerge in the tissues as they normalize to atmospheric pressure (Figure 4.2). The formation of gas bubble emboli in fish tissue can cause vascular occlusion, tissue necrosis and death (Smith, 1988). Increased morbidity and mortality are associated with higher levels of supersaturation and longer exposure times. However, the fish response to supersaturated water is not uniform: it can differ for fish of different sizes and ages. In a recent experiment we observed marked differences in morbidity between zebrafish strains held on a single rack with replicate tanks for each strain. ABs, followed by 5D fish, showed high mortality amongst the replicate tanks following a supersaturation event, whereas Casper and Tübingen (TU) fish on the same system showed no mortality (see the Zebrafish

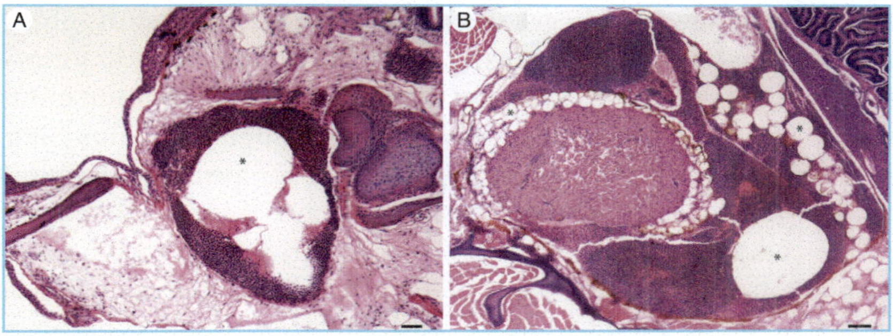

Figure 4.2 Gas bubble disease. Bubbles (*) in a large vessel in the head (A) and around the heart (B). Scale bars, 50 μm (A) and 100 μm (B). Sagittal sections stained with haematoxylin and eosin.

Information Network (ZFIN) https://zfin.org/ for details of wild-type lines). Moreover, even within and between populations of the same age and size, some fish will become moribund while others appear normal (Bouck, 1980; Weitkamp & Katz, 1980). Affected zebrafish generally hyperventilate at the bottom of the tank, unlike with other diseases of water quality that typically result in hyperventilation at the surface. Bubbles may be observed in fish tissues, especially in the fins and eyes where they can cause unilateral and bilateral exophthalmia (Lund & Heggberget, 1985; Machado et al., 1987). Finding bubbles in fish tissues is pathognomonic for gas bubble disease, but their absence does not rule out the diagnosis. Mortalities may occur several days after a supersaturation incident with no gross lesions. Tissue damage and stress may precipitate secondary infections (Speare, 1991) and growth rate may also be negatively affected (Batzios et al., 1998).

A saturometer can be used to determine whether total gas pressure is greater than 100% and water is supersaturated. However, the level of supersaturation in a recirculating water system can vary according to demands on the water system and may be isolated to a particular area of the system. Hence, measurements should be made at multiple locations and times throughout the day. Bubbles in the tissues are diagnostic but an inconsistent finding. Bubbles adherent to surfaces within the tank may suggest supersaturation. Therefore, diagnosis usually involves a careful evaluation of clinical signs and history. Vacuum and passive degassers will remove excess dissolved gas, as will aeration and allowing water to equilibrate with atmospheric pressure before sending it to fish (see Chapter 7 for more information). There is a risk that water trapped in unused plumbing could become supersaturated by pockets of trapped air. Draining these pipes before sending the water to fish may be advisable if time cannot be allowed for aeration or equilibration. We recommend turning off the incoming water and mixing the tank water with air stones to accelerate degassing if supersaturation is suspected in a given tank.

Nephrocalcinosis

Nephrocalcinosis refers to the accumulation of calcium deposits in renal tubules and ducts. In fish this is usually associated with high dissolved carbon dioxide (CO_2; see Chapter 7 for more information). Low oxygen and magnesium and high selenium and arsenic have also been associated with nephrocalcinosis, although these would likely have effects on tissues besides the kidney (Cockell et al., 1991; Cowey et al., 1977; Hilton et al., 1980). In zebrafish, nephrocalcinosis is generally subclinical and diagnosed on post-mortem microscopic examination. Often this is the only lesion present and, with no clear aetiology, it may be idiopathic. On the other hand, mineral deposits could accumulate secondary to systemic disease and impaired renal function.

Diagnosis of nephrocalcinosis is made by observation of basophilic deposits in renal tubules and ducts in histological sections stained with haematoxylin and eosin (Figure 4.3). Surrounding structures may exhibit mild to severe dilation. Chronic inflammation may be present as well. Granulomas occasionally form around the deposits, in which case acid-fast staining is recommended to look for evidence of mycobacteriosis (see Chapter 5 for more information). Treatment should focus on optimizing water quality to decrease CO_2: e.g. decreasing fish stocking density and increasing aeration and water exchange.

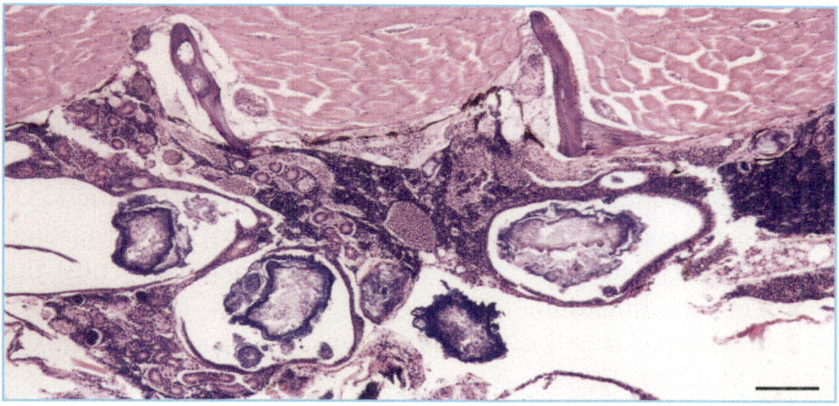

Figure 4.3 Nephrocalcinosis. Multiple renal tubules are dilated with basophilic, calcareous debris. Scale bar, 150 µm. Sagittal section stained with haematoxylin and eosin.

Metal toxicities

Copper, lead, zinc, mercury, cadmium, nickel, iron, cobalt and chromium have all been associated with disease in fishes (Dave & Xiu, 1991; Sorensen, 1991). Zebrafish may be exposed to metals via contaminated feed or water, usually by leaching from galvanized plumbing. Low alkalinity, high temperature and low pH all increase heavy metal solubility and hence toxicity. Most acute metal toxicities result from loss of total body sodium (Na+; Alsop & Wood, 2011), although iron toxicity is a result of precipitation of iron oxide on the gill surface (Abbas et al., 2002; Peuranen et al., 1994). Metal toxicities are typically associated with gill changes, including increased mucus production, epithelial hyperplasia, swelling, necrosis, separation from pillar cells, desquamation and increased chloride cells (Authman et al., 2015; Baker, 1969; Farrell & Iwama, 2010). Clinically these gill changes result in respiratory distress. Histological changes may be evident in muscle, liver and kidney. General depression of fecundity, growth and the immune system may occur (Authman et al., 2015). In zebrafish, decreased embryo hatching and survival were associated with metal exposures (Dave & Xiu, 1991). Feeding of chromium-contaminated decapsulated

Artemia spp. cysts resulted in orange-coloured zebrafish embryos with reduced survival (Tye et al., 2018).

Measurement of toxic levels of metal in fish water, feed or tissue is diagnostic for heavy metal toxicity. Other water parameters that should be considered when evaluating metal measurements, including hardness, alkalinity, pH, temperature and organic load, will affect metal solubility and toxicity. Treatment should include removing circulating metals and identifying and correcting the source of contamination.

Reverse osmosis and carbon filters will remove metals from the water. Metal chelators and ion-exchange filters are also effective. However, calcium and magnesium may need to be added back to the water because these essential metals will be removed along with toxic ones (see Chapter 7 for more information). Calcium has been shown to be protective by competing with other metals for binding sites in rainbow trout (Chowdhury et al., 2016).

Egg-associated inflammation

Inflammation in the ovary associated with degenerating egg material is a common finding in laboratory zebrafish. This lesion is believed to result from egg retention and degeneration and a subsequent inflammatory response. Egg development within females is continuous and zebrafish spawn very frequently. Some husbandry conditions and practices in the research setting may not be conducive to natural spawning, leading to egg retention. Females with egg retention will exhibit an excessively enlarged abdomen. Lesions range from small foci to locally extensive inflammation extending into adjacent tissues. A raft of eosinophilic debris, representing waxy material from atretic eggs, is often observed in the posterior ovary (Figure 4.4A). Often fish are subclinical and the lesion is appreciated only during histological examination of tissue sections. However, severe lesions have been associated with adhesions to the body wall and external ulcers (Kent et al., 2016). Large lesions likely impair spawning, although this has not been directly measured.

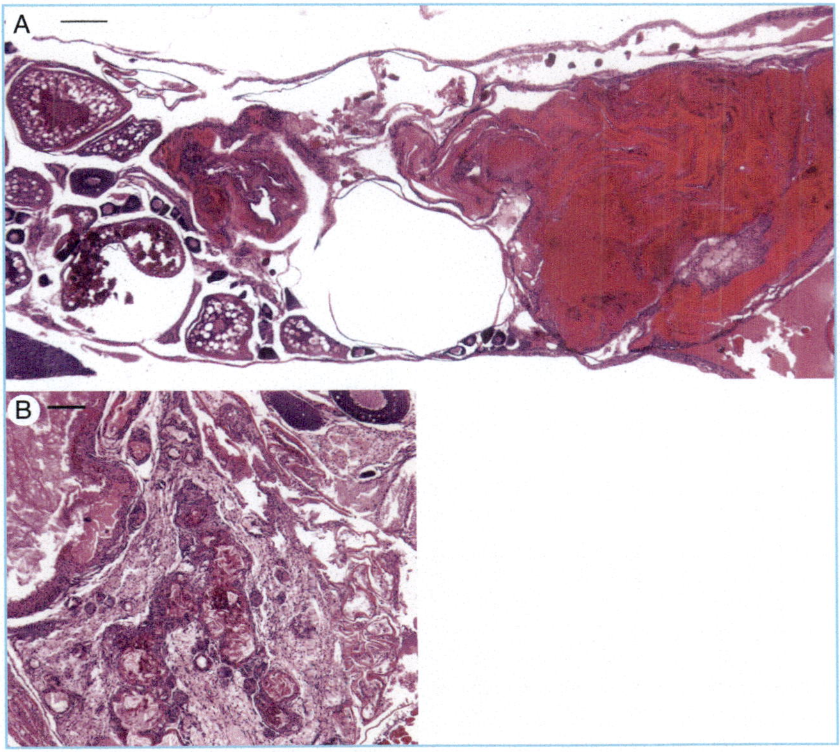

Figure 4.4 Egg-associated inflammation. (A) The ovary contains normal eggs and a large raft of degenerating egg material and inflammatory cells. Scale bar, 150 µm. (B) An area of inflammation in the ovary with granulomas. Scale bar, 100 µm. Sagittal sections stained with haematoxylin and eosin.

Presumptive diagnosis is made by observing an enlarged abdomen in females. Diagnosis is made by histopathology and observation of inflammation and degenerating egg material in the ovary and, in extreme cases, fibroplasia. Inflammation ranges from mild to severe and granulomas may be present (Figure 4.4B). Although lesions are usually sterile, acid-fast bacilli (i.e. *Mycobacterium* spp.) are occasionally observed within granulomas, which may represent secondary infection (see Chapter 5 for more information). Regular spawning may help prevent egg retention and subsequent disease. In a zebrafish toxicity study, degenerative, inflammatory lesions in the ovary were noted in 78% of control fish (n = 46; Rossteuscher et al.,

2008). The study authors concluded that the prevalence of these spontaneous lesions should be considered when determining sample size and in evaluations of the ovary as a study endpoint.

Hepatic megalocytosis

Hepatic megalocytosis describes the prominent enlargement of hepatocytes, involving both nuclear and cytoplasmic components (Figure 4.5). The distribution of megalocytes is generally diffuse but may be focally clustered in an area of the liver. Hepatic megalocytes may be multinucleated. In the English sole (*Parophrys vetulus*) this lesion has been described as an indicator of exposure to carcinogens (Myers et al., 2003). However, the actual affected hepatocytes are

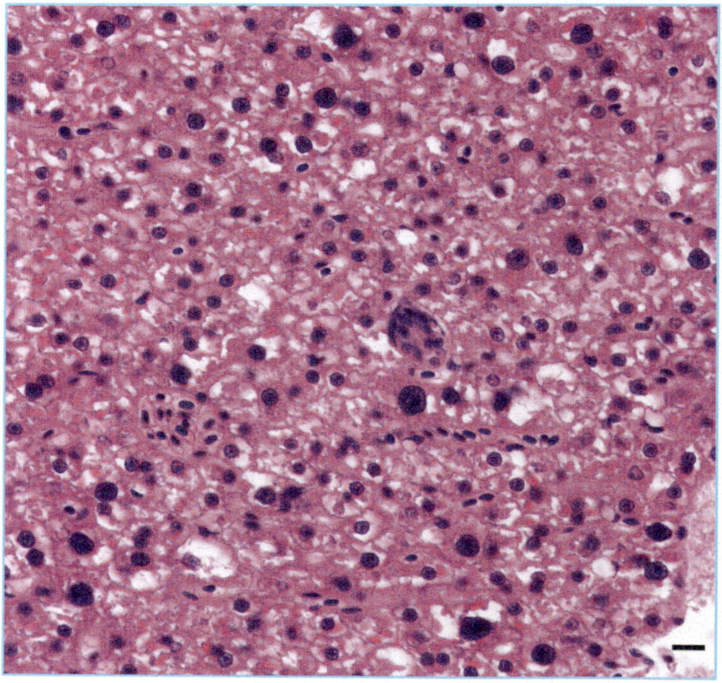

Figure 4.5 Hepatic megalocytosis. Multiple hepatocytes with enlarged nuclear and cytoplasmic components. Haematoxylin and eosin staining. Scale bar, 10 μm.

themselves not a pre-neoplastic lesion because the polyploid nature of the cells is caused by arrested cell division. Therefore, as seen in salmonid fishes with netpen liver disease (Kent, 1990), hepatic megalocytosis in zebrafish has not been linked to liver cancer. In addition to anthropogenic toxicants, hepatic megalocytosis has been described in other fishes associated with naturally occurring xenobiotics including pyrrolizidine alkaloids (Hendricks et al., 1981), microcystin (Andersen et al., 1993) and an unidentified food- or water-borne toxin (Kent, 1990). In these cases, hepatic megalocytosis was only part of a group of hepatic pathologies and fish were noticeably moribund. Observation of hepatic megalocytosis in adult fish with otherwise normal livers may indicate that either the liver previously had other, acute changes because these 'megalocytic hepatocytes' persist for many months, or that the only change was hepatic megalocytosis.

Hepatic megalocytosis in zebrafish often occurs in an otherwise normal liver, and it is diagnosed in fish that appear physically and behaviourally normal. A relationship between hepatic megalocytosis in zebrafish and recirculating water systems with fluidized sand filters has been suggested, although the lesion does not occur in all such systems, suggesting that additional factors are involved (Spitsbergen et al., 2012; Spitsbergen & Kent, 2003). In a new facility, hepatic megalocytosis was diagnosed in moribund zebrafish where it was believed to be the result of toxicants leaching from new plastics, and was ameliorated by the installation of a carbon filter on the recirculating water system (Kent et al., 2011). At the Zebrafish International Resource Centre (ZIRC) we frequently diagnose hepatic megalocytosis in AB fish but rarely see it in other wild-type lines on the same system, suggesting a possible genetic component to the aetiology (Murray et al., 2020).

Diagnosis is made by histological evaluation of the liver. For spontaneous occurrence, no treatment has been described. The addition of carbon filters may be preventative if leaching and recirculating toxins are suspected, particularly in systems with new plumbing (see Chapter 7 for more information). The prevalence of this lesion may be high in particular lines, and the effects on liver metabolism and physiology are unknown. Therefore, it should be considered when choosing a line for toxicity studies and when evaluating response to treatment.

Spinal deformities

Spinal curvature and shortening (platyspondyly) occur in zebrafish (Figure 4.6); dorsal curvature (kyphosis), ventral curvature (lordosis) and lateral curvature (scoliosis) are not uncommon findings. Spinal changes may be noted in individual fish or affecting tanks or stocks and the distribution likely reflects the underlying aetiology, of which there are many. Physical trauma, temperature, toxins and infectious agents can precipitate spinal deformities (Kent et al., 2016; Pohl, 1990; Sfakianakis et al., 2006; Stickland et al., 1988; see Chapter 5 for more information). Spinal deformities are often observed in a small percentage of fish within a population infected with *Pseudoloma neurophilia*, although spinal deformities also occur in fish with no exposure to the parasite. Dietary quality should be considered if large groups of fish are affected (see Chapter 3 for more information). Many vitamins and minerals have roles in skeletal development, in particular vitamins A, C, D, E and K along with calcium, phosphorus, manganese, selenium and zinc (Cahu et al., 2003; Lall & Lewis-McCrea, 2007). Spinal deformities may also result from degenerative changes in mature zebrafish (Hayes et al., 2013).

Diagnosis of spinal deformities is made by observation of spinal curvature, vertebral compression or alteration in the gross specimen and histological sections. Given the three-dimensional characteristics of the lesions, methods that involve visualization of the skeleton in whole fish, such as X-rays or tissue clearing, are more sensitive for detecting vertebral changes (Kent et al., 2004). Environmental, dietary, genetic and infectious parameters, along with prevalence, age of onset and distribution, should be evaluated when investigating spinal deformities.

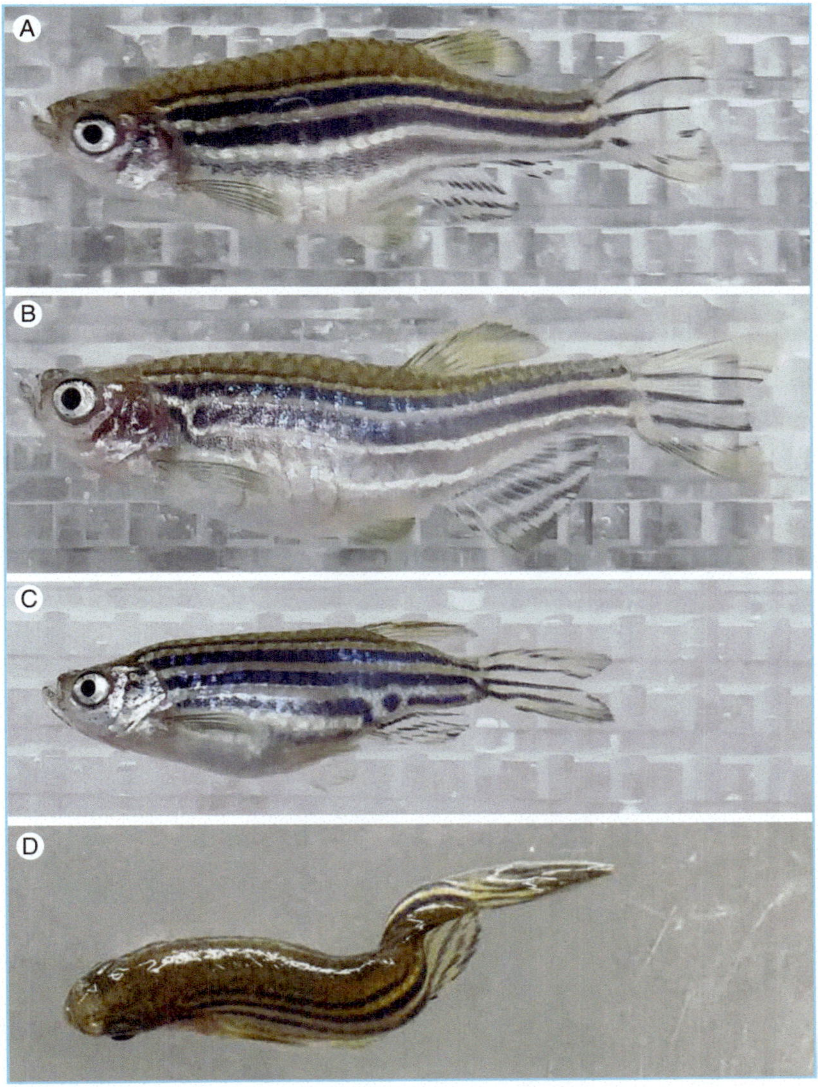

Figure 4.6 Spinal deformities. (A) Kyphosis, (B) lordosis, (C) platyspondyly, and (D) scoliosis.

Opercular malformations

Malformation of the opercular flap, especially eversion, can be common in some zebrafish lines. Usually both opercula are affected but the eversion can be unilateral. The malformation results in incomplete coverage of the gill cavity and exposure of the gills, which may impair growth rate and ability to cope with respiratory distress (Koumoundouros et al., 1997). Gilthead sea bream (*Sparus aurata* L.) recovered from unilateral inward curling of the operculum (Beraldo & Canavese, 2011), but regression and recovery of everted opercula has not been documented in zebrafish.

Diagnosis is made by gross observation of eversion of one or both of the opercula and exposure of the gills (Figure 4.7). Mild

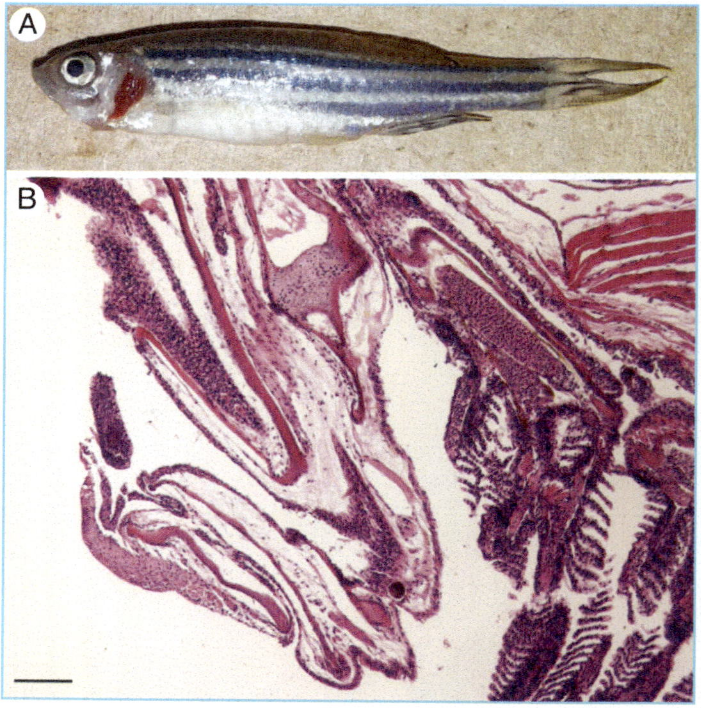

Figure 4.7 Opercular malformation. (A) Zebrafish with an everted operculum and exposed gills. (B) Sagittal section, stained with haematoxylin and eosin, of everted operculum. Scale bar, 100 μm.

malformations may be observed only when fish are anaesthetized or in histological sections. Change in the operculum is generally not associated with degeneration, inflammation or other skeletal changes in zebrafish. Malformations may be highly prevalent in particular lines, suggesting a possible heritable component. At ZIRC we performed in- and out-crosses to zebrafish with normal opercula and did not observe Mendelian inheritance, which may indicate either that multiple genes are involved or partial penetrance. Environmental factors likely also have a role and have been associated with opercular malformation in other fish (Hilomen-Garcia, 1997; Lindesjöö et al., 1994). Nutrition should be considered if other skeletal deformities are observed (see Chapter 3 for more information).

Bile duct proliferation

Bile duct hyperplasia is not uncommon in laboratory zebrafish and is typically diagnosed on post-mortem histological examination of clinically normal fish. Proliferation occurs throughout the biliary tree, and multiple foci of duct hyperplasia in tissue sections are likely connected by the three-dimensional duct tract rather than isolated foci (Spagnoli & Murray, 2020). Low cuboidal to squamous epithelial cells line ducts that are tortuous and diminished in size (Figure 4.8). Fibrous connective tissue often surrounds ductal structures. Granulomas formed around presumptive bile material may be present (Spagnoli et al., 2015).

A higher prevalence of biliary hyperplasia was noted in the wild-type line Tübingen Long Fin (TL; https://zfin.org/) compared with other wild-type lines (Spitsbergen et al., 2012). In another study, biliary hyperplasia and/or liver granulomas were noted in over half of 5D control and experimental fish (Spagnoli et al., 2015).

Bile duct hyperplasia is typically a post-mortem incidental finding and no treatment is recommended. Biliary hyperplasia is diagnosed by histologic examination and is distinguished from biliary neoplasia by the diffuse pattern and absence of cellular atypia

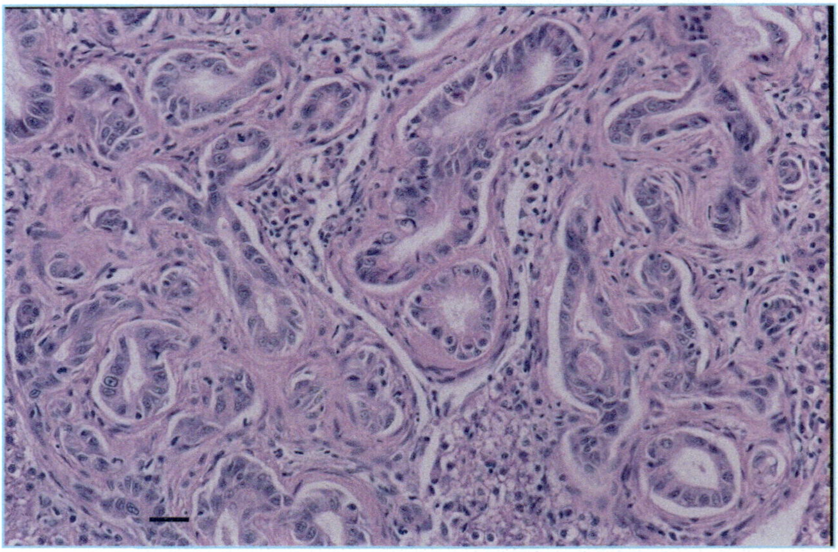

Figure 4.8 Bile duct hyperplasia. Scale bar, 25 μm.

(Spagnoli & Murray, 2020). Mitotic figures and areas of necrosis also indicate neoplasia rather than hyperplasia.

Neoplasia

Zebrafish are susceptible to a wide range of spontaneous neoplasms. In addition, studies of toxicology, genetic manipulation and xenotransplantation expand the range of tumours beyond the naturally occurring categories (Langenau, 2016). Neoplasms often result in nonspecific clinical signs, e.g. weight loss, oedema, exophthalmia, lethargy, externally visible tumours and abdominal swelling. Presumptive diagnosis can be made with the above, particularly the latter two changes. Confirmatory diagnosis is made by evaluation of histological sections. It should be noted that many of the most invasive and anaplastic neoplasms are difficult to describe precisely in small fish as it is often difficult to determine the organ of origin in histological sections of whole fish. A few of the more common tumours are briefly summarized here, but detailed descriptions of these and other tumours can be found elsewhere (Spagnoli & Murray, 2020; Spitsbergen et al., 2012).

As with higher vertebrates, neoplasms in zebrafish can be caused by chemical carcinogens (Spitsbergen et al., 2012). Also, numerous neoplastic diseases of fishes have been shown to be caused by transmissible agents (Anders & Yoshimizu, 1994) and these diseases are usually associated with viruses (Bowser & Casey, 1993; Coffee et al., 2013). To date, one common neoplasm of zebrafish, intestinal carcinoma, has been associated with a *Mycoplasma* spp. infection (Burns et al., 2018).

The most frequently diagnosed neoplasm in zebrafish is the seminoma. These masses can be quite large, filling most of the coelomic cavity. Histologically, most of these tumours consist of normal testicular cellular architecture and contain all stages of sperm maturation (Figure 4.9D). Therefore, it is questionable whether these masses represent hypertrophy and hyperplasia rather than neoplasia. The ultimobranchial gland, which is important for calcium regulation, is also known to undergo hyperplastic and neoplastic growth (Figure 4.9B). An enlarged but well-differentiated gland with normal cellular architecture is likely hyperplastic, whereas anaplastic cells and mitotic figures are more indicative of an adenoma and invasion of neighbouring tissues is typical of adenocarcinomas.

Peripheral nerve sheath tumours (spindle cell sarcomas) are the most common soft tissue sarcoma observed in zebrafish. These are dense, infiltrative masses of spindle-shaped cells usually in a storiform pattern (Figure 4.9C). They can be highly aggressive and destructive to affected tissues (Spitsbergen et al., 2012).

Lymphosarcomas are occasionally observed in zebrafish. Often there is diffuse epidermal invasion by lymphoid cells, which can result in skin pallor, and they may appear similar to scale pocket oedema (dropsy). Lymphoid cell accumulation in the retro-orbital space can result in exophthalmia. Lymphoid cells may also invade gills, head tissues, kidney and coelomic organs. Over the last 20 years we have seen two general types of lymphosarcoma; (1) massive, diffuse invasion of all organs and (2) epitheliotropic neoplasms that appear to originate in the thymus, invading the gills and skin (Figure 4.10B). In addition to just individual fish being affected,

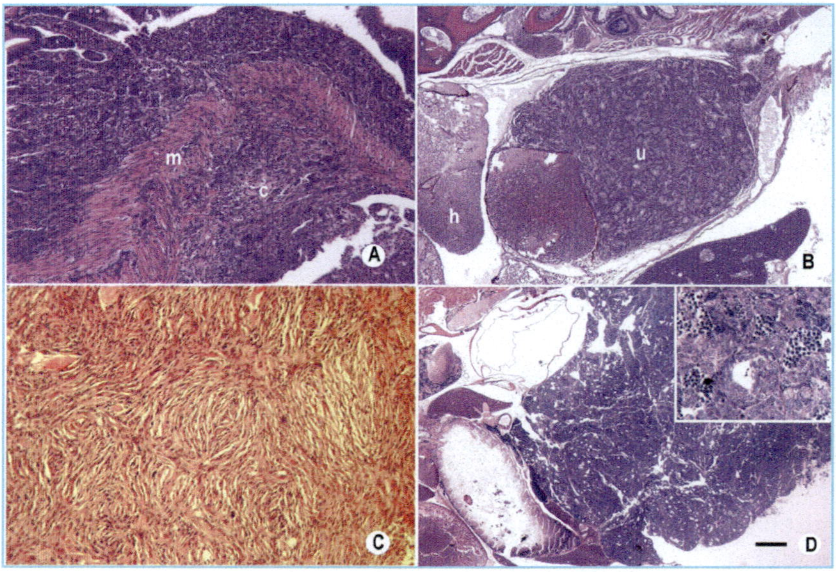

Figure 4.9 Some common neoplasms in zebrafish. Haematoxylin and eosin staining. (A) Small cell carcinoma of the intestine, with proliferation and invasion of neoplastic epithelial cells through the muscularis (m) of the intestine, and into the coelom (c). (B) Ultimobranchial tumour. The ultimobranchial gland (u), adjacent to the heart (h), is extremely enlarged and composed of simple to pseudostratified columnar epithelial cells surrounding small tubular lumina. (C) Peripheral nerve sheath (spindle cell) sarcoma. This tumour is composed of short, intersecting streams and bundles of spindle-shaped cells. (D) Well-differentiated seminoma. The testis is extremely enlarged, with a mass filling most of the coelomic cavity. Scale bar, 500 μm. Insert shows that the tumour is histologically consistent with normal testis at all stages of sperm development.

for the latter we have observed two outbreaks in which many fish at a facility were affected. Given that lymphoid neoplasms are often caused by oncogenic viruses (Coffee et al., 2013), an infectious aetiology should be considered, particularly for the epitheliotropic lymphosarcoma.

Interestingly, in zebrafish the most common location for chordoma is the intestinal tract (Cooper et al., 2015). Axial skeleton chordomas have also been reported, but at a lower frequency. Intestinal chordomas are generally subclinical, diagnosed on histopathological evaluation. These masses are often multi-lobulated and consist of vacuolated polygonal cells (Figure 4.10A).

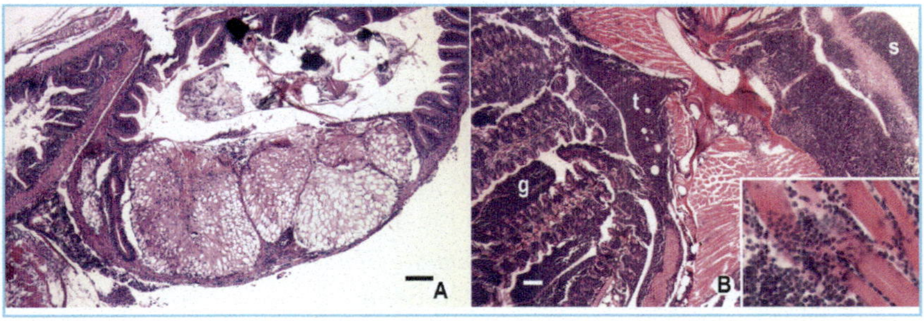

Figure 4.10 Two neoplasms occasionally observed in zebrafish. Haematoxylin and eosin staining. (A) Chordoma in the intestine. Note multiple lobules resembling notochord. Scale bar, 100 μm. (B) Epitheliotropic lymphosarcoma. Neoplastic cells primarily invade epithelial tissues, including gills (g) and skin (s); t, thymus. Scale bar, 100 μm. Insert shows neoplastic cells consistent with lymphocytes, having large nuclei and non-apparent cytoplasm.

The most frequently identified tumour in the zebrafish intestine is the intestinal small cell carcinoma. These tumours often involve multiple layers of the intestine, from epithelium, lamina propria, tunica muscularis to serosa (Figure 4.9A). Interpretation of neoplastic lesions can be complicated by inflammation (enteritis) and there may be a range of pathological changes, from hyperplasia and dysplasia to overt neoplasia, within an affected population (Paquette et al., 2013). While these tumours can be spontaneous, they have also been associated with infectious agents. The intestinal nematode *Pseudocapillaria tomentosa* has been associated with intestinal neoplasia and should be considered if chronic inflammation is consistently present (see Chapter 5 for more information; Kent et al., 2002; Schaaf et al., 2020). Some small cell carcinomas of the intestine are transmissible, and a single 16S ribosomal RNA sequence highly similar to *Mycoplasma penetrans* was enriched in fish with these tumours compared with controls (Burns et al., 2018). A recent study showed that laboratory exposure of *M. penetrans* caused intestinal changes, but the addition of *P. tomentosa* resulted in tumours (Kent et al., 2021). However, these tumours do occur in many laboratories in the absence of the worm (Schaaf et al., 2020), indicating that the worm is a strong promoter, but not a required initiator, of the neoplasms. Transmissible causes of intestinal cancer should be considered if multiple fish from a tank, stock or facility are diagnosed with intestinal neoplasia, whereas occasional diagnoses in single fish may represent spontaneous occurrence.

Conclusion

Non-infectious diseases are an important cause of acute to subclinical disease states in laboratory zebrafish. Hence, in addition to common and important pathogens, health surveillance and monitoring strategies should incorporate methodologies that capture non-infectious pathologic changes. The history of clinical and environmental changes, water chemistry measurements, diet, fish genetics and husbandry protocols complement diagnostic histopathology in investigations of non-infectious diseases.

Often diseases associated with suboptimal water quality, e.g. ammonia and nitrite toxicity and supersaturation, have obvious negative effects on fish health and welfare. Non-infectious tissue changes such as hepatic megalocytosis, bile duct hyperplasia, egg-associated inflammation and nephrocalcinosis may have undetermined effects on organ physiology, confound interpretation of research endpoints and are variables to consider when replicating experiments.

References

Abbas, H. H., Zaghloul, K. H., & Mousa, M. A. A. (2002). Effect of some heavy metal pollutants on some biochemical and histopathological changes in blue tilapia Oreochromis aureus. *Egyptian Journal of Agricultural Research*, *80*(3), 1395–1411. https://doi.org/10.21608/ejar.2002.312666

Alsop, D., & Wood, C. M. (2011). Metal uptake and acute toxicity in zebrafish: Common mechanisms across multiple metals. *Aquatic Toxicology*, *105*(3–4), 385–393. https://doi.org/10.1016/j.aquatox.2011.07.010

Anders, K., & Yoshimizu, M. (1994). Role of viruses in the induction of skin tumors and tumor-like proliferations of fish. *Diseases of Aquatic Organisms*, *19*(3), 215–232.

Andersen, R. J., Luu, H. A., Chen, D. Z., Holmes, C. F., Kent, M. L., Le Blanc, M., Taylor, F. J., & Williams, D. E. (1993). Chemical and biological evidence links microcystins to salmon "netpen liver disease". *Toxicon*, *31*(10), 1315–1323. https://doi.org/10.1016/0041-0101(93)90404-7

Authman, M. N. M., Zaki, M. S., Khallaf, E. A., & Abbas, H. H. (2015). Use of fish as bio-indicator of the effects of heavy metals pollution. *Journal of Aquaculture Research and Development*, *6*(4), 1–13.

Baker, J. T. P. (1969). Histological and electron microscopical observations on copper poisoning in winter flounder (Pseudopleuronectes americanus). *Journal of the Fisheries Research Board of Canada*, *26*(11), 2785–2793. https://doi.org/10.1139/f69-275

Batzios, C., Fotis, G., & Gavriilidou, I. (1998). Economic dimension of gas bubble disease effects on rainbow trout culture. *Aquaculture International*, *6*(6), 451–455. https://doi.org/10.1023/A:1009235518343

Beraldo, P., & Canavese, B. (2011). Recovery of opercular anomalies in gilthead sea bream, *Sparus aurata* L.: Morphological and morphometric analysis. *Journal of Fish Diseases*, *34*(1), 21–30. https://doi.org/10.1111/j.1365-2761.2010.01206.x

Bouck, G. R. (1980). Etiology of gas bubble disease. *Transactions of the American Fisheries Society*, *109*(6), 703–707. https://doi.org10.1577/1548-8659(1980)109<703:EOGBD>2.0.CO;2

Bowser, P. R., & Casey, J. W. (1993). Retroviruses of fish. *Annual Review of Fish Diseases*, *3*, 209–224. https://doi.org/10.1016/0959-8030(93)90035-A

Bowser, P. R., Falls, W. W., Vanzandt, J., Collier, N., & Phillips, J. D. (1983). Methemoglobinemia in channel catfish – Methods of prevention. *Progressive Fish-Culturist*, *45*(3), 154–158. https://doi.org/10.1577/1548-8659(1983)45[154:MICC]2.0.CO;2

Braun, M. H., Steele, S. L., & Perry, S. F. (2009). The responses of zebrafish (Danio rerio) to high external ammonia and urea transporter inhibition: Nitrogen excretion and expression of rhesus glycoproteins and urea transporter proteins. *Journal of Experimental Biology*, *212*(23), 3846–3856. https://doi.org/10.1242/jeb.034157

Burns, A. R., Watral, V., Sichel, S., Spagnoli, S., Banse, A. V., Mittge, E., Sharpton, T. J., Guillemin, K., & Kent, M. L. (2018). Transmission of a common intestinal neoplasm in zebrafish by cohabitation. Journal of Fish Diseases, 41(4), 569–579. https://doi.org/10.1111/jfd.12743.

Cahu, C., Zambonino Infante, J. Z., & Takeuchi, T. (2003). Nutritional components affecting skeletal development in fish larvae.

Aquaculture, 227(1–4), 245–258. https://doi.org/10.1016/S0044-8486(03)00507-6

Chowdhury, M. J., Girgis, M., & Wood, C. M. (2016). Revisiting the mechanisms of copper toxicity to rainbow trout: Time course, influence of calcium, unidirectional Na+ fluxes, and branchial Na+, K+ ATPase and V-type H+ ATPase activities. *Aquatic Toxicology, 177*, 51–62. https://doi.org/10.1016/j.aquatox.2016.05.009

Cockell, K. A., Hilton, J. W., & Bettger, W. J. (1991). Chronic toxicity of dietary disodium arsenate heptahydrate to juvenile rainbow-trout (Oncorhynchus mykiss). *Archives of Environmental Contamination and Toxicology, 21*(4), 518–527. https://doi.org/10.1007/BF01183873

Coffee, L. L., Casey, J. W., & Bowser, P. R. (2013). Pathology of tumors in fish associated with retroviruses: A review. *Veterinary Pathology, 50*(3), 390–403. https://doi.org/10.1177/0300985813480529

Colt, J. E., & Armstrong, D. A. (1979). *Nitrogen toxicity to fish, crustaceans, and molluscs.* Department of Civil Engineering, University of California, Davis.

Cooper, T. K., Murray, K. N., Spagnoli, S., & Spitsbergen, J. M. (2015). Primary intestinal and vertebral chordomas in laboratory zebrafish (Danio rerio). *Veterinary Pathology, 52*(2), 388–392. https://doi.org/10.1177/0300985814537531

Cowey, C. B., Knox, D., Adron, J. W., George, S., & Pirie, B. (1977). the production of renal calcinosis by magnesium-deficiency in rainbow trout (Salmo gairdneri). *British Journal of Nutrition, 38*(1), 127–135. https://doi.org/10.1079/bjn19770068

Daoust, P.-Y., & Ferguson, H. W. (1984). The pathology of chronic ammonia toxicity in rainbow trout, Salmo gairdneri Richardson. *Journal of Fish Diseases, 7*(3), 199–205. https://doi.org/10.1111/j.1365-2761.1984.tb00924.x

Dave, G., & Xiu, R. Q. (1991). Toxicity of mercury, copper, nickel, lead, and cobalt to embryos and larvae of zebrafish, Brachydanio rerio. *Archives of Environmental Contamination and Toxicology, 21*(1), 126–134. https://doi.org/10.1007/BF01055567

Farrell, A. P. A., & Iwama, G. K. (2010). Disorders of the cardiovascular and respiratory systems. In J. F. Leatherland & P. T. K. Woo (Eds.), *Fish diseases and disorders, volume 2: Non-infectious disorders* (2nd ed) (pp. 287–322). CABI Publishing.

Gaino, E., Arillo, A., & Mensi, P. (1984). Involvement of the gill chloride cells of trout under acute nitrite intoxication. *Comparative*

Biochemistry and Physiology Part A, *77*(4), 611–617. https://doi.org/10.1016/0300-9629(84)90171-3

Hayes, A. J., Reynolds, S., Nowell, M. A., Meakin, L. B., Habicher, J., Ledin, J., Bashford, A., Caterson, B., & Hammond, C. L. (2013). Spinal deformity in aged zebrafish is accompanied by degenerative changes to their vertebrae that resemble osteoarthritis. *PloS One*. Public Library of Science ONE, *8*(9), e75787. https://doi.org/10.1371/journal.pone.0075787

Hendricks, J. D., Sinnhuber, R. O., Henderson, M. C., & Buhler, D. R. (1981). Liver and kidney pathology in rainbow trout (Salmo gairdneri) exposed to dietary pyrrolizidine (Senecio) alkaloids. *Experimental and Molecular Pathology*, *35*(2), 170–183. https://doi.org/10.1016/0014-4800(81)90057-5

Hilomen-Garcia, G. V. (1997). Morphological abnormalities in hatchery-bred milkfish (Chanos chanos Forsskal) fry and juveniles. *Aquaculture*, *152*(1–4), 155–166. https://doi.org/10.1016/S0044-8486(96)01518-9

Hilton, J. W., Hodson, P. V., & Slinger, S. J. (1980). The requirement and toxicity of selenium in rainbow trout (Salmo gairdneri). *Journal of Nutrition*, *110*(12), 2527–2535. https://doi.org/10.1093/jn/110.12.2527

Jensen, G. (1989). *Handbook for common calculations in finfish aquaculture*. Louisiana State University Agricultural Center.

Kent, M. L. (1990). Netpen liver disease (NLD) of salmonid fishes reared in sea water: Species susceptibility, recovery, and probable cause. *Diseases of Aquatic Organisms*, *8*, 21–28. https://doi.org/10.3354/dao008021

Kent, M. L., Bishop-Stewart, J. K., Matthews, J. L., & Spitsbergen, J. M. (2002). Pseudocapillaria tomentosa, a nematode pathogen, and associated neoplasms of zebrafish (Danio rerio) kept in research colonies. *Comparative Medicine*, *52*(4), 354–358.

Kent, M. L., Buchner, C., Watral, V. G., Sanders, J. L., Ladu, J., Peterson, T. S., & Tanguay, R. L. (2011). Development and maintenance of a specific pathogen-free (SPF) zebrafish research facility for Pseudoloma neurophilia. *Diseases of Aquatic Organisms*, *95*(1), 73–79. https://doi.org/10.3354/dao02333

Kent, M. L., Harper, C., & Wolf, J. C. (2012). Documented and potential research impacts of subclinical diseases in zebrafish. *ILAR Journal*, *53*(2), 126–134. https://doi.org/10.1093/ilar.53.2.126

Kent, M. L., Spitsbergen, J. M., Matthews, J. L., Fournie, J. W., Murray, K. N., & Westerfield, M. (2016). Diseases of zebrafish in research facilities.

https://zebrafish.org/wiki/health/disease_manual/start Retrieved April 15, 2023

Kent, M. L., Wall, E. S., Sichel, S., Watral, V., Stagaman, K., Sharpton, T. J., & Guillemin, K. (2021). Pseudocapillaria tomentosa, Mycoplasma spp., and intestinal lesions in experimentally infected zebrafish Danio rerio. *Zebrafish*, *18*(3), 207–220. https://doi.org/10.1089/zeb.2020.1955

Kent, M. L., Watral, V. G., Whipps, C. M., Cunningham, M. E., Criscione, C. D., Heidel, J. R., Curtis, L. R., Spitsbergen, J., & Markle, D. F. (2004). A digenean Metacercaria (Apophallus sp.) and a Myxozoan (Myxobolus sp.) Associated with Vertebral Deformities in Cyprinid Fishes from the Willamette River, Oregon. *Journal of Aquatic Animal Health*, *16*(3), 116–129. https://doi.org/10.1577/H04-004.1

Koumoundouros, G., Oran, G., Divanach, P., Stefanakis, S., & Kentouri, M. (1997). The opercular complex deformity in intensive gilthead sea bream (Sparus aurata L.) larviculture. Moment of apparition and description. *Aquaculture*, *156*(1–2), 165–177. https://doi.org/10.1016/S0044-8486(97)89294-0

Kroupova, H., Machova, J., & Svobodova, Z. (2005). Nitrite influence on fish: A review. *Veterinární Medicína*, *50*(11), 461–471. https://doi.org/10.17221/5650-VETMED

Lall, S. P., & Lewis-McCrea, L. M. (2007). Role of nutrients in skeletal metabolism and pathology in fish – An overview. *Aquaculture*, *267*(1–4), 3–19. https://doi.org/10.1016/j.aquaculture.2007.02.053

Langenau, D. (2016). *Cancer and zebrafish: Mechanisms, techniques, and model.* Springer.

Lewis, W. M., & Morris, D. P. (1986). Toxicity of nitrite to fish – A review. *Transactions of the American Fisheries Society*, *115*(2), 183–195. https://doi.org/10.1577/1548-8659(1986)115<183:TONTF>2.0.CO;2

Lindesjöö, E., Thulin, J., Bengtsson, B. E., & Tjärnlund, U. (1994). Abnormalities of a gill cover bone, the operculum, in perch Perca fluviatilis from a pulp-mill effluent area. *Aquatic Toxicology*, *28*(3–4), 189–207. https://doi.org/10.1016/0166-445X(94)90033-7

Lund, M., & Heggberget, T. G. (1985). Avoidance-response of 2-year-old rainbow-trout, Salmo gairdneri Richardson, to air-supersaturated water – Hydrostatic compensation. *Journal of Fish Biology*, *26*(2), 193–200. https://doi.org/10.1111/j.1095-8649.1985.tb04256.x

Machado, J. P., Garling, D. L., Kevern, N. R., Trapp, A. L., & Bell, T. G. (1987). Histopathology and the pathogenesis of embolism (gas bubble disease) in rainbow trout (Salmo gairdneri). *Canadian Journal of Fisheries and Aquatic Sciences, 44*(11), 1985–1994. https://doi.org/10.1139/f87-243

Mocho, J. P., Collymore, C., Farmer, S. C., Leguay, E., Murray, K. N., & Pereira, N. (2022). FELASA-AALAS recommendations for monitoring and reporting of laboratory fish diseases and health status, with an emphasis on zebrafish (Danio rerio). *Comparative Medicine, 72*(3), 127–148. https://doi.org/10.30802/AALAS-CM-22-000034

Murray, K. N., Lains, D., & Spagnoli, S (2020) Water quality and idiopathic diseases of laboratory zebrafish. In S. C. Cartner et al. (Eds.), *The zebrafish in biomedical research: Biology, husbandry, diseases and research applications* (pp. 463–477). Elsevier.

Myers, M. S., Johnson, L. L., & Collier, T. K. (2003). Establishing the causal relationship between polycyclic aromatic hydrocarbon (PAH) exposure and hepatic neoplasms and neoplasia-related liver lesions in English sole (Pleuronectes vetulus). *Human and Ecological Risk Assessment, 9*(1), 67–94. https://doi.org/10.1080/713609853

Paquette, C. E., Kent, M. L., Buchner, C., Tanguay, R. L., Guillemin, K., Mason, T. J., & Peterson, T. S. (2013). A retrospective study of the prevalence and classification of intestinal neoplasia in zebrafish (Danio rerio). *Zebrafish, 10*(2), 228–236. https://doi.org/10.1089/zeb.2012.0828

Peuranen, S., Vuorinen, P. J., Vuorinen, M., & Hollender, A. (1994). The effects of iron, humic acids and low pH on the gills and physiology of brown trout (*Salmo trutta*). *Annales Zoologici Fennici, 31*, 389–396.

Pohl, C. (1990). Skeletal deformities and trace-metal contents of European smelt, Osmerus eperlanus, in the Elbe estuary. *Meeresforschung – Reports on Marine Research, 33*(1), 76–89.

Rosenberg, G. A. (2012). *Molecular physiology and metabolism of the nervous system: A clinical perspective*. Oxford University Press.

Rossteuscher, S., Schmidt-Posthaus, H., Schäfers, C., Teigeler, M., & Segner, H. (2008). Background pathology of the ovary in a laboratory population of zebrafish Danio rerio. *Diseases of Aquatic Organisms, 79*(2), 169–172. https://doi.org/10.3354/dao01893

Schaaf, R. M., Sharpton, T. J., Murray, K. N., Kent, A. D., & Kent, M. L. (2020). Retrospective analysis of the Zebrafish International Resource Center diagnostic data links Pseudocapillaria tomentosa to intestinal neoplasms in zebrafish Danio rerio (Hamilton 1822). *Journal of Fish Diseases, 43*(11), 1459–1462. https://doi.org/10.1111/jfd.13233

Schwedler, T. E., Tucker, C. S., & Beleau, M. H. (1985). Non-infectious diseases. In C. S. Tucker (Ed.), *Channel catfish culture* (pp. 497–541). Elsevier.

Sfakianakis, D. G., Georgakopoulou, E., Papadakis, I. E., Divanach, P., Kentouri, M., & Koumoundouros, G. (2006). Environmental determinants of haemal lordosis in European sea bass, *Dicentrarchus labrax* (Linnaeus, 1758). *Aquaculture, 254*(1–4), 54–64. https://doi.org/10.1016/j.aquaculture.2005.10.028

Smith, C. E. (1988). Histopathology of gas bubble disease in juvenile rainbow trout. *Progressive Fish-Culturist, 50*(2), 98–103. https://doi.org/10.1577/1548-8640(1988)050<0098:CHOGBD>2.3.CO;2

Sorensen, E. M. B. (1991). *Metal poisoning in fish*. CRC Press.

Spagnoli, S., & Murray, K. N. (2020). Nonexperimentally induced neoplastic and proliferative lesions in laboratory zebrafish. In S. C. Cartner et al. (Eds.), *The zebrafish in biomedical research: Biology, husbandry, diseases and research applications* (pp. 527–546). Elsevier.

Spagnoli, S., Xue, L., & Kent, M. L. (2015). The common neural parasite Pseudoloma neurophilia is associated with altered startle response habituation in adult zebrafish (Danio rerio): Implications for the zebrafish as a model organism. *Behavioural Brain Research, 291*, 351–360. https://doi.org/10.1016/j.bbr.2015.05.046

Speare, D. J. (1991). Endothelial lesions associated with gas bubble disease in fish. *Journal of Comparative Pathology, 104*(3), 327–335. https://doi.org/10.1016/s0021-9975(08)80044-8

Spitsbergen, J. M., Buhler, D. R., & Peterson, T. S. (2012). Neoplasia and neoplasm-associated lesions in laboratory colonies of zebrafish emphasizing key influences of diet and aquaculture system design. *ILAR Journal, 53*(2), 114–125. https://doi.org/10.1093/ilar.53.2.114

Spitsbergen, J. M., & Kent, M. L. (2003). The state of the art of the zebrafish model for toxicology and toxicologic pathology research—Advantages and current limitations. *Toxicologic Pathology, 31*, Suppl., 62–87. https://doi.org/10.1080/01926230390174959

Stickland, N. C., White, R. N., Mescall, P. E., Crook, A. R., & Thorpe, J. E. (1988). The effect of temperature on myogenesis in embryonic-development of the Atlantic salmon (Salmo salar L). *Anatomy and Embryology, 178*(3), 253–257. https://doi.org/10.1007/BF00318228

The Zebrafish Information Network (ZFIN). (1993). https://zfin.org Retrieved April 15, 2023

Tye, M. T., Montgomery, J. E., Hobbs, M. R., Vanpelt, K. T., & Masino, M. A. (2018). An adult zebrafish diet contaminated with chromium reduces the viability of progeny. *Zebrafish, 15*(2), 179–187. https://doi.org/10.1089/zeb.2017.1514

Walters, G. R., & Plumb, J. A. (1980). Environmental-stress and bacterial-infection in channel catfish, Ictalurus punctatus Rafinesque. *Journal of Fish Biology, 17*(2), 177–185. https://doi.org/10.1111/j.1095-8649.1980.tb02751.x

Wedemeyer, G. A. (1996). *Physiology of fish in intensive culture systems.* Springer.

Wedemeyer, G. A., & Yasutake, W. T. (1978). Prevention and treatment of nitrite toxicity in juvenile steelhead trout (Salmo gairdneri). *Journal of the Fisheries Research Board of Canada, 35*(6), 822–827. https://doi.org/10.1139/f78-132

Weitkamp, D. E., & Katz, M. (1980). A review of dissolved-gas supersaturation literature. *Transactions of the American Fisheries Society, 109*(6), 659–702. https://doi.org/10.1577/1548-8659(1980)109<659:ARODGS>2.0.CO;2

Chapter 5

Infectious diseases of zebrafish

Joanna Cambray-Young

Abstract

Zebrafish can be affected by pathogens or opportunistic microbes responsible for subclinical or clinical diseases when fish health is impacted by husbandry conditions, other health conditions or scientific procedures. These diseases may compromise fish welfare or experimental data. Here we review the main zebrafish pathogens that need to be considered. For example, *Pseudoloma neurophilia* is a common fungal parasite potentially introducing bias to behavioural studies; its elimination from zebrafish colonies can be particularly demanding. Another parasite is *Pseudocapillaria tomentosa*, which has been associated with increased mortality, oncogenesis and microbiome variation. Moreover, fish systems hold pathogens that can eventually induce disease to humans: zoonosis. For example, mycobacteria are a group of bacteria, usually present in all aquatic systems, that do not induce major morbidity or mortality. However, staff must protect themselves against the risk of mycobacterial infections. Some mycobacteria, such as *Mycobacterium marinum*, are particularly a risk to fish and staff health, and should be avoided in conventional aquatic facilities.

While the impact of some pathogens has been researched to some degree, there is a general dearth of information regarding the significance of many of the pathogens on the screening panels. This chapter will briefly summarize the current knowledge of each of these pathogens.

Claire Allen and Jean-Philippe Mocho (eds) **Zebrafish**
DOI: 10.1079/9781800629431.0005,

Keywords

zebrafish, *Danio rerio*, *Mycobacterium* spp., *Pseudoloma neurophilia*, *Pseudocapillaria tomentosa*, fish pathogens

Introduction

There are now numerous laboratories offering diagnostic tests for a long list of pathogens. The importance of each of these pathogens varies greatly, as they each have different effects on zebrafish health, welfare, reproductive success and a multitude of physiological parameters that may impact upon scientific data collected.

As well as the obvious welfare impact for fish displaying clinical signs, infectious diseases can be a confounding factor in scientific research. Even subclinical infections may introduce variables as the fish's immune system initiates cytokine cascades and cellular pathways to clear the pathogen. This has downstream effects on other body systems and energy partitioning. Immune responses, as well as the pathogen burden, will vary between individual animals, creating unknown variables within the experimental model. This intra- and inter-study variable can result in poor reproducibility of experiments and can be further exacerbated, especially if the health status of the aquarium changes in the interim period between repeats of studies or is compared with results from aquaria of differing health status.

If infection leads to overt disease with fish exhibiting clinical signs, this will impact the welfare of the fish. It may also reduce fecundity and reproductive success, with subsequent effects on maintaining transgenic lines and the availability of fish for experiments. If disease results in death, or the need to euthanize on welfare grounds, this can affect the statistical power of experimental studies.

Some fish pathogens are zoonotic, so have the potential to infect and cause disease in humans. Workers in aquaria should always assume that the water may contain pathogens such as *Mycobacterium* spp. and take appropriate caution (which will be greater for those with compromised immune systems). Where health screening for pathogens is undertaken, a negative result should not be interpreted as a guarantee that pathogens are not present in the system, merely that they are not abundant (Mocho et al., 2022a; see Chapters 7 and 9 for more information).

Bacterial pathogens

Mycobacterium *spp.*

These bacteria are acid-fast bacilli, requiring stains such as Ziehl–Neelsen for visualization in histological preparations (Figure 5.1). The unique composition of their cell wall renders them highly resistant to antibiotics and disinfectants (Chang et al., 2015), which is further augmented by their survival in biofilms (Chang et al., 2019b; Falkinham et al., 2001). Numerous species of *Mycobacterium* can be found in zebrafish aquaria, with polymerase chain reaction (PCR) tests commercially available for at least six of these plus a genus-level test: *M. abscessus*, *M. chelonae*, *M. fortuitum*, *M. haemophilum*, *M. marinum* and *M. peregrinum.*

There is some evidence that husbandry factors may play a role in the transmission of mycobacteria. Biofilms on tanks have been shown to include mycobacteria, which can then act as both a reservoir and source of mycobacterial infections in zebrafish (Figure 5.2). The live food *Paramecium caudatum* has been shown to enhance the transmission and infectivity of *M. marinum* and *M. chelonae* (Peterson et al., 2013). Recently it was demonstrated that mycobacterial transmission can also occur via brine shrimp (*Artemia franciscana*) and rotifers (*Branchionus* spp.), which highlights the potential biosecurity concerns of using live feeds (Chang et al., 2019b). Moreover, snails within the aquarium system have produced positive PCR results for unknown species of *Mycobacterium* (unpublished data). The significance of this to zebrafish is unknown.

The zoonotic potential of *M. marinum* is well documented (Decostere et al., 2004). These bacteria do not grow well at 37 °C, which limits the area affected to the skin of distal extremities, e.g. fingers/arms. Typically 'fish tank granuloma' results in nodules or ulcers forming on the affected

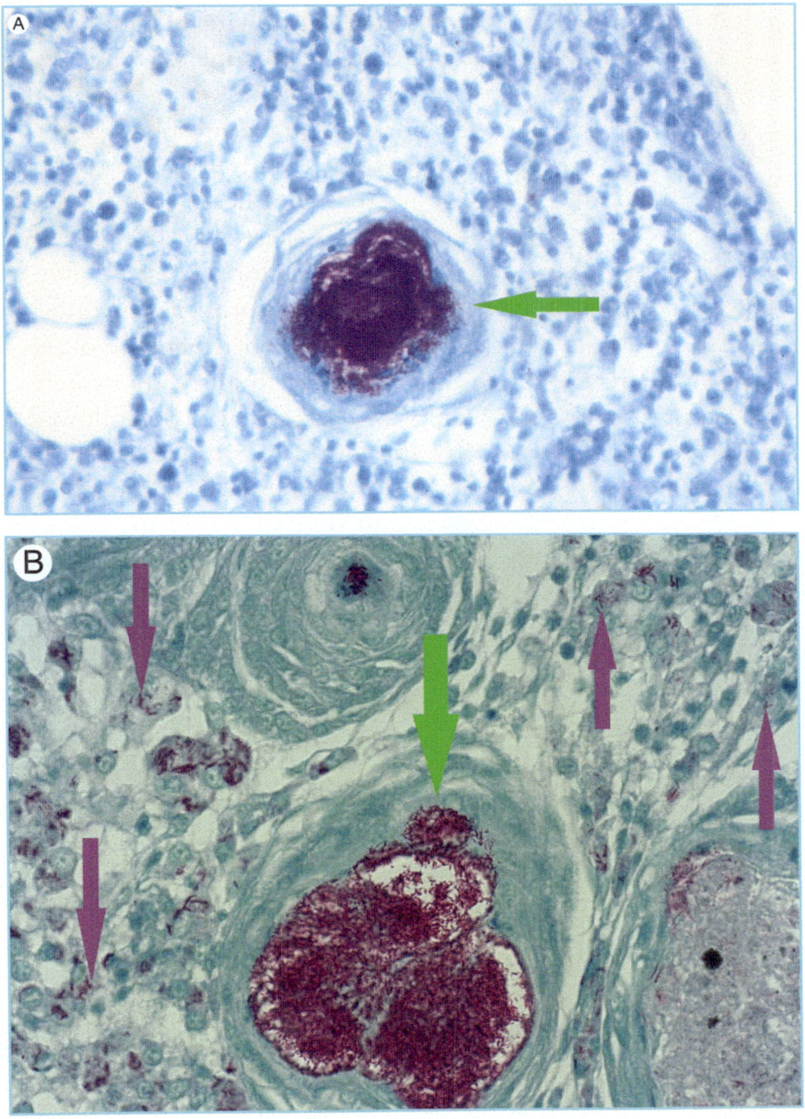

Figure 5.1 (A,B), Ziehl–Neelsen Fite staining of histological sections of an adult zebrafish highlighting acid-fast bacilli (stained red), forming dense extracellular aggregates, i.e. mycobacterial granulomas, but also present as sparse rods inside phagocytic cells (macrophages). Green arrows, granulomas; purple arrows, red-staining *Mycobacterium* spp. Original magnification 20× (A) and 40× (B). Credits: Nuno Pereira, NMS Research/Nova Medical School; Tânia Carvalho, Champalimaud Foundation.

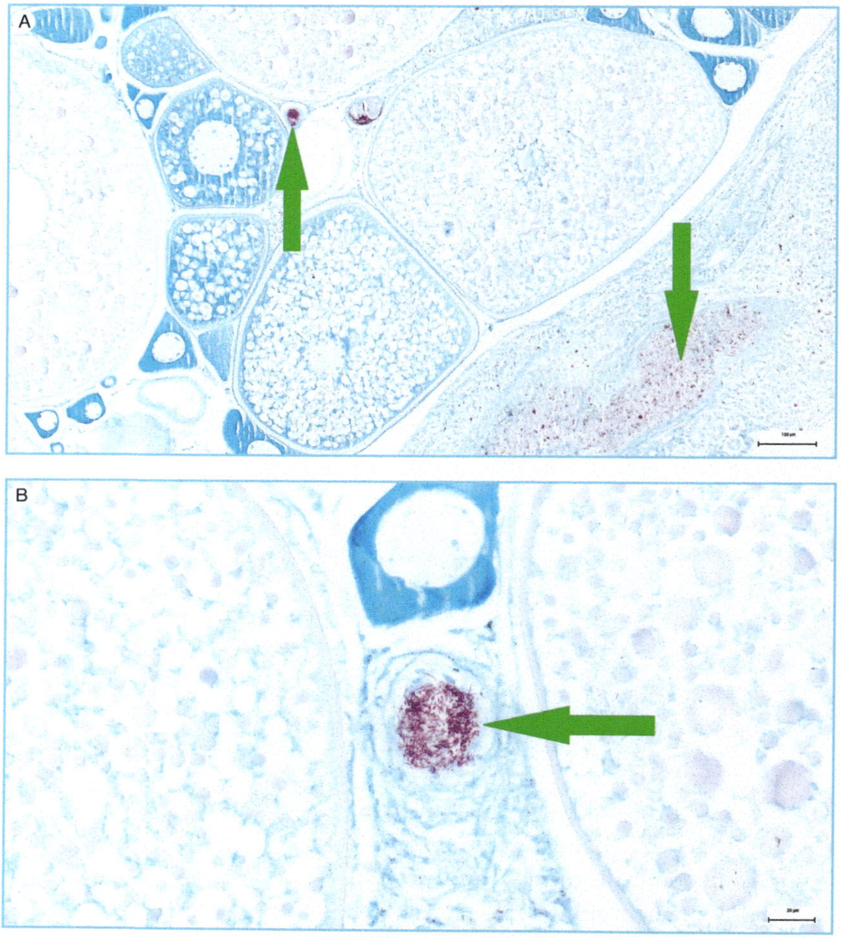

Figure 5.2 *Mycobacterium* spp. granulomas in histological sections of infected zebrafish. (A,B), mature mycobacterial granulomas, consisting of well-defined structures packed with Ziehl–Neelsen-positive bacilli, in the ovigerous stroma of an adult zebrafish. (C,D), large, multicentric granulomas in the stroma of the head kidney of the same zebrafish. Green arrows, granulomas; purple arrows, red-staining *Mycobacterium* spp. Staining: Ziehl–Neelsen Fite. Original magnification, 10× (A,C) and 40× (B,D). Credits: Tânia Carvalho, Champalimaud Foundation.

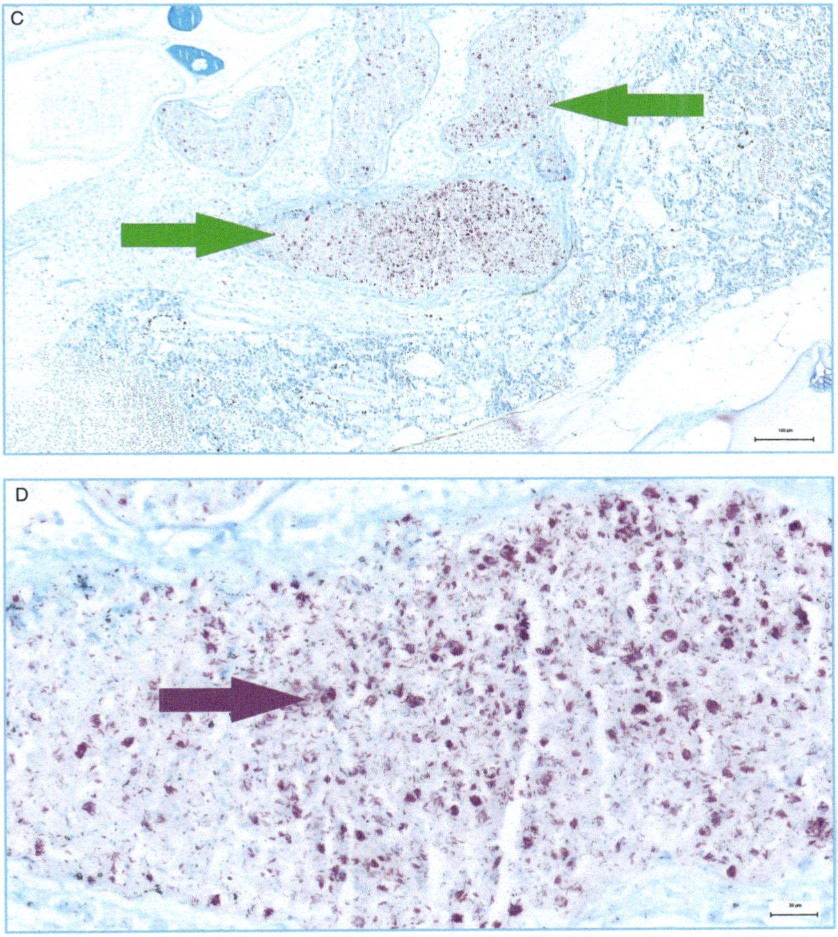

Figure 5.2 (continued)

skin, especially at the site of cuts or abrasions. Individuals with compromised immune systems may suffer more severe lesions. Untreated cases may take up to two years to resolve spontaneously. Treatment with antibiotics usually requires a prolonged course and multiple drugs. Less frequently, a more severe sporotrichoid form occurs and, in some cases, deeper tissue structures such as bone and tendons may become involved (Aubry et al., 2017). Other mycobacterial species have zoonotic potential but are less well characterized (Figure 5.3; Mocho et al., 2022b).

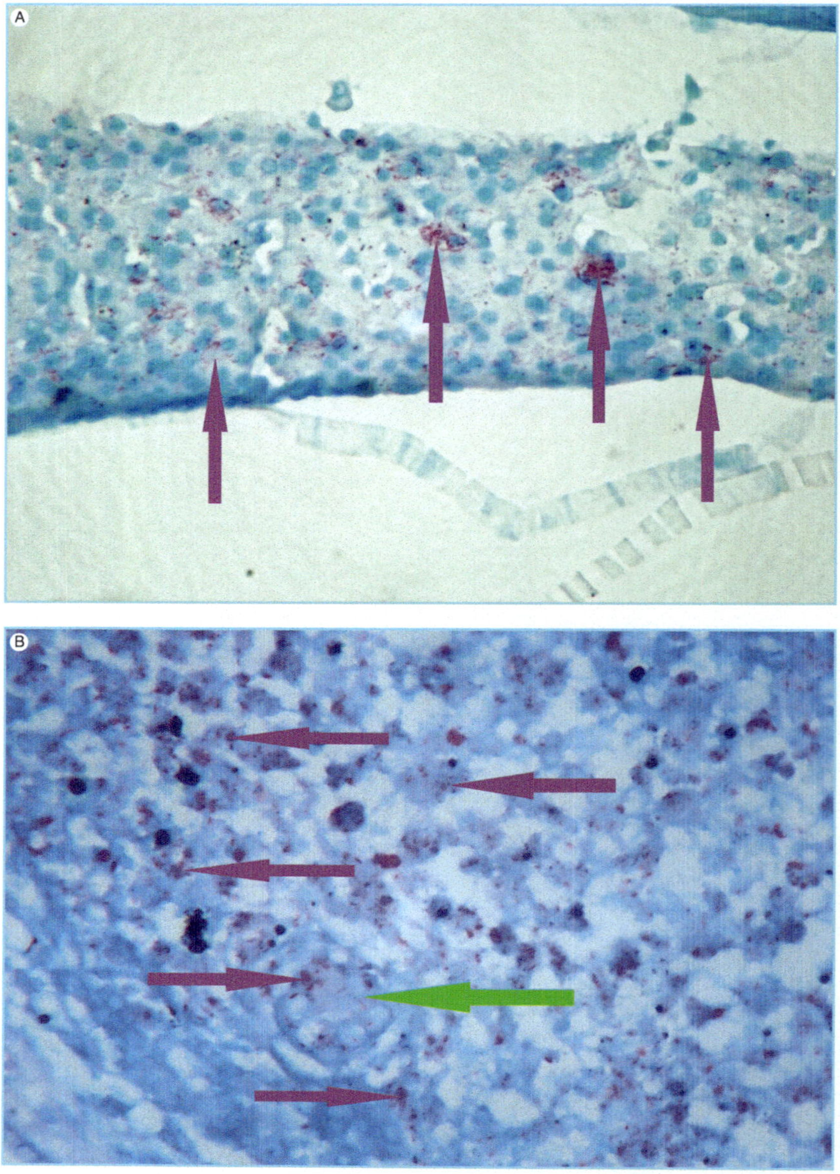

Figure 5.3 *Mycobacterium* spp. in the kidney of zebrafish, located both within and outside granulomas, coloured red by staining. Green arrow, granulomas; purple arrow, red-staining *Mycobacterium* spp. Staining: modified Fite–Faraco. Original magnification, 10× (A) and 40× (B). Credits: Vetofish.

The zoonotic nature of aquatic mycobacterial species has led to the use of zebrafish infection as a model for human tuberculosis (Cronan & Tobin, 2014).

In the zebrafish, *M. marinum* and *M. haemophilum* are the most pathogenic species and have been associated with severe disease (Figure 5.4; Kent et al., 2004; Rácz et al., 2019; Watral & Kent, 2007; Whipps et al., 2007). *M. peregrinum*, *M. abscessus*, *M. chelonae* and *M. fortuitum* are generally of lower virulence and can result in reduced survival and reproductive success. Many fish may be infected subclinically with mycobacteria (Murray et al., 2011). These asymptomatic infections can remain latent or could be stimulated to recrudesce, with the subsequent development of clinical disease following a stressor such as overcrowding (Ramsay et al., 2009a), exposure to gamma irradiation (Parikka et al., 2012) or concurrent infection (Mocho, 2016).

Mycobacteria shed from skin lesions and in faeces are ingested by zebrafish (Harriff et al., 2007) and then disseminate from the gastrointestinal tract systemically within macrophages, including via the pneumatic duct to the swim bladder (Whipps et al., 2012).

Mycobacteriosis is a disease of chronic inflammation. Granulomas may develop in the internal organs, detected as small white nodules at necropsy (Watral & Kent, 2007). Other clinical signs may be non-specific such as emaciation, skin ulceration, raised scales or enlarged abdomens. Behaviourally the fish may become lethargic or demonstrate abnormal swimming.

Mycobacterial infection, either overt or asymptomatic, has the potential to affect numerous types of research. Chronic inflammation will lead to the activation of inflammatory pathways with the release of cytokines, as suggested by the upregulation of genes coding for components of the inflammatory response observed in transcriptomic studies of *M. marinum* infection (Figure 5.5; Meijer et al., 2005). Behaviour abnormalities may confound studies that rely on behavioural measures as scientific outputs.

Even the relatively poorly pathogenic *M. chelonae* has been shown to adversely affect studies relying on the expression and visualization

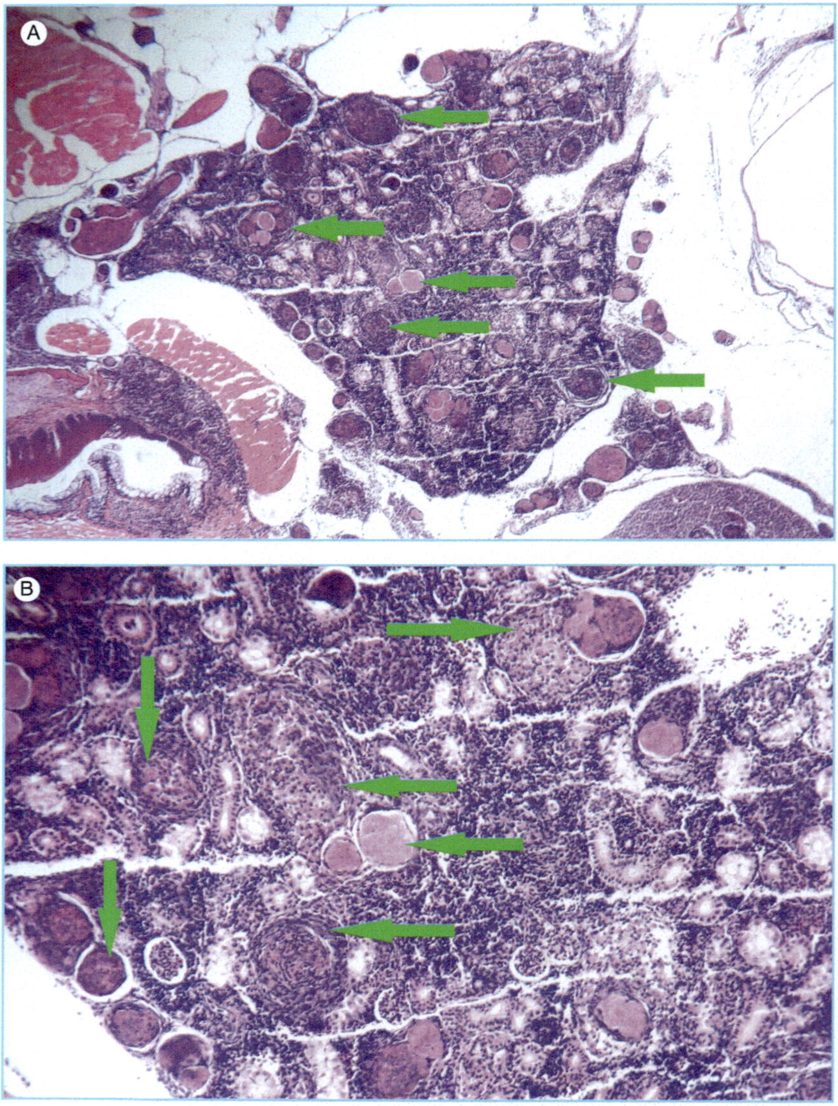

Figure 5.4 Severe granulomatous peritonitis due to *M. haemophilum* in a zebrafish. (A,B), kidney; (C), spleen, liver, pancreas and digestive tract; (D), testis. Green arrows, granulomas. Staining, haematoxylin and eosin. Original magnification 40× (A,C) and 100× (B,D). Credits: Vetofish.

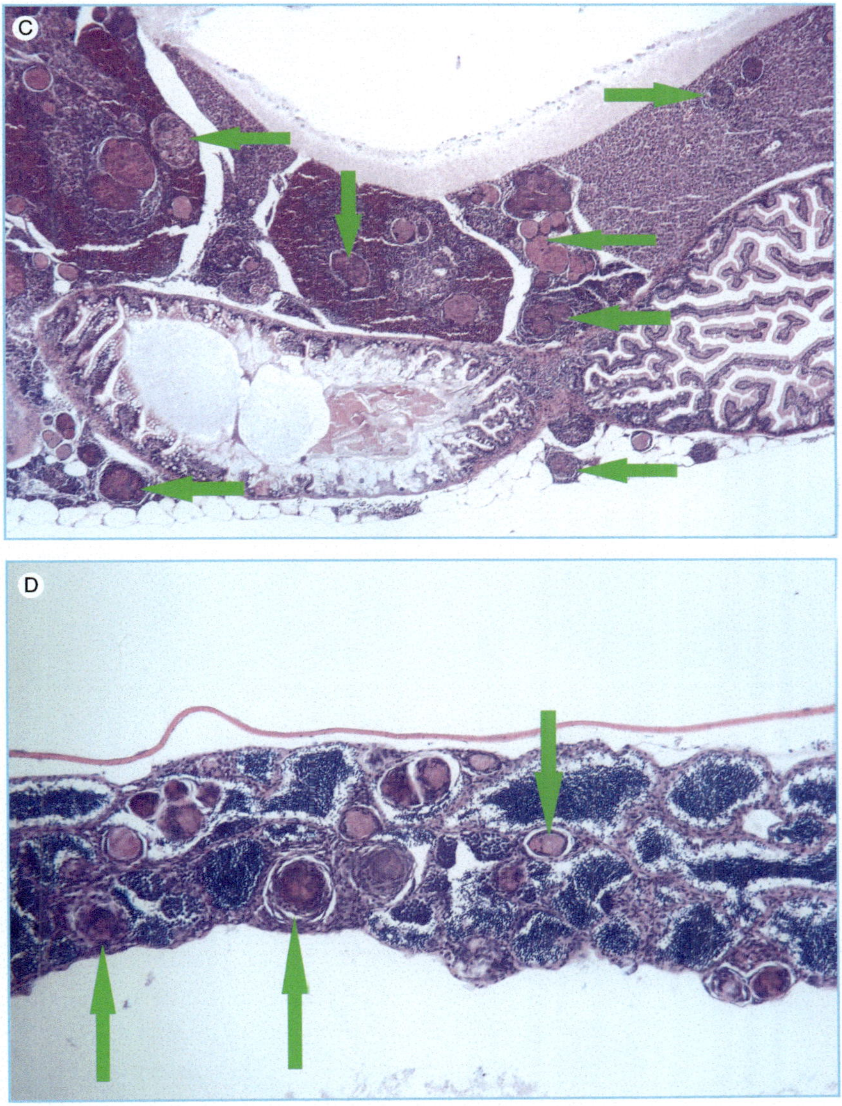

Figure 5.4 (continued)

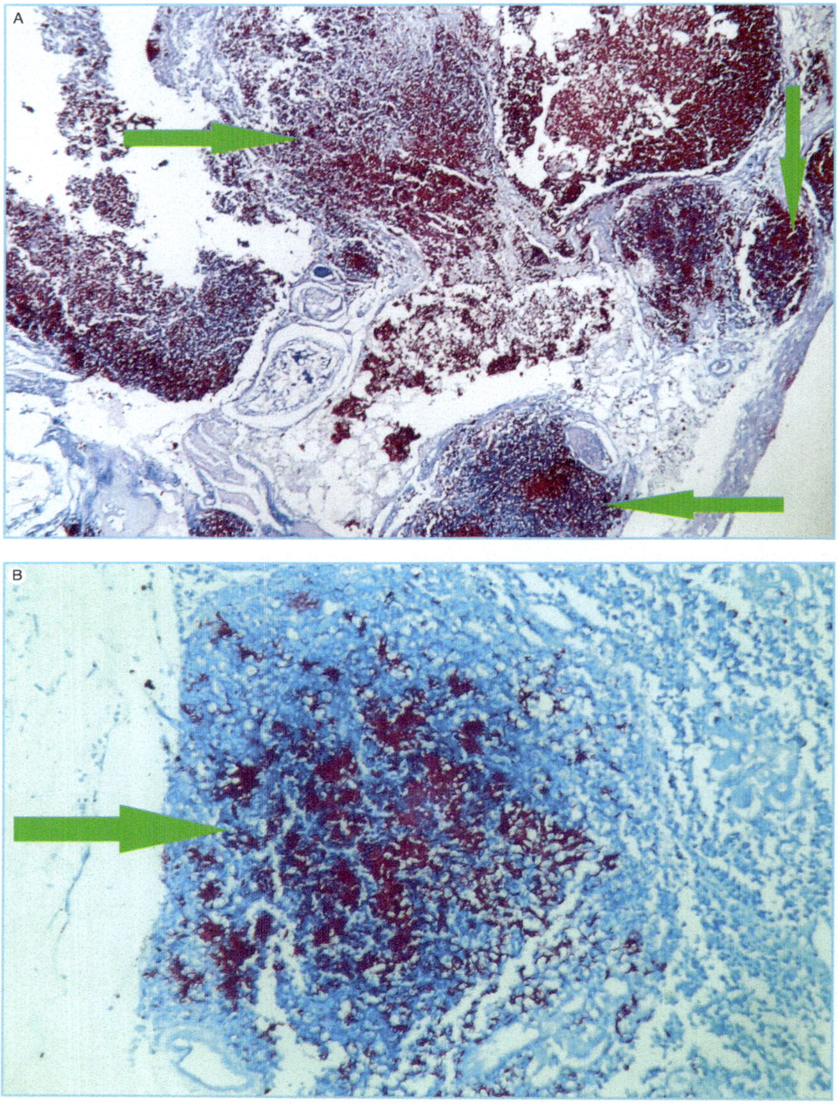

Figure 5.5 *M. marinum* coloured red by staining and located in both granulomas and other tissues of a zebrafish. Green arrows, granulomas. Staining, modified Fite–Faraco. Original magnification 40× (A) and 100× (B). Credits: Vetofish.

of endogenous fluorescent tags, due to the bacterium's autofluorescence (Whipps et al., 2014).

Edwardsiella ictaluri

This Gram-negative bacterium is able to infect many fish species, and edwardsiellosis causes high levels of mortality; in aquaculture it is the causal agent of enteric septicaemia of catfish (ESC), which is associated with losses on catfish farms worldwide. It is an obligate pathogen that is most likely transmitted between fish sharing the same water, via faecal shedding, and therefore potentially by vectors such as nets.

To date, five outbreaks have been described in research zebrafish systems, four of which were in quarantine facilities (Hawke et al., 2013). Each has been associated with high levels of mortality, with any remaining fish euthanized. Histopathological findings include widespread tissue necrosis, especially of the kidneys, spleen, nares and forebrain, but also involving the gills, heart and liver. Experimental infections have been induced to confirm that the bacterium induces disease as per Koch's postulates, and to explore zebrafish infection as a potential model for ESC (Petrie-Hanson et al., 2007).

Clinical signs include lethargy, raised scales, haemorrhage and/or ulceration of the skin and swollen abdomen due to ascites. Although short-term antibiotic treatment of a small number of zebrafish involved in an experiment was successfully attempted, it is not feasible to treat a large aquarium system. In the event of an outbreak, the most logical approach is to cull all remaining infected fish and disinfect the system. The bacteria do not survive for long in the environment, so decontamination should be relatively straightforward.

Opportunistic bacterial pathogens

Many bacteria exist within aquatic systems, often resident within biofilms. These are not primary pathogens but, if the fish becomes immunocompromised due to husbandry stress, such as over-stocking or poor water quality, or due to concurrent infection or wounds, these bacteria will seize the opportunity to colonize the fish.

Flavobacterium columnare *and* F. branchiophilum

These Gram-negative, filamentous bacteria, along with other bacteria, can cause columnaris or bacterial gill disease, in many species of fish (Declercq et al., 2013). This results in gill damage, which leads to respiratory problems and gasping. Skin lesions are also common, as either fin or tail rot, or ulcerative lesions and loss of pigmentation. Mouth ulcers have been reported in other fish species, which may lead to anorexia and weight loss. Microscopic analysis reveals fused gill lamellae and filamentous, gliding bacterial rods. This disease usually becomes apparent only when water quality decreases, for example during transport.

Aeromonas hydrophila, A. dhakhensis, Pseudomonas aeruginosa, P. fluorescens *and* Plesiomonas shigelloides

These Gram-negative bacteria live in the biofilm and may colonize gill lesions or skin wounds that fish sustain through fighting, poor netting, injecting or disease. Poor water parameters have been linked to mortality with involvement of these opportunistic bacteria (Pullium et al., 1999). Transgenic fish that are genetically altered to be immunocompromised are at especially high risk. Mixed bacterial infections may be found in bacterial aerocystitis (swim bladder infection), detected clinically by lethargy, weight loss and positioning at the bottom of the tank. Haemorrhaging at the base of fins and gills may be linked to aerocystitis but is also indicative of sepsis.

A positive PCR result for these bacteria is not necessarily indicative of a direct health risk to healthy fish. However, a heavy load of these bacteria in the system water may indicate that there is a problem with bacterial control systems, i.e. filtration and UV sterilization. Other disease problems experienced by the aquarium are likely to be exacerbated by the presence of these secondary infections.

Viral pathogens

Many aquatic viruses have a wide host range and can infect multiple fish species. To date there are few reports of naturally occurring viral infection in laboratory zebrafish aquaria (Kent et al., 2020). Health-screening laboratories commonly offer PCR tests for two viruses, discussed below. It is possible that other viruses could be a threat to zebrafish, which highlights the importance of good biosecurity and quarantine procedures especially in institutes where other fish species are housed (Mocho et al., 2022a,b). Unknown oncolytic viruses may contribute to increased neoplasia rates, which requires further research. Experimental exposure to viruses of other fish species has been shown to increase susceptibility to bacterial diseases, which could have relevance for infection and immunity studies in particular (Pakingking et al., 2003).

Infectious spleen and kidney necrosis virus (ISKNV)

Belonging to the genus *Megalocytivirus*, family Iridoviridae, this virus can infect a wide range of fish species and cause mortality in aquaculture systems. Until recently only experimental infections were described for zebrafish (Xu et al., 2008), but in 2015 a Spanish facility reported a naturally occurring outbreak in laboratory zebrafish (Bermúdez et al., 2018). Five fish were analysed following development of clinical signs: lethargy, inappetance, abnormal swimming, abdominal enlargement, respiratory distress, hyperaemic gills, scale protrusion and petechial haemorrhages at the base of fins. Enlarged spleen and kidneys have also been observed in experimental infections. Histologically, megalocytosis was seen in the kidney, spleen and intestine, associated with a degree of necrosis. Pathological changes were also detected in the gills, skin and fins. No other fish from the aquarium tested positive for ISKNV, which suggests that horizontal spread in a modern aquarium setting may be restricted.

However, the report reinforces the potential for zebrafish to be infected with this virus and the need for vigilance when importing wild-caught or petshop fish, of any species, into a zebrafish aquarium (Mocho et al., 2022b; see Chapter 7 for more information).

Infectious pancreatic necrosis virus (IPNV)

IPNV is in the Birnaviridae family and typically infects young salmon and trout, causing some mortality. Clinical signs seen in these species include abdominal distension, inappetance, abnormal swimming and skin darkening. Histologically, pancreatic necrosis and thick mucus in the intestine may be observed. There may be petechial haemorrhages in the pancreatic tissue and necrosis of renal haematopoietic tissue. Some fish will recover, and adult fish may become asymptomatic carriers. Natural infection of zebrafish with IPNV has never been reported, but experimental infection with the virus has been demonstrated to produce necrosis of the renal haematopoietic tissue (Crim & Riley, 2012; LaPatra et al., 2000).

Unicellular eukaryotic parasites

P. neurophilia

This single-celled, obligate intracellular microsporidian fungal parasite has a strong tropism for neural tissue including brain, spinal nerve roots and even single axons (Figure 5.6). It has also been observed in numerous tissues throughout the body of infected zebrafish. The most notable organ is the ovary, where it has the capacity to enter the eggs and undergo vertical transmission to the offspring (Sanders et al., 2013). The accumulation of microsporidian parasite clusters in the nervous system is documented to lead to clinical signs that may interfere with research models (West et al., 2014). Infections can also be subclinical, with no overt clinical signs of disease. A review of the Zebrafish International Resource Center (ZIRC) diagnostic service database (2006–2018; approximately

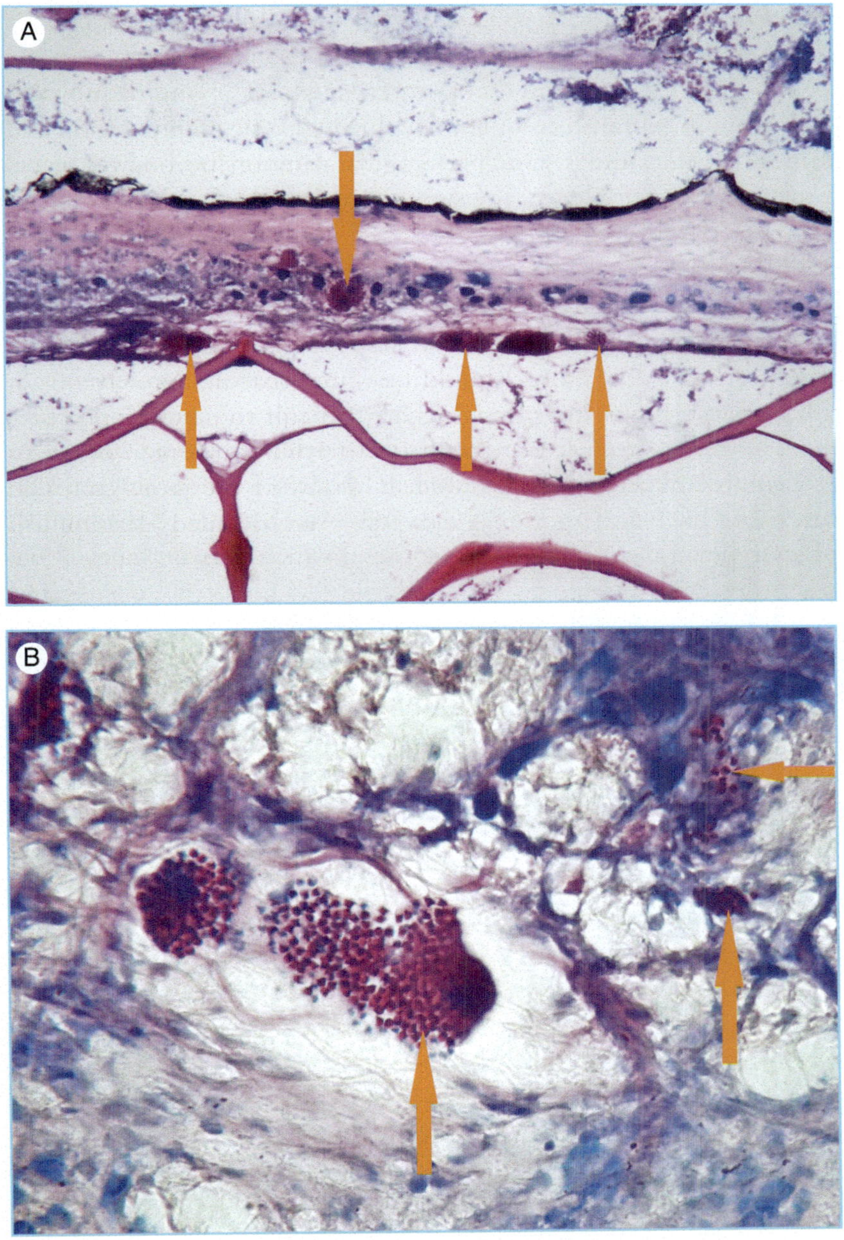

Figure 5.6 *P. neurophilia* xenoma (i.e. aggregates of microsporidia) in zebrafish spinal cord. Orange arrows, xenomas. Staining, modified Fite–Faraco. Original magnification 10× (A) and 40× (B). Credits: Vetofish.

10,000 fish) found that 45% of zebrafish facilities were positive for *P. neurophilia* but few displayed clinical signs of disease (Kent et al., 2020). This demonstrates that there is a significant level of subclinical disease in zebrafish colonies, which could confound the results of some scientific studies. One previous database review had suggested a male bias in prevalence, the significance of which requires further research (Chow et al., 2016).

Clinical signs that infected fish may display include altered swimming patterns, emaciation and scoliosis (bending) of the spine (Sanders et al., 2012). One recent study suggests that loss of body condition may be more pronounced in females (Sanders et al., 2020). Other signs and physiological effects of subclinical infection may, however, be elusive and difficult to detect from the tank-side.

Zebrafish are increasingly used as models for neurological diseases, and have undergone behavioural tests that aim to distinguish phenotypes and drug responses, some of which may be subtle. One study detected differences in the startle test between infected and non-infected zebrafish and observed that the former, even those with no apparent clinical signs, showed enhanced evasion to capture by net (Spagnoli et al., 2015). Furthermore, two recent studies have demonstrated that zebrafish infected with *P. neurophilia* exhibit different shoaling behaviour, with a decrease in inter-fish distance as well as elevated anxiety (freezing behaviour) and altered exploration (Midttun et al., 2020a; Spagnoli et al., 2017). Alterations in baseline behavioural parameters could confound studies of diseases such as schizophrenia, Asperger's disease and social anxiety, where behavioural tests of social interaction, shoaling, exploration and freezing might be used as scientific outcome measures. Fish are likely to have varying levels of infection and pathological change in the tissues, which creates an uncontrolled variable within the experiment. Indeed, transcriptomic analysis of infected zebrafish brains revealed that many genes were upregulated, with most of these associated with the immune response and circadian rhythm rather than being directly associated with neural function (Midttun et al., 2020b).

Disease may be exacerbated, as with many diseases, by the immunosuppressive effects of stress. Fish with previously subclinical disease may then present with clinical signs of disease and those with clinical signs may suffer a more severe disease. This was observed in a study that employed husbandry stressors and observed a decrease in fecundity and body weight, as well as an increase in severity of lesions observed, such as size of parasite cluster in the spinal cord and myositis (Figure 5.7; Ramsay et al., 2009b).

In a research environment there are likely to be many stressors that could precipitate disease, such as handling, poor water quality, vibrations from the pump, over- or under-stocking, bright light, anaesthesia or procedures such as imaging, injection or surgery. As research with zebrafish becomes more advanced and complex, the effects of pathogens are likely to become increasingly evident. For example, gamma irradiation (commonly used in mammalian research to ablate bone marrow) of zebrafish infected with *P. neurophilia* resulted in nearly 50% mortality at 6 weeks post irradiation compared with both non-irradiated, infected controls and irradiated, non-infected controls (Spagnoli et al., 2016).

There is the potential for many, if not all, research programmes to be affected by infection with this pathogen: not just the obvious experiments that study neurological or behavioural parameters, but potentially all studies. The variable effects on weight, fecundity, behaviour, anxiety, social interaction and underlying activation of the immune system and inflammatory pathways all have the potential to alter the physiology, with each individual fish responding differently, thus creating huge variation in baseline physiological parameters (Figure 5.8).

With this in mind, it may be desirable to report the health status of fish used in experiments and to eradicate *P. neurophilia* from laboratory aquarium systems. This has been achieved by laboratories such as Sinnhuber Aquatic Resource Laboratory (SARL), Oregon, USA and is described in their 2016 publication (Barton et al., 2016). Recommendations for how to eradicate microsporidia are also detailed in a 2011 publication by Murray et al.

Control is stymied by the resistant nature of the microsporidian spores that are excreted by the fish into the system water, in both the faeces and eggs (within eggs and externally). These have been shown to survive chlorine at standard egg-disinfection levels (25–50 ppm;

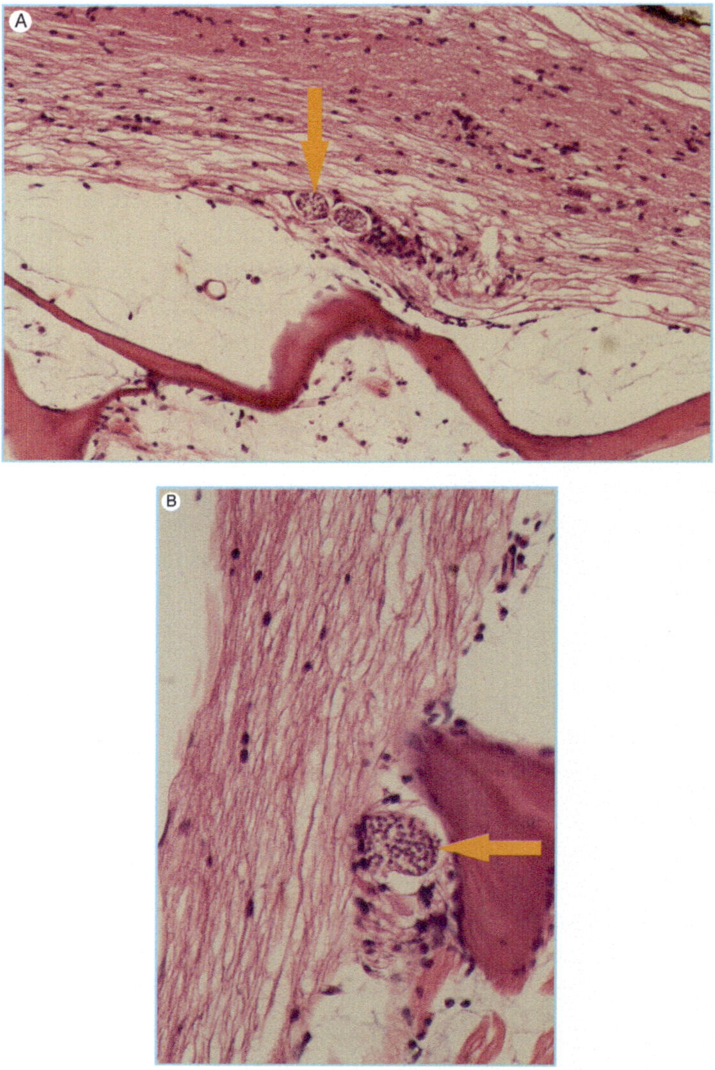

Figure 5.7 *P. neurophilia* spores in histological sections of zebrafish with different histological staining methods. (A,B), haematoxylin and eosin staining of aggregates of spores in the spinal cord, stained blue and pink, similar to the surrounding tissues. (C), Ziehl–Neelsen staining of an aggregate of spores in the muscle of an adult zebrafish, stained blue, similar to the surrounding tissues. (D), Luna staining of an aggregate of spores in the muscle of an adult zebrafish, differentially stained red and contrasting with the remaining tissue staining light blue. Orange arrows, xenomas. Original magnification 4× (A), 20× (B) and 40× (C,D). Credits: Nuno Pereira, NMS Research/Nova Medical School.

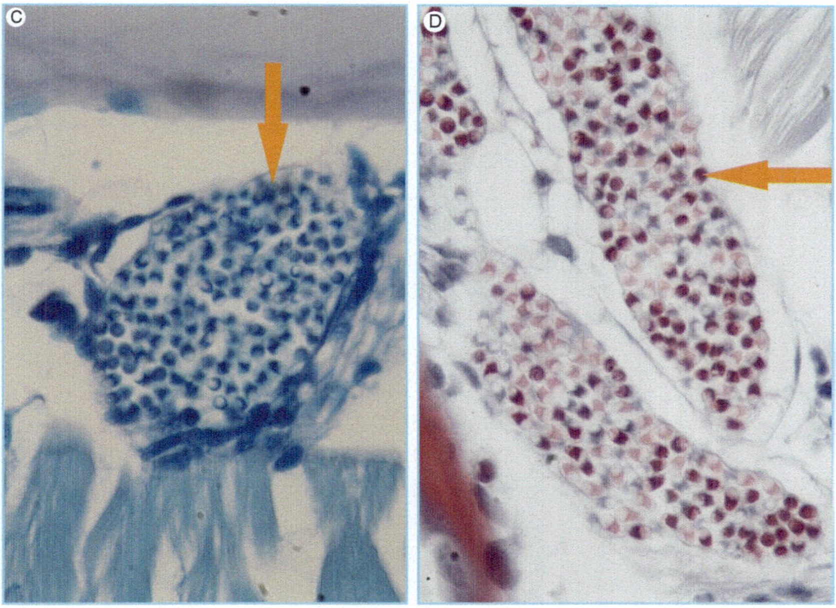

Figure 5.7 (continued)

Ferguson et al., 2007). Disinfection of spawning chambers with 100 ppm chlorine was shown to kill spores in one study, but such spores remained infectious in another. Breeding protocols using uninfected females paired with potentially infected males (ideally under 1 year old) may produce eggs that are free from spores (although there is a small risk of faecal contamination). Spores can survive being frozen at –80 °C for 3 years (Norris et al., 2018), so care must be taken when resurrecting past lines from cryopreservation, or from facilities of unknown health status.

To assess the initial prevalence and confirm subsequent eradication, the pathogen may be detected by testing a suitable number of fish by PCR or histologically (see Chapters 7 and 9 for more information), using either Fungi-Fluor or Luna (Figure 5.7) stain to optimize spore identification (or haematoxylin and eosin for large clusters). Autofluorescence of muscle granulomas containing spores may be useful to detect infection in live fish (West et al., 2014).

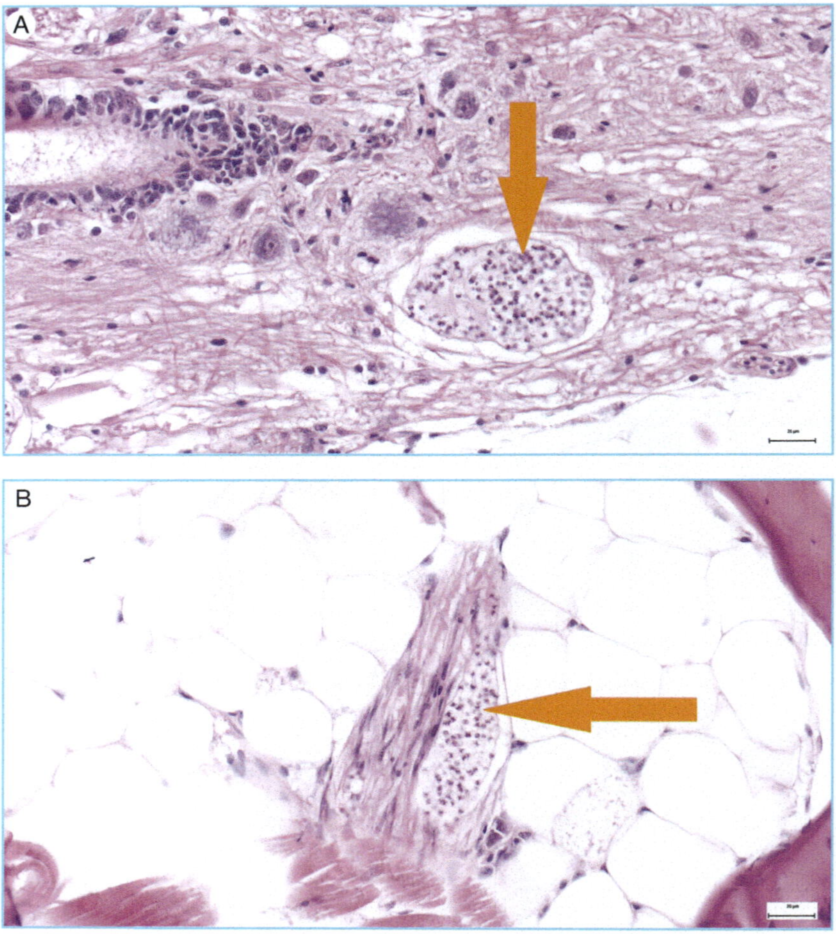

Figure 5.8 Spores of *P. neurophilia* in histological sections of infected zebrafish. (A,B), spores in the nervous system: spinal cord (A) and peripheral nerve (B) of an adult zebrafish, stained with haematoxylin and eosin. (C), spores in the ovigerous stroma of the same adult zebrafish, adjacent to developing follicles, stained with haematoxylin and eosin. (D), spore aggregate in the ovary associated with numerous inflammatory cells, namely macrophages. Orange arrows, xenomas; yellow arrows, macrophages. Staining, haematoxylin and eosin. Original magnification 10× (C) and 20× (A,B,D). Credits: Tânia Carvalho, Champalimaud Foundation.

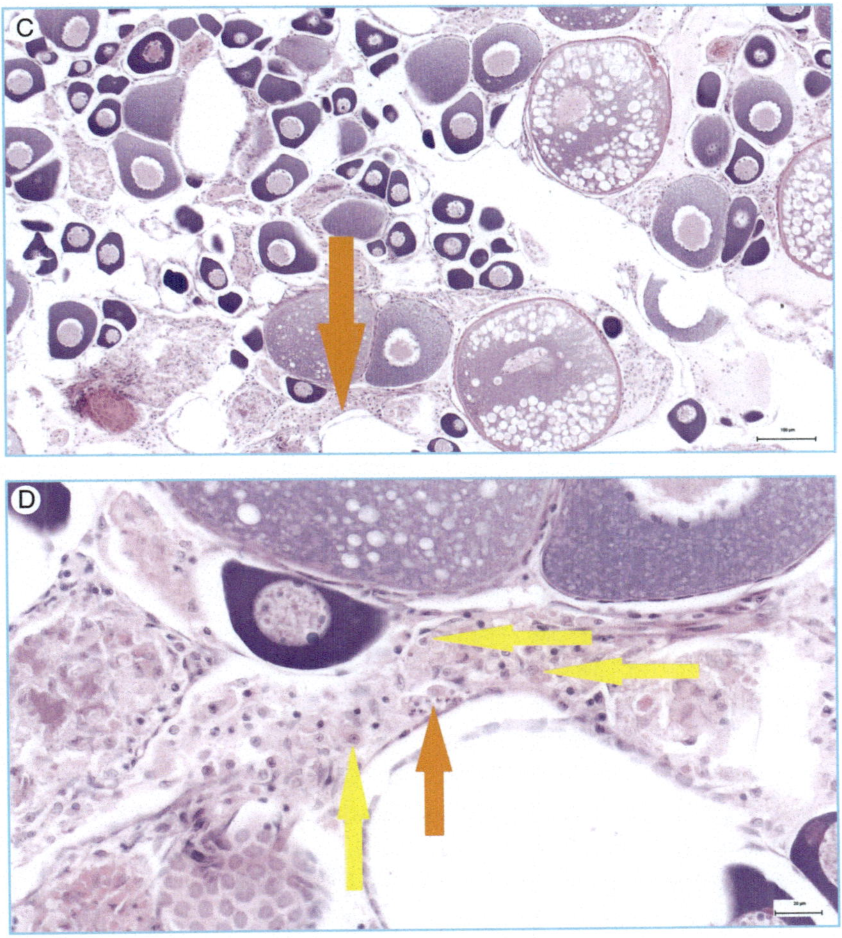

Figure 5.8 (continued)

Pleistophora hyphessobryconis

This Microsporidian parasite has a wide host range, its colloquial name being 'neon tetra disease' due to its prevalence in ornamental fish of certain species. *P. hyphessobryconis* has a tropism for skeletal muscle, where large numbers of spores are produced and cause muscle damage. Infected fish often appear lethargic and may exhibit weight loss, abnormal swimming, spinal curvature and distinctive deep-white nodules throughout the musculature. Asymptomatic fish have also been detected. Necropsy reveals skeletal muscle nodules that are soft, white and oedematous. Microscopically the skeletal muscle fibres may be disrupted by spores and associated with varying amounts of inflammation.

The three incidents of reported infections in laboratory zebrafish were associated with the mixing of pet store-bought fish into the aquarium, even after a period in quarantine (Sanders et al., 2010). Transmission is presumed to occur via ingestion of spores excreted or released from dead fish. The effectiveness of sporicidal action of regular bleaching is unknown, but spores may be as resistant as *P. neurophilia* and therefore difficult to eliminate.

Ichthyophthirius multifiliis

This ciliated protozoal parasite, known as 'Ich', infects the gills, fins and skin of a wide range of fish (von Gersdorff Jørgensen, 2017). No natural infections in zebrafish aquariums have been reported, but zebrafish have been successfully experimentally infected (predominantly for immunological studies; Wang et al., 2019). The life cycle includes a trophont phase that feeds on the fish's epithelial layer and mucus followed by a free-swimming phase, which progresses to produce numerous motile infective theronts that attach to the fish's skin. The disease is distinctive because small white nodules/trophonts form all over the fish's body. Damage to gills may cause respiratory difficulties and mortality. Lesions of the skin and mucosal layer integrity allow opportunistic pathogens to establish secondary bacterial or fungal infections. If the fish survives it may develop immunity. Treatments are described in the literature and could be considered if rare, genetically altered strains need to be preserved (Picón-Camacho et al., 2012).

Piscinoodinium pillulare

This motile dinoflagellate parasite causes 'velvet disease' in a number of fish species, especially from petshops, but has never been reported in laboratory zebrafish. It has free-living motile life stages that attach to the fish's gills and skin, especially around the dorsal fin. This is thought to cause irritation as fish may be seen trying to 'rub' themselves or flip around in corners of the tank. Fish may be lethargic and show reduced fecundity (Mocho et al., 2022a).

Multicellular parasites

P. tomentosa

Helminths affect all vertebrates, and many species are known to infect fish. There are numerous reports of laboratory zebrafish becoming infected with this intestinal nematode parasite. Transmission occurs via ingestion of excreted eggs and can occur directly, without the need for an intermediate host (Moravec et al., 1987). The helminths are invasive and cause pathological changes in the epithelium and lamina propria of the proximal intestine. This induces variable levels of inflammation, which can progress to epithelial hyperplasia, chronic granuloma formation and even peritonitis. Infection has also been associated with a possible increased incidence of intestinal neoplasia, especially at the intestinal–oesophageal junction (Kent et al., 2002, 2021). Clinically, infected fish may present as emaciated and lethargic although asymptomatic infections have been observed (Kent et al., 2012). Infection of zebrafish with *P. tomentosa* has been shown to cause alterations in the balance of the gut microbiome (Gaulke et al., 2019); the significance of this is currently unknown but could potentially impact disease models.

Due to damage and inflammation of the intestine, nutrient absorption is likely to be disrupted, which will therefore impact metabolic parameters and potentially behavioural aspects too, especially if the damage causes pain. Emaciated fish may show reduced fecundity, which is often used as a scientific output to indicate good welfare and health. The impact on the immune system (e.g. increases in intestinal eosinophil numbers that have been observed in infected fish) and microbiome, as well as potentially increased neoplasia risk, should be considered as potential confounds, especially for infection or toxicity studies. *P. tomentosa* infection therefore has the potential to impact numerous studies in an insidious manner that is difficult to quantify.

Diagnosis is typically by post-mortem and histological analysis of sections of intestines containing readily apparent cross-sections of worms and/or eggs. PCR can be used as a diagnostic test and for more rapid health screening of pooled fish samples. Environmental screening has also been shown to be useful for the detection of eggs in the tank sludge by microscopy or PCR (Mocho et al., 2017).

Bleaching embryos will not kill the nematode eggs. Because these are present only externally, their removal is aided by agitation during the bleaching process (see Chapter 7 for more information). Treatment of infected zebrafish colonies has been demonstrated using anthelmintic drugs such as fenbendazole and ivermectin, although adverse effects on transgenic lines and experimental parameters would need to be fully assessed before embarking on such a programme (Collymore et al., 2014; Maley et al., 2013). Several different methods have been proposed to disinfect a recirculating aquarium system following infection with capillarid worms: these involve heat, high levels of chlorine and/or desiccation (Martins et al., 2017). Cryopreservation kills the nematode eggs (Norris et al., 2018), so cryopreserved sperm and eggs can be used to restock the aquarium (after screening for other pathogens such as *P. neurophilia*).

Myxidium streisingeri

This is an obligate intracellular Myxozoan microbe (multicellular) commonly found in zebrafish colonies (Kent et al., 2020; Whipps et al., 2015). Most Myxozoa species require an intermediate host for transmission; this is often a worm but can be another fish species (e.g. the Myxozoan agent of proliferative kidney disease, which causes high mortality in salmonids, uses several fish species as hosts, but direct transmission between salmon has not been observed). The species and habitat of the intermediate host in a recirculating laboratory zebrafish system is unknown. From 2006 to 2018 the

ZIRC diagnostic service found that 22% of facilities had positive results for this microbe (Kent et al., 2020). Spores have been found in the ducts of zebrafish kidneys, but no pathology is reported. More research is required to determine whether *M. streisingeri* is a commensal or potential pathogen and what, if any, the impacts on research may be.

Fungi

Fungal infections observed in zebrafish are usually opportunistic, occurring secondarily to a primary disease or stressor.

Lecythophora mutabilis

A research zebrafish colony in the USA reported an incident of high mortality in young fry (5–24 days post hatching; Dykstra et al., 2001). Fish had material protruding from the opercula, mouth and anus that was found to contain fungal hyphae, mixed bacterial populations and protozoa. Affected fish may have starved or died from asphyxiation. The epizootic was controlled by improving the aquarium's water filtration and sterilization systems. Rarely, *L. mutabilis* has been reported as an opportunistic pathogen in human patients.

Saprolegnia *spp.*

Several species of *Saprolegnia* have occasionally been detected in zebrafish colonies and have been used to infect zebrafish experimentally (Kent et al., 2020; Shin et al., 2017). The hyphae can be observed as a fluffy white layer around affected fish, which are likely to have a pre-existing disease and/or wound.

Conclusion

In all zebrafish colonies there are likely to be a number of pathogens present unless significant effort has been made to exclude these to create specific-pathogen-free (SPF) or germ-free colonies. These measures come at a high cost and are normally not achievable by many facilities. Many of these pathogens exist subclinically, with overt clinical disease occurring in large numbers of fish only when there are husbandry issues, or in individual fish with concurrent health issues. The presence of subclinical disease may result in a lack of standardization between individual fish and may act as an unknown variable and hence confound studies that do not take it into consideration. The increasing use of genetically altered immunocompromised strains and more advanced surgical models increases the risk that the incidence of clinical disease may increase. For these models that are at increased risk, it is worth considering keeping these fish in stand-alone, isolated, flow-through racks that can more easily be maintained as SPF (see Chapter 7 for more information).

References

Aubry, A., Mougari, F., Reibel, F., & Cambau, E. (2017). Mycobacterium marinum. *Microbiology Spectrum*, *5*(2). https://doi.org/10.1128/microbiolspec.TNMI7-0038-2016

Barton, C. L., Johnson, E. W., & Tanguay, R. L. (2016). Facility Design and Health Management Program at the Sinnhuber Aquatic Research Laboratory. *Zebrafish*, *13*, Suppl. 1, S39–S43. https://doi.org/10.1089/zeb.2015.1232

Bermúdez, R., Losada, A. P., de Azevedo, A. M., Guerra-Varela, J., Pérez-Fernández, D., Sánchez, L., Padrós, F., Nowak, B., & Quiroga, M. I. (2018). First description of a natural infection with spleen and kidney necrosis virus in zebrafish. *Journal of Fish Diseases*, *41*(8), 1283–1294. https://doi.org/10.1111/jfd.12822

Chang, C. T., Colicino, E. G., DiPaola, E. J., Al-Hasnawi, H. J., & Whipps, C. M. (2015). Evaluating the effectiveness of common disinfectants at preventing the propagation of Mycobacterium spp. isolated from zebrafish. *Comparative Biochemistry and Physiology. Toxicology and Pharmacology*, 178, 45–50. https://doi.org/10.1016/j.cbpc.2015.09.008

Chang, C. T., Benedict, S., & Whipps, C. M. (2019a). Transmission of Mycobacterium chelonae and Mycobacterium marinum in laboratory zebrafish through live feeds. *Journal of Fish Diseases, 42*(10), 1425–1431. https://doi.org/10.1111/jfd.13071

Chang, C. T., Lewis, J., & Whipps, C. M. (2019b). Source or sink: Examining the role of biofilms in transmission of Mycobacterium spp. in laboratory zebrafish. *Zebrafish, 16*(2), 197–206. https://doi.org/10.1089/zeb.2018.1689

Chow, F. W., Xue, L., & Kent, M. L. (2016). Retrospective study of the prevalence of Pseudoloma neurophilia shows male sex bias in zebrafish Danio rerio (Hamilton-Buchanan). *Journal of Fish Diseases, 39*(3), 367–370. https://doi.org/10.1111/jfd.12328

Collymore, C., Watral, V., White, J. R., Colvin, M. E., Rasmussen, S., Tolwani, R. J., & Kent, M. L. (2014). Tolerance and efficacy of emamectin benzoate and ivermectin for the treatment of Pseudocapillaria tomentosa in laboratory zebrafish (Danio rerio). *Zebrafish, 11*(5), 490–497. https://doi.org/10.1089/zeb.2014.1021

Crim, M. J., & Riley, L. K. (2012). Viral diseases in zebrafish: What is known and unknown. *ILAR Journal, 53*(2), 135–143. https://doi.org/10.1093/ilar.53.2.135

Cronan, M. R., & Tobin, D. M. (2014). Fit for consumption: Zebrafish as a model for tuberculosis. *Disease Models and Mechanisms, 7*(7), 777–784. https://doi.org/10.1242/dmm.016089

Declercq, A. M., Haesebrouck, F., Van den Broeck, W., Bossier, P., & Decostere, A. (2013). Columnaris disease in fish: A review with emphasis on bacterium-host interactions. *Veterinary Research, 44*(1), 27. https://doi.org/10.1186/1297-9716-44-27

Decostere, A., Hermans, K., & Haesebrouck, F. (2004). Piscine mycobacteriosis: A literature review covering the agent and the disease it causes in fish and humans. *Veterinary Microbiology, 99*(3–4), 159–166. https://doi.org/10.1016/j.vetmic.2003.07.011

Dykstra, M. J., Astrofsky, K. M., Schrenzel, M. D., Fox, J. G., Bullis, R. A., Farrington, S., Sigler, L., Rinaldi, M. G., & McGinnis, M. R. (2001). High mortality in a large-scale zebrafish colony (Brachydanio rerio Hamilton and Buchanan, 1822) associated with Lecythophora mutabilis (van Beyma) W. Gams and McGinnis. *Comparative Medicine, 51*(4), 361–368.

Estes, J. M., Altemara, M. L., Crim, M. J., Fletcher, C. A., & Whitaker, J. W. (2021). Behavioral and Reproductive Effects of Environmental Enrichment and Pseudoloma neurophilia infection on Adult zebrafish (Danio rerio). *Journal of the American Association for*

Laboratory Animal Science, *60*(3), 249–258. https://doi.org/10.30802/AALAS-JAALAS-20-000113

Falkinham, J. O., 3rd, Norton, C. D., & LeChevallier, M. W. (2001). Factors influencing numbers of Mycobacterium avium, Mycobacterium intracellulare, and other Mycobacteria in drinking water distribution systems. *Applied and Environmental Microbiology*, *67*(3), 1225–1231. https://doi.org/10.1128/AEM.67.3.1225-1231.2001

Ferguson, J. A., Watral, V., Schwindt, A. R., & Kent, M. L. (2007). Spores of two fish Microsporidia (Pseudoloma neurophilia and Glugea anomala) are highly resistant to chlorine. *Diseases of Aquatic Organisms*, *76*(3), 205–214. https://doi.org/10.3354/dao076205

Gaulke, C. A., Martins, M. L., Watral, V. G., Humphreys, I. R., Spagnoli, S. T., Kent, M. L., & Sharpton, T. J. (2019). A longitudinal assessment of host–microbe-parasite interactions resolves the zebrafish gut microbiome's link to Pseudocapillaria tomentosa infection and pathology. *Microbiome*, *7*(1), 10. https://doi.org/10.1186/s40168-019-0622-9

Harriff, M. J., Bermudez, L. E., & Kent, M. L. (2007). Experimental exposure of zebrafish, Danio rerio (Hamilton), to Mycobacterium marinum and Mycobacterium peregrinum reveals the gastrointestinal tract as the primary route of infection: A potential model for environmental mycobacterial infection. *Journal of Fish Diseases*, *30*(10), 587–600. https://doi.org/10.1111/j.1365-2761.2007.00839.x

Hawke, J. P., Kent, M., Rogge, M., Baumgartner, W., Wiles, J., Shelley, J., Savolainen, L. C., Wagner, R., Murray, K., & Peterson, T. S. (2013). Edwardsiellosis caused by Edwardsiella ictaluri in laboratory populations of zebrafish Danio rerio. *Journal of Aquatic Animal Health*, *25*(3), 171–183. https://doi.org/10.1080/08997659.2013.782226

Jørgensen, L. V. G. (2017). The fish parasite Ichthyophthirius multifiliis – Host immunology, vaccines and novel treatments. *Fish and Shellfish Immunology*, *67*, 586–595. https://doi.org/10.1016/j.fsi.2017.06.044

Jørgensen, L. V. G., Korbut, R., Jeberg, S., Kania, P. W., & Buchmann, K. (2018). Association between adaptive immunity and neutrophil dynamics in zebrafish (Danio rerio) infected by a parasitic ciliate. *PloS One*. Public Library of Science ONE, *13*(9), e0203297. https://doi.org/10.1371/journal.pone.0203297

Ke, X., Wang, J., Gu, Z., Li, M., & Gong, X. (2009). Saprolegnia brachydanis, a new oomycete isolated from zebra fish. *Mycopathologia*, *167*(2), 107–113. https://doi.org/10.1007/s11046-008-9150-z

Kent, M. L., Bishop-Stewart, J. K., Matthews, J. L., & Spitsbergen, J. M. (2002). Pseudocapillaria tomentosa, a nematode pathogen, and associated neoplasms of zebrafish (Danio rerio) kept in research colonies. *Comparative Medicine*, *52*(4), 354–358.

Kent, M. L., Whipps, C. M., Matthews, J. L., Florio, D., Watral, V., Bishop-Stewart, J. K., Poort, M., & Bermudez, L. (2004). Mycobacteriosis in zebrafish (Danio rerio) research facilities. *Comparative Biochemistry and Physiology. Toxicology and Pharmacology*, 138(3), 383–390. https://doi.org/10.1016/j.cca.2004.08.005

Kent, M. L., Spitsbergen, J. M., Matthews, J. L., Fournie, J. W., Murray, K. N., & Westerfield, M. (2016). Diseases of zebrafish in research facilities. https://zebrafish.org/wiki/health/disease_manual/start Retrieved April 15, 2023

Kent, M. L., Harper, C., & Wolf, J. C. (2012). Documented and potential research impacts of subclinical diseases in zebrafish. *ILAR Journal*, *53*(2), 126–134. https://doi.org/10.1093/ilar.53.2.126

Kent, M. L., Sanders, J. L., Spagnoli, S., Al-Samarrie, C. E., & Murray, K. N. (2020). Review of diseases and health management in zebrafish Danio rerio (Hamilton 1822) in research facilities. *Journal of Fish Diseases*, *43*(6), 637–650. https://doi.org/10.1111/jfd.13165

Kent, M. L., Wall, E. S., Sichel, S., Watral, V., Stagaman, K., Sharpton, T. J., & Guillemin, K. (2021). Pseudocapillaria tomentosa, Mycoplasma spp., and intestinal lesions in experimentally infected zebrafish Danio rerio. *Zebrafish*, *18*(3), 207–220. https://doi.org/10.1089/zeb.2020.1955

LaPatra, S. E., Barone, L., Jones, G. R., & Zon, L. I. (2000). Effects of infectious hematopoietic necrosis virus and infectious pancreatic necrosis virus infection on hematopoietic precursors of the zebrafish. *Blood Cells, Molecules and Diseases*, *26*(5), 445–452. https://doi.org/10.1006/bcmd.2000.0320

Maley, D., Laird, A. S., Rinkwitz, S., & Becker, T. S. (2013). A simple and efficient protocol for the treatment of zebrafish colonies infected with parasitic nematodes. *Zebrafish*, *10*(3), 447–450. https://doi.org/10.1089/zeb.2013.0868

Martins, M. L., Watral, V., Rodrigues-Soares, J. P., & Kent, M. L. (2017). A method for collecting eggs of Pseudocapillaria tomentosa (Nematoda: Capillariidae) from zebrafish Danio rerio and efficacy of heat and chlorine for killing the nematode's eggs. *Journal of Fish Diseases*, *40*(2), 169–182. https://doi.org/10.1111/jfd.12501

Meijer, A. H., Verbeek, F. J., Salas-Vidal, E., Corredor-Adámez, M., Bussman, J., van der Sar, A. M., Otto, G. W., Geisler, R. & Spaink, H. P. (2005). Transcriptome profiling of adult zebrafish at the late stage of chronic tuberculosis due to *Mycobacterium marinum* infection. *Molecular Immunology, 42*(10), 1185–1203. doi: 10.1016/j.molimm.2004.11.014

Midttun, H. L. E., Vindas, M. A., Nadler, L. E., Øverli, Ø., & Johansen, I. B. (2020a). Behavioural effects of the common brain-infecting parasite Pseudoloma neurophilia in laboratory zebrafish (Danio rerio). *Scientific Reports, 10*(1), 8083. https://doi.org/10.1038/s41598-020-64948-8

Midttun, H. L. E., Vindas, M. A., Whatmore, P. J., Øverli, Ø., & Johansen, I. B. (2020b). Effects of Pseudoloma neurophilia infection on the brain transcriptome in zebrafish (Danio rerio). *Journal of Fish Diseases, 43*(8), 863–875. https://doi.org/10.1111/jfd.13198

Mocho, J. P. (2016). Three-dimensional screen: A comprehensive approach to the health monitoring of zebrafish. *Zebrafish, 13*, Suppl. 1, S132–S137. https://doi.org/10.1089/zeb.2015.1200

Mocho, J. P., Martin, D. J., Millington, M. E., & Saavedra Torres, Y. (2017). Environmental screening of Aeromonas hydrophila, Mycobacterium spp., and Pseudocapillaria tomentosa in zebrafish systems. *Journal of Visualized Experiments: JoVE*, (130), 55306. https://doi.org/10.3791/55306

Mocho, J. P., Collymore, C., Farmer, S. C., Leguay, E., Murray, K. N., & Pereira, N. (2022a). FELASA-AALAS recommendations for monitoring and reporting of laboratory fish diseases and health status, with an emphasis on zebrafish (Danio rerio). *Comparative Medicine, 72*(3), 127–148. https://doi.org/10.30802/AALAS-CM-22-000034

Mocho, J. P., Collymore, C., Farmer, S. C., Leguay, E., Murray, K. N., & Pereira, N. (2022b). FELASA-AALAS recommendations for biosecurity in an aquatic facility, including prevention of zoonosis, introduction of new fish colonies, and quarantine. *Comparative Medicine, 72*(3), 149–168. https://doi.org/10.30802/AALAS-CM-22-000042

Moravec, F., Prokopic, J., & Shlikas, A. V. (1987). The biology of nematodes of the family Capillariidae Neveu-Lemaire, 1936. *Folia Parasitologica, 34*(1), 39–56.

Murray, K. N., Bauer, J., Tallen, A., Matthews, J. L., Westerfield, M., & Varga, Z. M. (2011). Characterization and management of asymptomatic Mycobacterium infections at the Zebrafish International Resource Center. *Journal of the American Association for Laboratory Animal Science, 50*(5), 675–679.

Murray, K. N., Dreska, M., Nasiadka, A., Rinne, M., Matthews, J. L., Carmichael, C., Bauer, J., Varga, Z. M., & Westerfield, M. (2011). Transmission, diagnosis, and recommendations for control of Pseudoloma neurophilia infections in laboratory zebrafish (Danio rerio) facilities. *Comparative Medicine*, *61*(4), 322–329.

Norris, L. J., Watral, V., & Kent, M. L. (2018). Survival of bacterial and parasitic pathogens from zebrafish (Danio rerio) after cryopreservation and thawing. *Zebrafish*, *15*(2), 188–201. https://doi.org/10.1089/zeb.2017.1528

Pack, M., Belak, J., Boggs, C., Fishman, M., & Driever, W. (1995). Intestinal capillariasis in zebrafish. *Zebrafish Science Monitor*, *3*(4), 1–3.

Pakingking, R., Takano, R., Nishizawa, T., Mori, K., Iida, Y., Arimoto, M., & Muroga, K. (2003). Experimental coinfection with Aquabirnavirus and viral hemorrhagic septicemia virus (VHSV), Edwardsiella tarda or Streptococcus iniae in Japanese flounder Paralichthys olivaceus. *Fish Pathology*, *38*(1), 15–21. https://doi.org/10.3147/jsfp.38.15

Parikka, M., Hammarén, M. M., Harjula, S. K., Halfpenny, N. J., Oksanen, K. E., Lahtinen, M. J., Pajula, E. T., Iivanainen, A., Pesu, M., & Rämet, M. (2012). Mycobacterium marinum causes a latent infection that can be reactivated by gamma irradiation in adult zebrafish. *PLoS Pathogens*, *8*(9), e1002944. https://doi.org/10.1371/journal.ppat.1002944

Peterson, T. S., Ferguson, J. A., Watral, V. G., Mutoji, K. N., Ennis, D. G., & Kent, M. L. (2013). Paramecium caudatum enhances transmission and infectivity of Mycobacterium marinum and M. chelonae in zebrafish Danio rerio. *Diseases of Aquatic Organisms*, *106*(3), 229–239. https://doi.org/10.3354/dao02649

Petrie-Hanson, L., Romano, C. L., Mackey, R. B., Khosravi, P., Hohn, C. M., & Boyle, C. R. (2007). Evaluation of zebrafish Danio rerio as a model for enteric septicemia of catfish (ESC). *Journal of Aquatic Animal Health*, *19*(3), 151–158. https://doi.org/10.1577/H06-026.1

Picón-Camacho, S. M., Marcos-Lopez, M., Bron, J. E., & Shinn, A. P. (2012). An assessment of the use of drug and non-drug interventions in the treatment of Ichthyophthirius multifiliis Fouquet, 1876, a protozoan parasite of freshwater fish. *Parasitology*, *139*(2), 149–190. https://doi.org/10.1017/S0031182011001867

Pullium, J. K., Dillehay, D. L., & Webb, S. (1999). High mortality in zebrafish (Danio rerio). *Contemporary Topics in Laboratory Animal Science*, *38*(3), 80–83.

Rácz, A., Dwyer, T., & Killen, S. S. (2019). Overview of a disease outbreak and introduction of a step-by-step protocol for the eradication of Mycobacterium haemophilum in a zebrafish system. *Zebrafish, 16*(1), 77–86. https://doi.org/10.1089/zeb.2018.1628

Ramsay, J. M., Watral, V., Schreck, C. B., & Kent, M. L. (2009a). Husbandry stress exacerbates mycobacterial infections in adult zebrafish, Danio rerio (Hamilton). *Journal of Fish Diseases, 32*(11), 931–941. https://doi.org/10.1111/j.1365-2761.2009.01074.x

Ramsay, J. M., Watral, V., Schreck, C. B., & Kent, M. L. (2009b). Pseudoloma neurophilia infections in zebrafish Danio rerio: Effects of stress on survival, growth, and reproduction. *Diseases of Aquatic Organisms, 88*(1), 69–84. https://doi.org/10.3354/dao02145

Sanders, J. L., Lawrence, C., Nichols, D. K., Brubaker, J. F., Peterson, T. S., Murray, K. N., & Kent, M. L. (2010). Pleistophora hyphessobryconis (Microsporidia) infecting zebrafish Danio rerio in research facilities. *Diseases of Aquatic Organisms, 91*(1), 47–56. https://doi.org/10.3354/dao02245

Sanders, J. L., Watral, V., & Kent, M. L. (2012). Microsporidiosis in zebrafish research facilities. *ILAR Journal, 53*(2), 106–113. https://doi.org/10.1093/ilar.53.2.106

Sanders, J. L., Watral, V., Clarkson, K., & Kent, M. L. (2013). Verification of intraovum transmission of a microsporidium of vertebrates: Pseudoloma neurophilia infecting the Zebrafish, Danio rerio. PloS One, 8(9), e76064. https://doi.org/10.1371/journal.pone.0076064

Sanders, J. L., Monteiro, J. F., Martins, S., Certal, A. C., & Kent, M. L. (2020). The Impact of Pseudoloma neurophilia Infection on Body Condition of zebrafish. *Zebrafish, 17*(2), 139–146. https://doi.org/10.1089/zeb.2019.1812

Shin, S., Kulatunga, D. C. M., Dananjaya, S. H. S., Nikapitiya, C., Lee, J., & De Zoysa, M. (2017). Saprolegnia parasitica isolated from rainbow trout in Korea: Characterization, anti-Saprolegnia activity and host pathogen interaction in zebrafish disease model. *Mycobiology, 45*(4), 297–311. https://doi.org/10.5941/MYCO.2017.45.4.297

Spagnoli, S., Xue, L., & Kent, M. L. (2015). The common neural parasite Pseudoloma neurophilia is associated with altered startle response habituation in adult zebrafish (Danio rerio): Implications for the zebrafish as a model organism. *Behavioural Brain Research, 291*, 351–360. https://doi.org/10.1016/j.bbr.2015.05.046

Spagnoli, S., Sanders, J., & Kent, M. L. (2017). The common neural parasite Pseudoloma neurophilia causes altered shoaling behaviour in adult laboratory zebrafish (Danio rerio) and its implications for neurobehavioural research. *Journal of Fish Diseases*, *40*(3), 443–446. https://doi.org/10.1111/jfd.12512

Spagnoli, S. T., Sanders, J. L., Watral, V., & Kent, M. L. (2016). Pseudoloma neurophilia Infection Combined with gamma Irradiation Causes Increased Mortality in Adult zebrafish (Danio rerio) Compared to Infection or Irradiation Alone: New Implications for Studies Involving immunosuppression. *Zebrafish*, *13*, Suppl. 1, S107–S114. https://doi.org/10.1089/zeb.2015.1223

Wang, Q., Yu, Y., Zhang, X., & Xu, Z. (2019). Immune responses of fish to Ichthyophthirius multifiliis (Ich): A model for understanding immunity against protozoan parasites. *Developmental and Comparative Immunology*, *93*, 93–102. https://doi.org/10.1016/j.dci.2019.01.002

Watral, V., & Kent, M. L. (2007). Pathogenesis of Mycobacterium spp. in zebrafish (Danio rerio) from research facilities. *Comparative Biochemistry and Physiology. Toxicology and Pharmacology*, *145*(1), 55–60. https://doi.org/10.1016/j.cbpc.2006.06.004

West, K., Miles, R., Kent, M. L., & Frazer, J. K. (2014). Unusual fluorescent granulomas and myonecrosis in Danio rerio infected by the microsporidian pathogen Pseudoloma neurophilia. *Zebrafish*, *11*(3), 283–290. https://doi.org/10.1089/zeb.2013.0933

Whipps, C. M., Dougan, S. T., & Kent, M. L. (2007). Mycobacterium haemophilum infections of zebrafish (Danio rerio) in research facilities. *FEMS Microbiology Letters*, *270*(1), 21–26. https://doi.org/10.1111/j.1574-6968.2007.00671.x

Whipps, C. M., Matthews, J. L., & Kent, M. L. (2008). Distribution and genetic characterization of Mycobacterium chelonae in laboratory zebrafish Danio rerio. *Diseases of Aquatic Organisms*, *82*(1), 45–54. https://doi.org/10.3354/dao01967

Whipps, C. M., Lieggi, C., & Wagner, R. (2012). Mycobacteriosis in zebrafish colonies. *ILAR Journal*, *53*(2), 95–105. https://doi.org/10.1093/ilar.53.2.95

Whipps, C. M., Moss, L. G., Sisk, D. M., Murray, K. N., Tobin, D. M., & Moss, J. B. (2014). Detection of autofluorescent Mycobacterium chelonae in living zebrafish. *Zebrafish*, *11*(1), 76–82. https://doi.org/10.1089/zeb.2012.0863

Whipps, C. M., Murray, K. N., & Kent, M. L. (2015). Occurrence of a myxozoan parasite Myxidium streisingeri n. sp. in laboratory zebrafish Danio rerio. *Journal of Parasitology, 101*(1), 86–90. https://doi.org/10.1645/14-613.1

Xu, X., Zhang, L., Weng, S., Huang, Z., Lu, J., Lan, D., Zhong, X., Yu, X., Xu, A., & He, J. (2008). A zebrafish (Danio rerio) model of infectious spleen and kidney necrosis virus (ISKNV) infection. *Virology, 376*(1), 1–12. https://doi.org/10.1016/j.virol.2007.12.026

Chapter 6

Adult zebrafish as a vertebrate model of ageing

Pam S. Ellis, Raquel R. Martins and Catarina M. Henriques

Abstract

Age is the greatest risk factor for disease. Until recently, ageing research was focused on studying specific diseases of ageing. However, most people in old age suffer from multimorbidity. Understanding the common primary drivers of disease in ageing is therefore an urgent need. For this we need animal models that share fundamental hallmarks of ageing with humans. One such hallmark is restricted telomerase expression and consequent shortening and dysfunction of telomeres with ageing. Zebrafish have been shown to have human-like telomere length and, accordingly, telomerase-deficient zebrafish (*tert*$^{-/-}$) age and die prematurely, providing a valuable accelerated model of ageing that allows researchers to identify telomerase-dependent pathways that contribute to disease. Here we describe the fundamental pathological, cellular and molecular phenotypes of natural ageing and point towards those that have been identified as being telomerase dependent.

Keywords

zebrafish, ageing, telomeres, telomerase, DNA damage, senescence

Claire Allen and Jean-Philippe Mocho (eds) **Zebrafish**
DOI: 10.1079/9781800629431.0006, © CAB International, 2024

Introduction

Age is the greatest risk factor for chronic diseases. These pose serious medical and socio-economic challenges, already comprising about 80% of healthcare costs in the developed world. Because the number of people over 65 is predicted to double in the next 50 years (European Commission, 2015), developing effective therapies for debilitating chronic diseases of ageing, improving quality of life and reducing healthcare costs is an urgent goal (Rae et al., 2010). Recognizing this, the European Commission has highlighted ageing research as a priority in the Horizon 2020 programme. To date, ageing research has mainly focused on single diseases. However, this approach does not address the fact that approximately 60% of people over 65 years suffer from multiple diseases at the same time, termed multimorbidity (Fabbri et al., 2015; Tinetti et al., 2012). There is, therefore, an urgent need to understand the fundamental hallmark mechanisms that go awry with ageing and underlie multiple chronic diseases of ageing. To face this challenge and determine and test new therapeutic targets and interventions, we need to use model organisms that have hallmarks of ageing similar to those of humans. Here we describe the key aspects of zebrafish natural ageing, highlight the best-described premature model of zebrafish ageing, the telomerase (*tert*$^{-/-}$) mutant, and refer to key methodologies from both published and unpublished sources (methods available upon request).

The zebrafish as a telomere-dependent vertebrate model for ageing

One important hallmark of ageing in humans is telomere dysfunction. Critically short telomeres cause cells to stop dividing and subsequently they either die by apoptosis or enter a state of 'dormancy', termed senescence. Importantly, senescent cells accumulate aberrantly with ageing in humans (Dimri et al., 1995), mice (Krishnamurthy et al., 2004) and zebrafish (Carneiro et al., 2016b; Henriques et al., 2013). There is strong evidence implicating telomere dysfunction and senescence in age-associated pathologies such as atherosclerosis

(Minamino and Komuro, 2007), arthritis (Price et al., 2002), liver cirrhosis (Wiemann et al., 2002), chronic obstructive pulmonary disease (COPD) (Amsellem et al., 2011) and cancer (Campisi, 2013). Research suggests that this accumulation is likely to happen because senescent cells secrete several tissue remodelling factors including matrix metalloproteinases (MMPs), pro-inflammatory cytokines and chemokines, termed the senescence-associated secretory phenotype (SASP) (Coppe et al., 2010), creating a chronic inflammatory state in the elderly (Ohtani & Hara, 2013). Progressive telomere dysfunction is therefore a major player in the impairment of tissue homeostasis with age in a cell-autonomous manner, by impairing cell proliferation, and in a non-cell-autonomous manner, by contributing to the accumulation of senescent cells in tissues (Atzmon et al., 2010; Campisi, 2013; Henriques & Ferreira, 2012; Njajou et al., 2007, 2009; Pereira & Ferreira, 2013; Shay & Wright, 2007).

The zebrafish has recently emerged as a powerful complementary model to investigate the fundamental mechanisms of ageing that underlie disease (Carneiro et al., 2016a). Like humans, and unlike most laboratory mice, both naturally aged (Carneiro et al., 2016b) and telomerase-mutant zebrafish (*tert*$^{-/-}$) (Anchelin et al., 2013; Henriques et al., 2013) accumulate telomere-dependent replicative senescent cells and inflammation in aged tissues (Carneiro et al., 2016b; Henriques et al., 2013). The progressive shortening of telomeres, accumulation of senescence and inflammation in *tert*$^{-/-}$ tissues over a relatively short period of time (6–12 months as compared with 18–36 in wild type (WT)) (Carneiro et al., 2016b; Henriques et al., 2013) allows an exceptional temporal analysis from young to old *tert*$^{-/-}$ throughout the life course of the animal.

Previous work showed that zebrafish tissue ageing, as assessed by tissue homeostasis markers including proliferation, DNA damage, cell death and senescence, does not occur in all tissues at the same time. Zebrafish age-dependent tissue degeneration occurs in a time- and tissue-dependent manner and, for most tissues, it is anticipated and exacerbated in the absence of telomerase (Anchelin et al., 2013; Carneiro et al., 2016b; Henriques et al., 2013). The authors showed that, in zebrafish, some critical tissues age before others, such as the intestine and muscle, and that this is tightly accompanied by telomere shortening and DNA damage in natural, WT ageing (Carneiro et al., 2016b). In fact, DNA damage at telomeres tightly predicted cellular damage, gut inflammation and myocyte degeneration (Carneiro et al., 2016b). Importantly, this is reminiscent of the human scenario in which telomerase loss-of-function mutations or mutations affecting telomere stability lead to premature ageing syndromes (progerias), with particular impact on highly proliferative tissues (Alter et al., 2012; Hofer et al., 2005). This type of study benefited from the ease of imaging whole adult zebrafish on a slide, allowing the authors to take an unbiased approach, performing different immunostainings across whole adult zebrafish paraffin sections, and imaging and quantifying most of the tissues. In parallel, the authors strengthened their conclusions by quantifying the same parameters at the gene expression level using real-time quantitative polymerase chain reaction (RT-QPCR; Carneiro et al., 2016b; Henriques et al., 2013). Given that these were the key techniques used to establish the kinetics of ageing across the different zebrafish tissues, and that they are highly comparable with what has been used to characterize human and mice ageing, it is a good place to start when studying ageing in any model. Below we highlight the key discoveries in zebrafish.

Importantly, these analyses were performed in animals of varying age, comparing WT with telomerase mutant siblings derived from a heterozygous cross, to ensure these were the first generation of mutants (Henriques et al., 2013). The authors also ensured that the mutant had been outcrossed to WT unrelated animals at least five times, to minimize the possibility of other deleterious mutations. By comparing prematurely aged *tert*$^{-/-}$ side by side with naturally ageing WT siblings, the authors were able to determine which features of WT ageing were being anticipated in the *tert*$^{-/-}$, highlighting the naturally ageing phenotypes that are likely to depend on telomerase (Carneiro et al., 2016b). We propose that a similar strategy should be used to determine whether other mutants also age prematurely, and, if so, which cellular and tissue-ageing phenotypes are anticipated, and thereby likely to be dependent on the particular mutated or deleted gene. It is important to realize, however, that any natural age-associated pathologies are unlikely to have 100%

penetrance, so studying natural, WT ageing requires larger numbers than genetically modified, premature/accelerated models of ageing. As in humans, not every single aged individual will develop cataracts, diabetes, heart failure, Alzheimer's disease or cancer. It is the increased probability of developing age-associated diseases that is important to quantify, and the challenge is to determine and directly test whether the cellular and molecular changes accompanying this degeneration are causal and whether they can be prevented or reversed therapeutically.

Fundamental observations in WT zebrafish ageing – establishing the equivalent of the 'frailty index'

As in humans, zebrafish display various age-associated phenotypes that become more frequent the longer the animals live. Examples of these are: spinal curvature, development of cataracts, cancer, infections and an overall loss of body mass (Carneiro et al., 2016b) (Figure 6.1A). Zebrafish also display altered behaviour, and this has been published (Gilbert et al., 2014), and protocols are described elsewhere in this book (see Chapter 12 for more information). The loss of body mass translates into a decreased width/length body ratio (Carneiro et al., 2016b; Henriques et al., 2013). Measuring this ratio, rather than zebrafish weight per se, ensures that changes in size, which can happen due to different fish densities in the tank, for example, are not a confounding issue. This ratio can be easily measured using a simple picture of the fish and any software ruler. No scale is required because the resulting number will be a ratio. Importantly, previous studies show that decreased body width/length ratio is an accurate predictor of mortality (Carneiro et al., 2016b; Henriques et al., 2013), suggesting that this may offer an equivalent to the human 'frailty index' (Mitnitski et al., 2017; Rockwood et al., 2015), commonly used in the clinic to estimate patient morbidity. Previous work established that a width/length body ratio approaching 0.15 correlates very strongly with mortality (Carneiro et al., 2016b; Henriques et al., 2013) (Figure 6.2A). Therefore we suggest that, in future studies aimed at determining whether a particular strain is displaying premature ageing phenotypes, it is key to closely monitor this width/length body ratio. Importantly, this cachectic/frail phenotype is underpinned by a significant decrease in subcutaneous adipose tissue and muscle fibre thickness (Carneiro et al., 2016b) (Figures 6.2B, 6.3A–C and 6.4A,B), as previously described for human frailty.

Fundamental HP alterations of WT zebrafish ageing and key methodologies

An important step in quantifying tissue ageing is histopathological (HP) analysis. This requires excellent tissue processing, sectioning and staining techniques (Ellis et al., 2022; methods available on request) to minimize tissue destruction artefacts, which can be easily confused with degeneration and it is important to learn to distinguish between these (Menke et al., 2011; Mumford et al., 2007). Qualitative and quantitative measurements should be performed with the help of a histopathologist. Zebrafish HP alterations with ageing are well described (Carneiro et al., 2016b; Henriques et al., 2013). In general, most HP changes in zebrafish natural ageing become significant from the age of 18–24 months (Carneiro et al., 2016b). Key tissues showing significant HP alterations in zebrafish WT ageing are, in approximate chronological order, subcutaneous adipose tissue, the haematopoietic system (in zebrafish this is located in the head kidney), gut, liver, gonads (with the testes being best described) and muscle (Carneiro et al., 2016b). These changes include a general degeneration of the tissue, increased inflammation and thickening of mucosa (Figure 6.4) as well as increased incidence of cancer (Carneiro et al., 2016b) (Figure 6.1).

Many, but not all, of these changes are anticipated and exacerbated in the absence of telomerase in the *tert*$^{-/-}$ premature ageing model (Figures 6.1–6.4) (Carneiro et al., 2016b). This type of study, comparing WT ageing with *tert*$^{-/-}$, allowed the authors to identify potential natural ageing phenotypes that are likely to depend on telomere shortening (Carneiro et al., 2016b). A similar strategy,

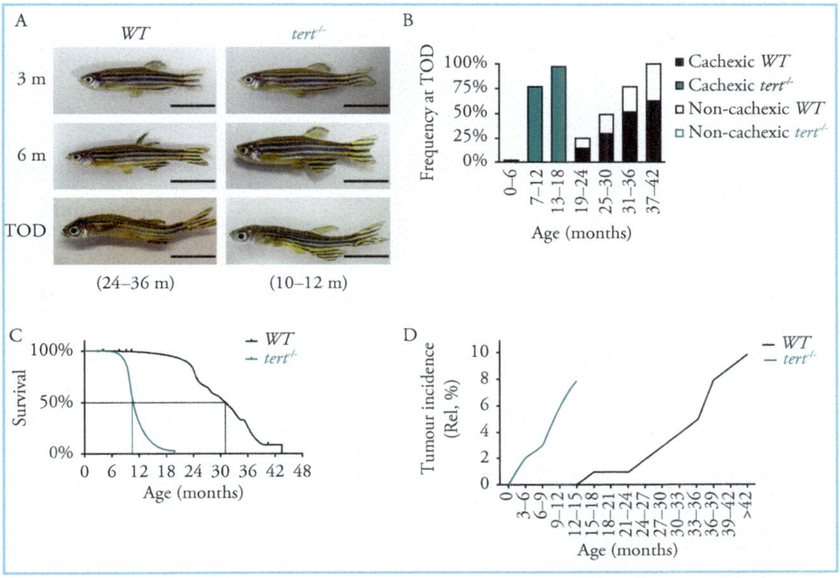

Figure 6.1 *tert*$^{-/-}$ mutants accelerate the onset of age-associated cachexia and neoplasia. (A) Representative images of WT and *tert*$^{-/-}$ mutants show that, at time-of-death (TOD), zebrafish exhibit signs of cachexia and deformation of the spine, with very low body mass indices. (B) In the WT population the incidence of these alterations increases with age while in *tert*$^{-/-}$ mutants, at TOD, 100% of the population is affected. (C) Sustained cachexia in older zebrafish is accompanied by an increase in mortality, as shown by Kaplan–Meier curves of WT and *tert*$^{-/-}$ zebrafish. WT zebrafish have a half-life of 30.8 m, threefold greater than that of the 10.6 m of *tert*$^{-/-}$ ($P < 0.0001$). $N = 426$ WT, $N = 98$ *tert*$^{-/-}$ for Figure 6.1A–C. TOD corresponds to the interval comprising the second and third quartiles of survival (25–75%). Scale bars, 1 cm. (D) Tumourigenesis exponentially accelerates with age in both WT and *tert*$^{-/-}$ zebrafish, reaching cumulative incidences of around 10 and 8%, respectively. $N = 23/238$ WT, $N = 5/66$ *tert*$^{-/-}$. *tert*$^{-/-}$ mutants have an earlier onset of neoplasia starting at 4 months of age ($P = 0.003$). (E) Tumours in WT male zebrafish affect mainly the reproductive system, following the appearance of Periodic Acid-Shiff (PAS$^+$)-stained cells in the tissue stroma (F) (quantified in G). WT females show an even distribution of tumours between the reproductive and haematopoietic systems, followed by liver, intestine and pancreas. $N = 118$ WT males, $N = 120$ WT females, $N = 58$ *tert*$^{-/-}$ males, $N = 8$ *tert*$^{-/-}$ females. Quantification of the percentage of PAS$^+$ cells was performed in three fields of view for each individual in the graph. Scale bars, 50 μm. Data are represented as mean ± standard error of the mean (SEM) (Carneiro et al., 2016b).

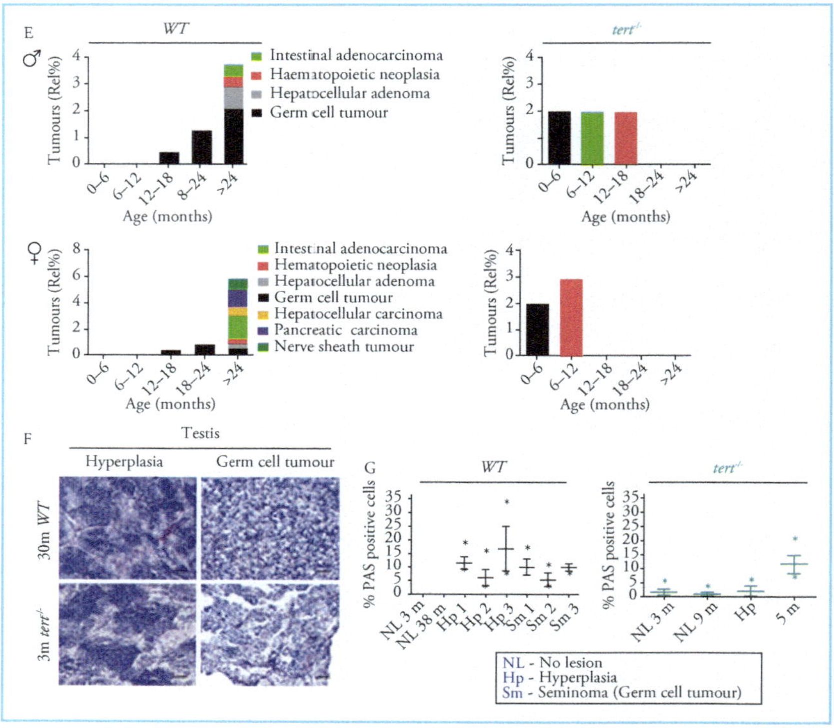

Figure 6.1 (continued)

comparing with other mutants, can be used to determine other likely mechanisms involved in age-associated diseases, as mentioned in the introduction.

Fundamental cellular damage in WT zebrafish ageing and key methodologies

The time- and tissue-dependent degeneration that occurs in natural zebrafish ageing is accompanied by cellular alterations that are conserved amongst multiple organisms, regardless of being triggered by critically short telomeres or not (López-Otín et al., 2013; Rossiello et al., 2014; Victorelli and Passos, 2017). These hallmarks of ageing include accumulation of DNA damage, peaks of cellular death and

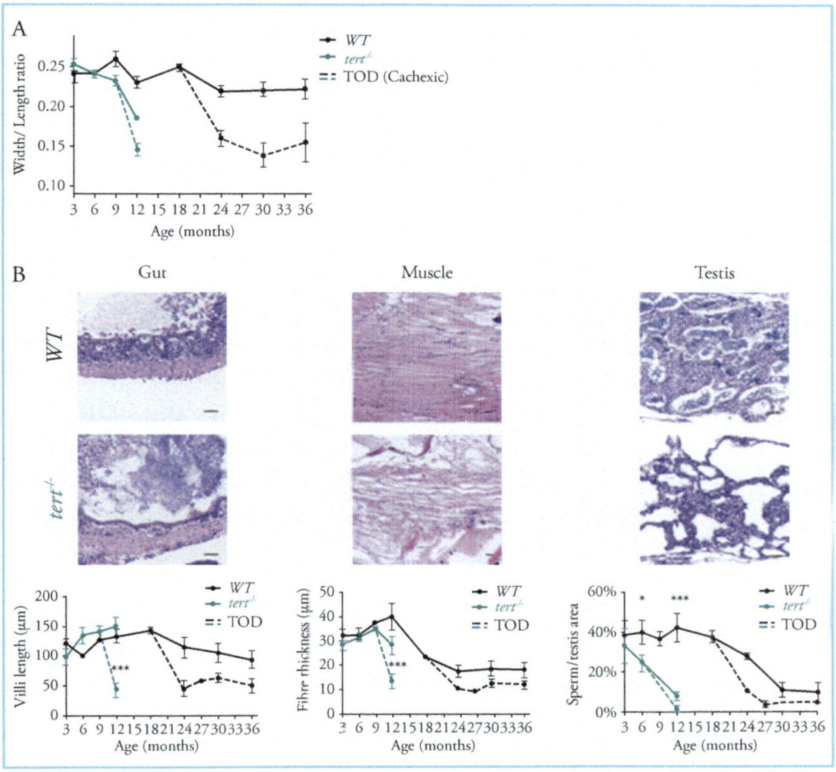

Figure 6.2 Cachexia at TOD is associated with lower body mass, gut villi shortening and marked muscle and testicular atrophy. (A) Quantification of width/length ratios shows that, at TOD, both WT and *tert*$^{-/-}$ zebrafish have significantly lower body mass indices when compared with non-cachexic siblings (N = 4–17 per time point for WT zebrafish and N = 3–7 for *tert*$^{-/-}$ mutants). (B) Top, representative haematoxylin and eosin-stained sections of gut, muscle and testis from WT and *tert*$^{-/-}$ siblings at TOD. Bottom, cachexia is associated with shorter villi (vil.) and/or villous atrophy (defined as flattening of the gut mucosal surface, N = 3–6 per genotype per time point), and severe myocyte atrophy and degeneration (N = 3–6 per genotype per time point), to a higher degree than found for non-cachexic siblings. Testis also shows pronounced atrophy, with reduced germ cell compartment associated with cachexia (N = 3–4 per genotype per time point). TOD corresponds to the interval comprising the second and third quartiles of survival (25–75%). Quantifications were performed in at least three different fields of view for each individual. Scale bars, 50 µm. Data are represented as mean ± SEM (Carneiro et al., 2016b).

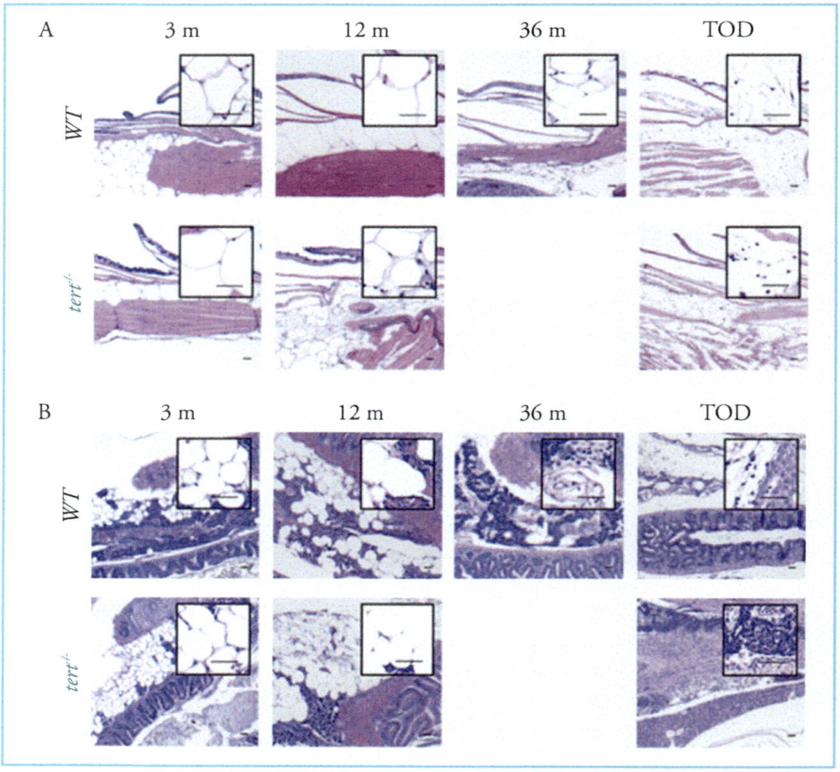

Figure 6.3 Subcutaneous adipose tissue progressively disappears with ageing whereas visceral reserves are maintained until more advanced ages. Representative haematoxylin and eosin-stained sections of subcutaneous (A) and visceral (B) adipose tissue depots of WT (3, 12 and 36 months and TOD) and *tert*$^{-/-}$ siblings (3 and 12 months and TOD). (C) quantification of adipocyte vacuole diameter at different ages. (A) WT zebrafish show progressive loss of the subcutaneous depot with age (C), accompanied by a reduction in adipocyte vacuole diameter (adipocytes are ~3.3 times smaller at 36 months versus 3 months). *tert*$^{-/-}$ anticipates this phenotype by 12 months of age. With cachexia, both WT and *tert*$^{-/-}$ zebrafish show complete exhaustion of the subcutaneous adipose tissue depot (at TOD). (B) Visceral adipose tissue (peri-pancreatic) is reduced or absent at 36 months in WT zebrafish (when adipocytes are ~3 times smaller than at 3 months) and by 12 months in tert mutants; and similarly to the subcutaneous depot, exhaustion of visceral adipose reserves is associated with cachexia. *N* = 3–7 zebrafish per genotype per time point. TOD corresponds to the interval comprising the second and third quartiles of survival (25–75%). Scale bars, 50 µm. Data are represented as mean ± SEM (Carneiro et al., 2016b).

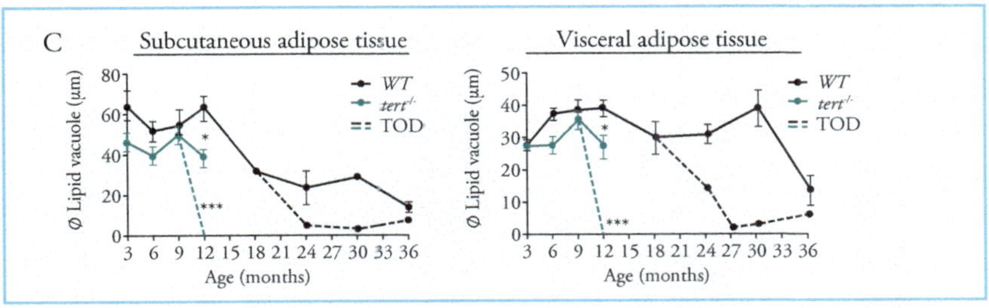

Figure 6.3 (continued)

ultimately an accumulation of senescent cells, which we now know contribute to tissue degeneration in multiple ways (Childs et al., 2015). DNA damage at telomeres has been shown to accumulate with ageing, in some situations even irrespectively of critically short telomere length (Fumagalli et al., 2012; Hewitt et al., 2012; Victorelli & Passos, 2017). This is because DNA damage at telomeres is largely irreparable, contributing to cell cycle arrest and senescence with ageing (Fumagalli et al., 2012; Rossiello et al., 2014).

In zebrafish, critically short telomeres measured by telomere restriction fragment analysis (TRF) using southern blots (methods available upon request) have been shown to anticipate the accumulation of DNA damage response (DDR) markers, namely 53BP1 (Henriques et al., 2013) and γH2AX (Carneiro et al., 2016b). These markers can be visualized in tissue sections using optimized immunofluorescence techniques (Ellis et al., 2022; methods available upon request). One can quantify levels of DDR by either quantifying the mean intensity of fluorescence (MIF) per cell nucleus across the tissue, comparing young versus old versus mutant, for example. An alternative method involves quantifying the percentage of cells with more than five clear γH2AX foci, as previously described (Carneiro et al., 2016b; Henriques et al., 2013). Combining γH2AX immunostaining with a telomere PNA-fluorescent probe allows for the quantification of telomere-associated DNA damage foci (TAFs) (Hewitt et al., 2012) in individual cells within tissues. Note that this is not the correct way to measure telomere length. In tissues, relative telomere length can be measured by DNA in situ hybridization of telomeres (Telo-FISH). This needs to be combined with a centromeric FISH

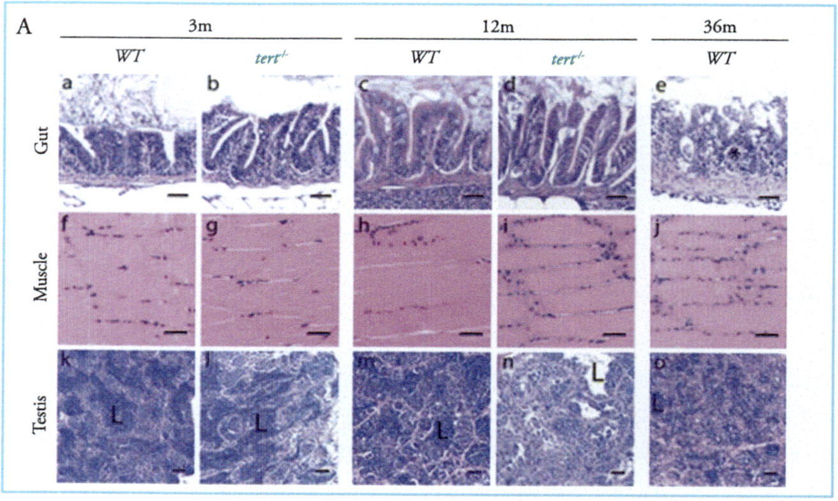

Figure 6.4 Telomere shortening culminates in tissue defects associated with ageing. (A) Representative haematoxylin and eosin-stained sections of gut, muscle and testis from WT and *tert*$^{-/-}$ siblings. By 12 months of age, *tert*$^{-/-}$ mutants (N = 3) show inflammatory cell infiltration of the lamina propria of the gut (b, d), myocyte atrophy and degeneration (characteristic of sarcopenia [i]) and reduced numbers of mature spermatozoa (l, n). Similarly, by 36 months, WT shows similar lesions in the gut (N = 5 [e]) and muscle (N = 5 [j]); and with ageing, WT testis shows a gradual decrease in the number of spermatozoa (k, m, o, n = 5). (B) Among these age-related histological lesions, intestinal inflammation and sarcopenia correlate with shortening of mTL (P = 0.002 and 0.028, respectively), while no correlation is found for reduction in mature spermatozoa numbers. Grey shaded area identifies the median telomeric length at which significant intestinal inflammation, sarcopenia and defects in production of mature spermatozoa are observed in *tert*$^{-/-}$ mutants. WT and *tert*$^{-/-}$ age groups are indicated in each graph by black- and red-coloured numbers, respectively. (C) In accordance with lower levels of spermatozoa production at later ages, WT shows reduced ability to successfully fertilize female eggs from 24 months of age onwards (N = 4). *tert*$^{-/-}$ has impairment of reproductive ability by 6 months and complete lack of function by 12 months (N = 3). (D) Erythrocyte levels (N = 5) decrease with ageing in WT and *tert*$^{-/-}$ mutants' kidney marrow, indicative of anaemia, but this is not predicted by mTL decline in this tissue. Grey shaded area represents the median telomeric length at which a significant decrease in erythrocyte levels is observed in *tert*$^{-/-}$ mutants. Scale bars, 50 µm. N = 3–6 for tissue mTL quantifications per genotype per time point (x axis in B,D). mTL, median telomere length. Data are represented as mean ± SEM (Carneiro et al., 2016b).

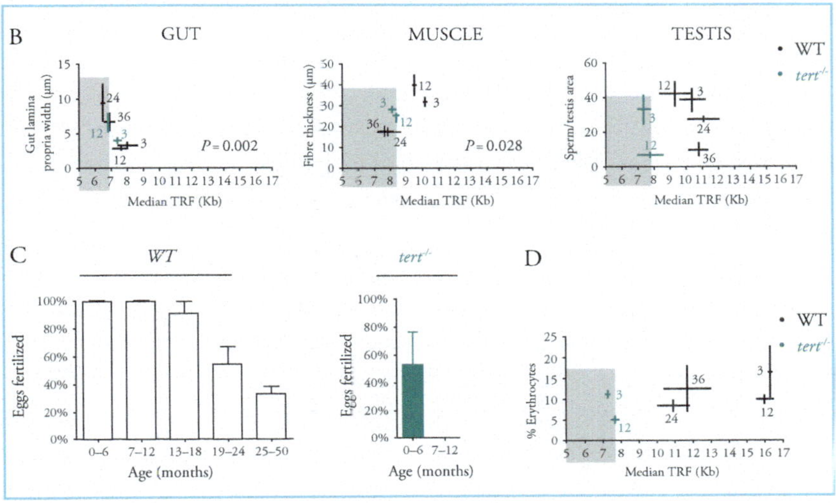

Figure 6.4 (continued)

(Junko Aida et al., 2014) probe that can be ordered specifically for zebrafish, since the human centromeric probe does not work for zebrafish. The centromeric probe we used (our unpublished data) was based on previously published work (Phillips & Reed, 2000), specifically the RFAL1 probe. A custom PNA probe can be made to order, specifically FAM-OO-TAGACAACATTTCATGCA (our unpublished data; method available upon request). The MIF of the telomere fluorescent signal should then be divided by the centromeric signal, in each cell, to give a relative telomere/centromere ratio. This is not an absolute telomere length measurement, but it can be used to compare relative length between different cells in the same tissue. The accumulation of telomere-dependent DDR, quantified as TAFs and expressed as percentage of γH2AX foci that co-localize with telomere foci in each cell nucleus (Hewitt et al., 2012), has been shown to increase in zebrafish ageing. This is more significant in the ageing gut than in many other tissues, and correlates with increased cellular senescence (Carneiro et al., 2016b).

Because cellular senescence is a complex phenomenon, there is still no single marker that can be used to qualify a cell as senescent. Instead, to identify a cell as senescent one should aim to identify at least the following characteristics in the same cell using a combination of immunofluorescence and enzymatic techniques (de Jesus & Blasco, 2012; Hernandez-Segura et al., 2017; Lawless et al., 2010) (methods available upon request): increased DDR (e.g. more than five strong foci of γH2AX); absence of proliferation markers (e.g. PCNA); absence of apoptotic markers (e.g. TUNEL or cleaved caspase 3); and increased senescence-associated β-galactosidase (SA-β-gal). Ideally one would confirm these results by RT-QPCR of the same markers and also others, such as cell cycle arrest markers *p21* and *p16* (in zebrafish, P16-like gene), DDR response markers such as *p53* and apoptotic markers such as *puma*, as previously described in zebrafish (Carneiro et al., 2016b; Henriques et al., 2013). Alternatively, RNA in situ hybridization of key genes can be combined with immunofluorescence in zebrafish tissue sections by protocols carefully developed by us (Ellis et al., 2022; methods available upon request).

Fundamental functional alterations and key methodologies

Even though there is still much to be explored in zebrafish ageing, there have been key reports showing that tissue function and general behaviour are affected with ageing and can be quantified.

A key example of impaired tissue function with ageing is the gut. As described above, the gut is one of the first tissues to display hallmarks of ageing in a telomerase-dependent fashion in zebrafish. Concurrently, we and others have shown that gut permeability is increased in zebrafish ageing, adapting the widely described SMURF assay (Dambroise et al., 2016; Martins et al., 2018). Gut permeability has been shown to increase with ageing and likely contribute to the spreading of inflammation and possibly infection to other parts of the body. Importantly, increased gut permeability has been shown to be a good biomarker to predict mortality in a variety of organisms, both invertebrates such as *Drosophila* and the vertebrate zebrafish (Dambroise et al., 2016; Martins et al., 2018; Tricoire & Rera, 2015).

Zebrafish have been widely used as a model for regeneration studies due to their ability to regenerate many tissues (reviewed in Marques et al., 2019). However, zebrafish show ageing-associated phenotypes suggesting that this regenerative capacity declines with age or is insufficient to restore healthy tissues in aged fish. Heart regeneration is impaired in the absence of telomerase (Bednarek et al., 2015), so it is likely, albeit yet to be shown, that it is also impaired with natural ageing. Furthermore, the zebrafish retina undergoes ageing-associated neurodegeneration. We have shown, however, that neuronal regeneration in the retina can be triggered following acute injury, even in the aged zebrafish (Martins et al., 2022). This highlights the zebrafish as a potential model to study mechanisms of reactivation of regeneration, particularly in the context of age-related degenerative diseases.

Neurodegeneration and associated diseases are also an important aspect of ageing in humans and have a significant negative impact on the quality of life (Niccoli & Partridge, 2012). Developing new model organisms amenable to genetic manipulation and high/semi-high-throughput drug screening is therefore an urgent endeavour. The zebrafish has emerged as a promising model for neurodegeneration and behavioural alterations with ageing (Brock, 2017; Van Houcke et al., 2015).

Special husbandry and procedure notes for maintaining and caring for ageing and aged fish

Scoring ageing phenotypes

Ageing phenotypes can be divided into early and late stages. Early phenotypes occur before the half-life of these animals (around 9 months of age for the *tert*$^{-/-}$ mutant and around 32 months for naturally aged WT fish); late phenotypes occur nearer the end point of their lives (roughly 12–15 months for the *tert*$^{-/-}$ mutant and roughly 40–43 months in WT fish). It is for this reason that one should use defined end-point criteria, such as width/length body ratio of about 0.15 (Carneiro et al., 2016b; Henriques et al., 2013). Indicators such as inability to feed or swim should prompt sacrifice of the animals to measure specific phenotypes, avoiding death as the end point. Regular, if not daily, checks, scoring sheets and imaging by standard photography should be used to monitor end points.

Crosses

It is important to note that fecundity decreases with ageing (Carneiro et al., 2016b) and premature ageing mutants such as the *tert*$^{-/-}$ are often infertile (Anchelin et al., 2013; Henriques et al., 2013). It is therefore important to keep breeding stocks of these mutant lines maintained in heterozygosity and, ideally, to generate these stocks from outcrosses. Heterozygous stocks need to be crossed every time mutants and respective WT siblings are required for experiments. To minimize variability, one should try to use segregated couples to generate siblings (e.g. three different couples to create the progeny for analysis).

Managing stress

To alleviate any possible distress, aged (WT or prematurely aged mutants) should be kept in separate tanks from younger WT fish to prevent competition and aggression from stronger fish. If the fish are at very low density, environmental enrichments such as artificial plants should be used (see Chapter 2 for more information).

If you are planning to grow large numbers of fish to age, consider keeping them in a quieter location in your fish facility and standardize feeding (see Chapter 7 chapter for more information). Keep the tanks clean but don't disturb the fish unnecessarily by frequent netting.

Anaesthesia

Aged fish may be more sensitive to anaesthesia and regeneration assays in general and adverse effects are more likely; however, death should not exceed 10%. Anaesthesia should be carefully monitored throughout the procedures and assessments of severity made. It is advisable to perform a pilot test under the advice of facility technicians/designated veterinarian the first time a procedure is performed, to establish the best anaesthesia method and welfare care for the particular procedure (see Chapter 13 for more information).

Conclusions

Zebrafish offer a valuable complementary model to study the fundamental mechanisms that drive ageing. Importantly, due to its size, genetics and cost, zebrafish offer great potential for high and medium throughput drug screening, which is critical for bridging the gap between in vitro and in vivo murine models used for therapeutic translation. Indeed, pharmaceutical companies are looking for such models to improve attrition rate when screening compounds, better predicting toxicity and also allowing testing of the effects on multiple tissue systems. Here we describe the fundamental pathological, cellular and molecular phenotypes of natural ageing in zebrafish. We hope this will provide fundamental support for all researchers interested in using this model for ageing research.

Acknowledgements

We would like to thank members of the Biological Services Aquarium (University of Sheffield) for all their support in maintaining our fish lines and for providing valued input towards experimental planning. In addition we thank members of the Skelet.AL histology unit (Mellanby Centre for Bone Research, University of Sheffield Medical School) for their expertise in tissue processing and sectioning; and the Wolfson Light Microscope facility for imaging support.

References

Alter, B. P., Rosenberg, P. S., Giri, N., Baerlocher, G. M., Lansdorp, P. M., & Savage, S. A. (2012). Telomere length is associated with disease severity and declines with age in dyskeratosis congenita. *Haematologica, 97*(3), 353–359. https://doi.org/10.3324/haematol.2011.055269

Amsellem, V., Gary-Bobo, G., Marcos, E., Maitre, B., Chaar, V., Validire, P., Stern, J. B., Noureddine, H., Sapin, E., Rideau, D., Hue, S., Le Corvoisier, P., Le Gouvello, S., Dubois-Randé, J. L., Boczkowski, J., & Adnot, S. (2011). Telomere dysfunction causes sustained inflammation in chronic obstructive pulmonary disease. *American Journal of Respiratory and Critical Care Medicine, 184*(12), 1358–1366. https://doi.org/10.1164/rccm.201105-0802OC

Anchelin, M., Alcaraz-Pérez, F., Martínez, C. M., Bernabé-García, M., Mulero, V., & Cayuela, M. L. (2013). Premature aging in telomerase-deficient zebrafish. *Disease Models and Mechanisms, 6*(5), 1101–1112. https://doi.org/10.1242/dmm.011635

Atzmon, G., Cho, M., Cawthon, R. M., Budagov, T., Katz, M., Yang, X., Siegel, G., Bergman, A., Huffman, D. M., Schechter, C. B., Wright, W. E., Shay, J. W., Barzilai, N., Govindaraju, D. R., & Suh, Y. (2010). Evolution in health and medicine Sackler colloquium: Genetic variation in human telomerase is associated with telomere length in Ashkenazi centenarians. *Proceedings of the National Academy of Sciences of the United States of America, 107*, Suppl. 1, 1710–1717. https://doi.org/10.1073/pnas.0906191106

Bednarek, D., González-Rosa, J. M., Guzmán-Martínez, G., Gutiérrez-Gutiérrez, Ó., Aguado, T., Sánchez-Ferrer, C., Marques, I. J., Galardi-Castilla, M., De Diego, I., Gómez, M. J., Cortés, A., Zapata, A., Jiménez-Borreguero, L. J., Mercader, N., & Flores, I. (2015). Telomerase is essential for zebrafish heart regeneration. *Cell Reports, 12*(10), 1691–1703. https://doi.org/10.1016/j.celrep.2015.07.064

Brock, A. J., Sudwarts, A., Parker, M. O., & Brennan, C. H. (2017). Zebrafish behavioral models of ageing. In A. V. Kalueff (Ed.), *The rights and wrongs of zebrafish: Behavioral phenotyping of zebrafish* (pp. 241–258). Springer.

Campisi, J. (2013). Aging, cellular senescence, and cancer. *Annual Review of Physiology, 75*, 685–705. https://doi.org/10.1146/annurev-physiol-030212-183653

Carneiro, M. C., De Castro, I. P., & Ferreira, M. G. (2016a). Telomeres in aging and disease: Lessons from zebrafish. *Disease Models and Mechanisms*, *9*(7), 737–748. https://doi.org/10.1242/dmm.025130

Carneiro, M. C., Henriques, C. M., Nabais, J., Ferreira, T., Carvalho, T., & Ferreira, M. G. (2016b). Short telomeres in key tissues initiate local and systemic aging in zebrafish. *PLOS Genetics*, *12*(1), e1005798. https://doi.org/10.1371/journal.pgen.1005798

Childs, B. G., Durik, M., Baker, D. J., & Van Deursen, J. M. (2015). Cellular senescence in aging and age-related disease: From mechanisms to therapy. *Nature Medicine*, *21*(12), 1424–1435. https://doi.org/10.1038/nm.4000

Coppé, J. P., Desprez, P. Y., Krtolica, A., & Campisi, J. (2010). The senescence-associated secretory phenotype: The dark side of tumor suppression. *Annual Review of Pathology*, *5*, 99–118. https://doi.org/10.1146/annurev-pathol-121808-102144

Dambroise, E., Monnier, L., Ruisheng, L., Aguilaniu, H., Joly, J. S., Tricoire, H., & Rera, M. (2016). Two phases of aging separated by the Smurf transition as a public path to death. *Scientific Reports*, *6*, 23523. https://doi.org/10.1038/srep23523

de Jesus, B. B., & Blasco, M. A. (2012). Assessing cell and organ senescence biomarkers. *Circulation Research*, *111*(1), 97–109. https://doi.org/10.1161/CIRCRESAHA.111.247866

Dimri, G. P., Lee, X., Basile, G., Acosta, M., Scott, G., Roskelley, C., Medrano, E. E., Linskens, M., Rubelj, I., & Pereira-Smith, O. (1995). A biomarker that identifies senescent human cells in culture and in aging skin in vivo. *Proceedings of the National Academy of Sciences of the United States of America*, *92*(20), 9363–9367. https://doi.org/10.1073/pnas.92.20.9363

Ellis, P. S., Martins, R. R., Thompson, E. J., Farhat, A., Renshaw, S. A., & Henriques, C. M. (2022). A subset of gut leukocytes has telomerase-dependent "hyper-long" telomeres and require telomerase for function in zebrafish. Immunity and Ageing, 19(1), 31. https://doi.org/10.1186/s12979-022-00287-8

European Commission, Directorate-general for Economic and Financial Affairs, & Economic Policy Committee. (2015). *The 2015 ageing report: Economic and budgetary projections for the 28 EU Member States (2013–2060)*. Publications Office.

Fabbri, E., Zoli, M., Gonzalez-Freire, M., Salive, M. E., Studenski, S. A., & Ferrucci, L. (2015). Aging and multimorbidity: New tasks, priorities, and frontiers for integrated gerontological and clinical research. *Journal of the American Medical Directors Association, 16*(8), 640–647. https://doi.org/10.1016/j.jamda.2015.03.013

Fumagalli, M., Rossiello, F., Clerici, M., Barozzi, S., Cittaro, D., Kaplunov, J. M., Bucci, G., Dobreva, M., Matti, V., Beausejour, C. M., Herbig, U., Longhese, M. P., & d'Adda di Fagagna, F. (2012). Telomeric DNA damage is irreparable and causes persistent DNA-damage-response activation. *Nature Cell Biology, 14*(4), 355–365. https://doi.org/10.1038/ncb2466

Gilbert, M. J., Zerulla, T. C., & Tierney, K. B. (2014). Zebrafish (Danio rerio) as a model for the study of aging and exercise: Physical ability and trainability decrease with age. *Experimental Gerontology, 50,* 106–113. https://doi.org/10.1016/j.exger.2013.11.013

Henriques, C. M., Carneiro, M. C., Tenente, I. M., Jacinto, A., & Ferreira, M. G. (2013). Telomerase is required for zebrafish lifespan. *PLOS Genetics, 9*(1), e1003214. https://doi.org/10.1371/journal.pgen.1003214

Henriques, C. M., & Ferreira, M. G. (2012). Consequences of telomere shortening during lifespan. *Current Opinion in Cell Biology, 24*(6), 804–808. https://doi.org/10.1016/j.ceb.2012.09.007

Hernandez-Segura, A., De Jong, T. V., Melov, S., Guryev, V., Campisi, J., & Demaria, M. (2017). Unmasking transcriptional heterogeneity in senescent cells. *Current Biology, 27*(17), 2652–2660.e4. https://doi.org/10.1016/j.cub.2017.07.033

Hewitt, G., Jurk, D., Marques, F. D., Correia-Melo, C., Hardy, T., Gackowska, A., Anderson, R., Taschuk, M., Mann, J., & Passos, J. F. (2012). Telomeres are favoured targets of a persistent DNA damage response in ageing and stress-induced senescence. *Nature Communications, 3,* 708. https://doi.org/10.1038/ncomms1708

Hofer, A. C., Tran, R. T., Aziz, O. Z., Wright, W., Novelli, G., Shay, J., & Lewis, M. (2005). Shared phenotypes among segmental progeroid syndromes suggest underlying pathways of aging. *Journals of Gerontology. Series A, Biological Sciences and Medical Sciences, 60*(1), 10–20. https://doi.org/10.1093/gerona/60.1.10

Junko Aida, N. I.-S., Nakamura, K.-I., Ishikawa, N., Terai, M., Matsuda, Y., Aida, S., Arai, T., & Takubo, K. (2014). Determination of telomere length by the quantitative fluorescence in situ hybridization (Q-FISH) Method. *American Journal of Analytical Chemistry, 5.*

Krishnamurthy, J., Torrice, C., Ramsey, M. R., Kovalev, G. I., Al-Regaiey, K., Su, L., & Sharpless, N. E. (2004). Ink4a/Arf expression is a biomarker of aging. *Journal of Clinical Investigation, 114*(9), 1299–1307. https://doi.org/10.1172/JCI22475

Lawless, C., Wang, C., Jurk, D., Merz, A., Zglinicki, T. V., & Passos, J. F. (2010). Quantitative assessment of markers for cell senescence. *Experimental Gerontology, 45*(10), 772–778. https://doi.org/10.1016/j.exger.2010.01.018

López-Otín, C., Blasco, M. A., Partridge, L., Serrano, M., & Kroemer, G. (2013). The hallmarks of aging. *Cell, 153*(6), 1194–1217. https://doi.org/10.1016/j.cell.2013.05.039

Marques, I. J., Lupi, E., & Mercader, N. (2019). Model systems for regeneration: Zebrafish. *Development, 146*(18), dev167692. https://doi.org/10.1242/dev.167692

Martins, R. R., McCracken, A. W., Simons, M. J. P., Henriques, C. M., & Rera, M. (2018). How to catch a Smurf? in vivo assessment of intestinal permeability in multiple model organisms. *Bio-Protocol, 8*(3), e2722. https://doi.org/10.21769/BioProtoc.2722

Martins, R. R., Zamzam, M., Tracey-White, D., Moosajee, M., Thummel, R., Henriques, C. M., & MacDonald, R. B. (2022). Muller Glia maintain their regenerative potential despite degeneration in the aged zebrafish retina. *Aging Cell, 21*(4), e13597. https://doi.org/10.1111/acel.13597

Menke, A. L., Spitsbergen, J. M., Wolterbeek, A. P. M., & Woutersen, R. A. (2011). Normal anatomy and histology of the adult zebrafish. *Toxicologic Pathology, 39*(5), 759–775. https://doi.org/10.1177/0192623311409597

Minamino, T., & Komuro, I. (2007). Vascular cell senescence: Contribution to atherosclerosis. *Circulation Research, 100*(1), 15–26. https://doi.org/10.1161/01.RES.0000256837.40544.4a

Mitnitski, A. B., Rutenberg, A. D., Farrell, S., & Rockwood, K. (2017). Aging, frailty and complex networks. *Biogerontology, 18*(4), 433–446. https://doi.org/10.1007/s10522-017-9684-x

Mumford, S., Heidel, J., Smith, C., Morrison, J., MacConnell, B., & Blaze, V. (2007). *Fish histology and histopathology, 4.* United States Fish and Wildlife Service.

Niccoli, T., & Partridge, L. (2012). Ageing as a risk factor for disease. *Current Biology, 22*(17), R741–R752. https://doi.org/10.1016/j.cub.2012.07.024

Njajou, O. T., Cawthon, R. M., Damcott, C. M., Wu, S. H., Ott, S., Garant, M. J., Blackburn, E. H., Mitchell, B. D., Shuldiner, A. R., &

Hsueh, W. C. (2007). Telomere length is paternally inherited and is associated with parental lifespan. *Proceedings of the National Academy of Sciences of the United States of America, 104*(29), 12135–12139. https://doi.org/10.1073/pnas.0702703104

Njajou, O. T., Hsueh, W. C., Blackburn, E. H., Newman, A. B., Wu, S. H., Li, R., Simonsick, E. M., Harris, T. M., Cummings, S. R., Cawthon, R. M., & Health ABC study. (2009). Association between telomere length, specific causes of death, and years of healthy life in health, aging, and body composition, a population-based cohort study. *Journals of Gerontology. Series A, Biological Sciences and Medical Sciences, 64*(8), 860–864. https://doi.org/10.1093/gerona/glp061

Ohtani, N., & Hara, E. (2013). Roles and mechanisms of cellular senescence in regulation of tissue homeostasis. *Cancer Science, 104*(5), 525–530. https://doi.org/10.1111/cas.12118

Pereira, B., & Ferreira, M. G. (2013). Sowing the seeds of cancer: Telomeres and age-associated tumorigenesis. *Current Opinion in Oncology, 25*(1), 93–98. https://doi.org/10.1097/CCO.0b013e32835b6358

Phillips, R. B., & Reed, K. M. (2000). Localization of repetitive DNAs to zebrafish (Danio rerio) chromosomes by fluorescence in situ hybridization (FISH). *Chromosome Research, 8*(1), 27–35. https://doi.org/10.1023/a:1009271017998

Price, J. S., Waters, J. G., Darrah, C., Pennington, C., Edwards, D. R., Donell, S. T., & Clark, I. M. (2002). The role of chondrocyte senescence in osteoarthritis. *Aging Cell, 1*(1), 57–65. https://doi.org/10.1046/j.1474-9728.2002.00008.x

Rae, M. J., Butler, R. N., Campisi, J., De Grey, A. D., Finch, C. E., Gough, M., Martin, G. M., Vijg, J., Perrott, K. M., & Logan, B. J. (2010). The demographic and biomedical case for late-life interventions in aging. *Science Translational Medicine, 2*(40), 40cm21. https://doi.org/10.1126/scitranslmed.3000822

Rockwood, K., Mitnitski, A., & Howlett, S. E. (2015). Frailty: Scaling from cellular deficit accumulation? *Interdisciplinary topics in gerontology and geriatrics, 41*, 1–14.

Rossiello, F., Herbig, U., Longhese, M. P., Fumagalli, M., & d'Adda di Fagagna, F. (2014). Irreparable telomeric DNA damage and persistent DDR signalling as a shared causative mechanism of cellular senescence and ageing. *Current Opinion in Genetics and Development, 26*, 89–95. https://doi.org/10.1016/j.gde.2014.06.009

Shay, J. W., & Wright, W. E. (2007). Hallmarks of telomeres in ageing research. *Journal of Pathology, 211*(2), 114–123. https://doi.org/10.1002/path.2090

Tinetti, M. E., Fried, T. R., & Boyd, C. M. (2012). Designing health care for the most common chronic condition—Multimorbidity. *JAMA, 307*(23), 2493–2494. https://doi.org/10.1001/jama.2012.5265

Tricoire, H., & Rera, M. (2015). A new, discontinuous 2 phases of aging model: Lessons from Drosophila melanogaster. *PLOS ONE, 10*(11), e0141920. https://doi.org/10.1371/journal.pone.0141920

Van Houcke, J., De Groef, L., Dekeyster, E., & Moons, L. (2015). The zebrafish as a gerontology model in nervous system aging, disease, and repair. *Ageing Research Reviews, 24*(B), 358–368. https://doi.org/10.1016/j.arr.2015.10.004

Victorelli, S., & Passos, J. F. (2017). Telomeres and cell senescence – Size matters not. *EBiomedicine, 21*, 14–20. https://doi.org/10.1016/j.ebiom.2017.03.027

Wiemann, S. U., Satyanarayana, A., Tsahuridu, M., Tillmann, H. L., Zender, L., Klempnauer, J., Flemming, P., Franco, S., Blasco, M. A., Manns, M. P., & Rudolph, K. L. (2002). Hepatocyte telomere shortening and senescence are general markers of human liver cirrhosis. *FASEB Journal.* Federation of American Societies for Experimental Biology, *16*(9), 935–942. https://doi.org/10.1096/fj.01-0977com

Chapter 7

Organization of the zebrafish facility

Biosecurity, water quality, health monitoring and environmental impact

Jean-Philippe Mocho

Abstract

The organization of a zebrafish facility requires numerous operational and practical considerations. Because a sound biosecurity programme is essential to the control of colonies' health status, biosecurity barriers, triage of imports and quarantine process are discussed here. These processes are evaluated by an exhaustive health monitoring, which involves recording morbidity and mortality and screening both fish and environmental samples. These data constitute the health status report and, with the addition of a thorough description of husbandry and biosecurity practices, should be shared transparently between laboratories to facilitate fish exchange. To preserve fish health and model reproducibility, many husbandry practices focus on water treatment. The means to control chemical and microbiological water quality are detailed within this chapter, including methods to standardize feed distribution. Finally we consider the environmental impact of zebrafish facilities and propose routes towards carbon footprint reduction for some key resource-consuming processes.

Claire Allen and Jean-Philippe Mocho (eds) **Zebrafish**
DOI: 10.1079/9781800629431.0007, © CAB International, 2024

Keywords

zebrafish, biosecurity, diet distribution, water quality, health monitoring, environmental impact

Introduction

In this chapter we discuss what is detailed in previous chapters and propose practical applications for the organization of the zebrafish facility. First we consider how to achieve a sound biosecurity programme to control microbiological populations and potential chemical pollutants across the animal facility. Biosecurity depends on barriers set to contain or protect colonies, and on the containment of hazardous incoming material such as fish, biological material, water and feed. Triage of imports and quarantine are therefore pillars of the biosecurity programme. Next, spreading of hazards within the facility must be controlled by designing flows for fish, staff and other material. This may require assigning equipment to specific flows, with provision of dedicated space and budget. For example, incoming water and recirculating water must be treated separately, and we propose practical means to control chemical and microbiological risks. To preserve water quality it is essential to control water pollution due to feed distribution. Indeed, feeding fish is the central activity to the daily routine of the animal technician (see Chapter 3 for more information). A mistake in feed distribution can induce acute or long-term ill health, and this can be challenging when different people feed zebrafish. There are practical solutions to prevent mistakes in feed distribution. The organization of the technicians' tasks also includes daily care and monitoring of fish health, primarily to remove suffering individuals. Moreover, the recording of clinical signs detected daily is key to monitoring of disease incidence in colonies (see Chapters 5 and 4 for more information), and to detection of adverse effects and harmful phenotypes (see Chapter 14 for more information). The screening for diseases must be completed with sample numbers, frequency and types to produce meaningful data (see Chapter 9 for more information) and support the safe exchange of fish between laboratories. This overview of the organization in a zebrafish facility cannot avoid discussing how the facility can be organized to reduce its environmental impact. First we discuss how consumable management and organizational decisions influence the facility's environmental impact. We detail carbon footprint evaluation, with the example of water supply and renewal. Noticeably, biosecurity considerations such as cleaning and disinfection processes may be resource intensive, and a sound organization may contribute to significant carbon footprint reductions.

Biosecurity

Barriers and epidemiological units EUs

The aim of a biosecurity programme is to prevent introduction and spreading of contaminants (Mocho et al., 2022a). This is in order to protect fish health and welfare, staff health and preserve the reproducibility of animal models (see Chapter 5 for more information). Biosecurity should be at the heart of the facility organization from the design stage to the daily flows of fish, material, staff and visitors. Indeed, the facility can be divided in different zones containing imported fish (i.e. quarantine), main fish colonies, fish with higher microbiological requirements (e.g. immune-deficient fish) and fish presenting higher hazards (e.g. ageing fish or those contaminated by procedure). The other important zones include storage space, food preparation room, injection suites, machinery, corridors, offices, entrances/exits etc. Any fish-holding zone with defined microbiological characteristics is called an epidemiological unit (EU), and barriers are set around the EU to control incoming staff, fish, water and material. Barriers can be a water-decontamination process, donning personal protective equipment (PPE) such as gloves or a physical separation between flows to prevent cross-contamination. This is why the definition of the EUs and their barriers is central to the design of the animal facilities. Corridors, storage and flows are difficult to adapt once walls and doors are set and space is allocated. The absence of reliable barriers means that fish are likely to be contaminated with the same microbes (i.e. viruses, bacteria, parasites, fungi etc.) and, therefore, to share the same EU. For example, fish in the same room, rack or tank are considered part of the same EU because droplets of water can spread contamination across the room. Logically, quarantine must be set in a dedicated room, with independent and dedicated flow and equipment, surrounded by barriers and protocols to create a quarantine EU (Figure 7.1; Mocho et al., 2022a).

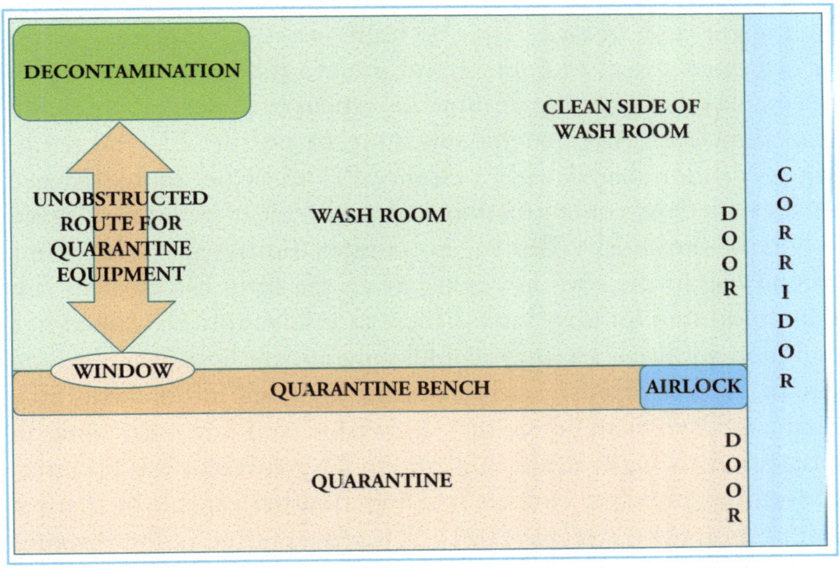

Figure 7.1 Quarantine equipped with a barrier box (airlock) to the corridor and a hatch to the washroom disinfection area. The barrier box shown here acts as the airlock to ensure negative pressure in quarantine and positive pressure in the corridor. The barrier box allows decontamination of small items taken out of quarantine, such as egg surface sanitation. Clean items can be delivered through the barrier box to quarantine without entering quarantine, e.g. feed. The hatch to the washroom creates a direct route between equipment decontamination and dirty equipment on the quarantine bench. There is no bench on the washroom side of the hatch, to avoid obstruction and contamination.

Flows between EUs and quarantine

Whether the aim is to contain a quarantine area or to separate two EUs – for example, holding different fish species – the definition of flows must first consider where dedicated equipment is stored. Bottlenecks between flows must be avoided. For example, microscopes, refrigerators (e.g. for feed distribution items), incubators, water-testing kits, recording systems, maintenance kits and other equipment should not be shared between different EUs. Space for an EU must include fish-holding plus other necessary areas for equipment and storage. Thus, defining an EU represents a cost in dedicated space, equipment and time to follow the barrier processes, whether this is consumable expenses or decontamination (e.g. autoclave). Any material taken out of a dirtier EU and reaching a corridor shared with a cleaner EU must be double bagged to prevent cross-contamination in transit. First, items are bagged inside the dirtier EU. This bag is considered dirty, and it is set in a second bag in the corridor, as it crosses the dirty EU barrier. This is a typical step for dirty tanks. These cannot simply be washed in a common washer since they would contaminate both machine and circuit. A safer practice is to clean them by hand in the dirtier EU, then autoclave them before they are washed and join other tanks in circulation (Mocho et al., 2022a). An alternative to the resource-consuming autoclave consists in using a washer capable of thermal disinfection (Mocho et al., 2021). When quarantine and washrooms are adjacent, to avoid double bagging of dirty tanks, a hatch can be set between the two rooms to create a direct route away from corridors and flows dedicated to clean equipment. The washroom technician can then reach for dirty tanks on the quarantine bench, turn around and set the dirty tanks directly into the decontamination chamber (e.g. autoclave, washer with thermal disinfection). To optimize quarantine flows, a lobby area or a barrier box between quarantine and common corridors is particularly useful to mitigate the risk of cross-contamination. It can create an ideal space to deliver clean items without entering quarantine (e.g. feed) and to decontaminate items taken out of quarantine (e.g. egg surface disinfection). Figure 7.1 shows an example of a quarantine equipped to ease flows in a small facility. The lobby and barrier box can act as airlocks, with negative pressure in quarantine and positive pressure in the corridor. Decontamination can also be adapted to a specific hazard – for instance, desiccation and cryopreservation are efficient processes to eliminate pathogens such as *Pseudocapillaria tomentosa* (Kent et al., 2019; Mocho et al., 2022a,b; Norris et al., 2018; see Chapter 5 for more information). The robustness of a barrier depends on the buy-in from the stakeholders: the biosecurity programme must be designed in agreement with users and higher management. Training and communication on this matter are essential, and stakeholders' compliance with the biosecurity programme is a valuable indicator to the local culture of care. The weakness of a barrier is often linked to its popularity. The more people and/or materials are processed through the barrier, the greater the risk of contamination. This can be due to human error or the sheer law of averages. Many disinfection processes rely on dilution or a percentage of reduction. The higher the number, the higher the risk of falling within the small percentage of decontamination failure.

PPE

Staff must protect themselves to mitigate the risk of zoonotic disease, i.e. infection from fish or water pathogens (e.g. *Mycobacterium* spp.; Mocho et al., 2022a; see Chapter 5 for more information). Therefore, PPE must be worn when touching water or contaminated material or handling fish. Gloves are an absolute minimum requirement, and jewellery and wristwatches should be avoided. Long-sleeved gloves are useful when the tasks involve reaching deeply into a reservoir, e.g. sump cleaning. Goggles should be considered when water droplets could reach the face. It is safer and more comfortable to wear dedicated scrubs when performing demanding and lasting cleaning duties, and this also avoids staff wearing wet and contaminated civil clothes on public transport. Long hair should be tied and/or kept in a hairnet. When working in dirtier EUs such as quarantine, dedicated PPE must be worn and changed when exiting, and this includes overalls. Otherwise, in cleaner EUs, each facility usually sets their own rules regarding wearing a gown or other body protection when performing non-invasive procedures (Mocho & von Krogh, 2022). To ensure better compliance with biosecurity protocols and identify flows, quarantine PPE can have a dedicated colour. For example, orange gloves could be worn in the quarantine room and blue gloves in the main fish EU. In this case, no one should wear orange gloves in the common corridor. Masks are not usually deemed necessary to preserving fish microbiological status. Other PPE is sometimes used to protect fish: for example, many facilities require overshoes to be worn. The justification for this remains unclear, because fish falling to the floor are more likely to be contaminated by chemicals than by live aquatic pathogens. This is often an avoidable consumption of single-use plastic that could be replaced by either proposing dedicated shoes when entering the EU or deciding to euthanize or quarantine any fish in contact with the floor and to disinfect any item that fell to the floor.

Imports

Cross-contamination within the facility aside, imports carry the main risk for pathogen introduction into the facility. Imports must be triaged before and after shipment (in quarantine; Mocho et al., 2022a). Facility stakeholders must define the list of microbes they want to exclude from each EU and from quarantine. This would vary depending on context and bearing in mind the short-, mid- and long-term objectives of the facility. European legislation does not permit the importation of *Danio rerio* into research facilities from pet shops, or from other non-laboratory suppliers. This significantly reduces the risk of importing unexpected pathogens that are rare in laboratory zebrafish but more common in other fish species (Mocho et al., 2022a,b). Ideally, fish should be imported from only resource centres (e.g. European Zebrafish Resource Centre, Zebrafish International Resource Centre) or establishments with strict biosecurity rules and robust historical screening data and where excluded pathogens are not detected. This is particularly important in regard to refreshing the genetic pool of wild-type colonies (Crim & Lawrence, 2021). In all cases, when import candidates are suspected of being contaminated with a pathogen excluded from quarantine (e.g. *Mycobacterium marinum*, due to its zoonotic potential and virulence), the import must be declined and the fish not shipped (Mason et al., 2016). If this first triage step is cleared, options to import eggs in quarantine should be favoured. If adults are received, screening can be performed from the quarantine tank within one week of arrival (Mocho et al., 2017). When a pathogen is detected in quarantine, a decision can be taken to euthanize the colony and disinfect quarantine. Alternatively, treatment can be attempted or the colony re-derived (e.g. *P. tomentosa*; Mocho et al., 2022a,b). Ideally, all fish are imported into quarantine and never leave quarantine. Only surface-disinfected (i.e. 'bleached') eggs from quarantined fish are accepted into the main holding EU (Mocho et al., 2022a).

Egg surface sanitation

First, eggs must be thoroughly washed by strongly squirting embryo media and treatment solutions on the eggs held in a sieve (mesh of 400 µm; ZIRC, 2019). This is facilitated by creating mesh baskets with PVC pipes. Hot glue can be used to stick the mesh to the PVC pipe (ZIRC, 2019), in which case the mesh tends to detach when the sieves are washed with hot water between uses. Ideally, the mesh is instead mechanically fixed within the PVC pipe to keep eggs slightly raised when immersed. This will create a space between the bottom of the treatment solution container and the treated eggs, allowing egg immersion and preventing flushing of dirt back through the mesh when eggs are immersed in treatment solutions. The wash phase of the sanitation process is absolutely key and essential to elimination of pathogens that are in the clutch, outside eggs and resistant to chemical treatment (e.g. *P. tomentosa* eggs).

Disinfection will then help to further reduce microbial contamination. Various protocols of disinfection utilizing bleach and/or iodine are popular in zebrafish laboratories (e.g. 35 ppm sodium hypochlorite for 10 min; ZIRC, 2019). Hydrogen peroxide is more commonly used for some other fish species. There is a great variation in egg surface disinfection protocols across zebrafish facilities worldwide, depending on the quality of available chemicals, treatment solution concentrations and exposure time. The only common point is that the procedure is referred to as 'bleaching', whether bleach is used or not. There is no optimal protocol; indeed, microbial sensitivity to chemicals varies depending on microbial strain (Chang et al., 2015). If egg surface sanitation is performed to eradicate a specific pathogen cultured from the fish colonies, it is therefore recommended to first identify the best protocol for that pathogen (Mason et al., 2016). On the other hand, when eggs are blindly bleached out of quarantine, no specific information on microbial sensitivity is available because a broad antiseptic efficacy is sought.

The determining factor may then be egg survival rate following sanitation, which can be kept above 90% (e.g. with the help of 0.5 g/l sodium thiosulfate for 5 s to prevent further effects of sodium hypochlorite; Peneyra et al., 2020; ZIRC, 2019). Caution must be taken when increasing disinfectant contact time and/or concentration to increase microbial elimination (Mendoza et al., 2022). Eggs may survive but an overly strong reduction of microbial flora on the egg surface may impair larval development in the couple of weeks following fertilization and induce high mortality and welfare impairment when fish have reached sentience. This can be a particular problem with stronger exposure to iodine. Regarding timing, disinfection must be performed before hatching otherwise larvae would not survive. Some fish strains may show a higher survival rate if bleached on the day of fertilization rather than the following day, or the opposite (Peneyra et al., 2020). Following disinfection the chorion may become too robust; in this regard, pronase (e.g. 40.8 ppm protease from *Streptomyces griseus* for 1 min; ZIRC, 2019) has been used with great success to ease hatching (Peneyra et al., 2020). In the absence of pronase, each egg must be manually dechorionated with watchmakers' tweezers or ophthalmic forceps to liberate embryos.

The main limitation of egg surface sanitation is its inability to reach pathogens within the egg (i.e. intra ovum, vertical transmission); this prevents elimination of parasites such as *Pseudoloma neurophilia*. However, some facilities have been unable to detect *P. neurophilia* in their main holding system following a two-step bleaching introduction of contaminated fish. In this case, a first generation of bleached eggs is raised in quarantine until sexual maturity. These spawn and produce a second generation of eggs that are bleached and introduced into the main holding system (Mocho et al., 2022a). The risk of raising a contaminated egg is diluted with each disinfection. Furthermore, breeding the subsequent generation when they are young, i.e. less likely to be heavily infested, reduces the risk of clutch contamination. This is not a total biosecurity barrier, but it may represent a statistically good chance of obtaining parasite-free animals. Finally, a potential caveat of egg surface disinfection is the impact on the egg microbiome. Bacterial flora is a growing topic of interest in zebrafish research, and routine attacks on the fish microbiome may not be an appropriate challenge for some models.

Water

Water supply

Water entering the animal facility may carry hazardous microbes and chemicals, including water delivered by taps in fish rooms. It may be too cold for safe addition to the fish system and have an unsuitable chemical composition, e.g. too hard. To address these issues it is common for the water to be treated by reverse osmosis (RO) before it enters the fish system. RO relies on two filtrations: first (usually) a carbon filter and then a RO membrane with which water molecules form hydrogen bonds and are forced through under pressure. Organic molecules and microbes are sieved out. The resulting RO water is free of most pollutants and disinfected (but neither pure nor sterilized). Smaller molecules, such as salt ions of low valence, are not rejected as efficiently (>85%; FDA, 2014). In practice, although RO is often deemed sufficient to produce microbiologically and chemically acceptable water for fish water systems, experience shows that carbon filtration is often the weak point of RO devices. When incoming water is contaminated with pollutants, carbon filters can rapidly be overwhelmed. The inclusion of an additional and reliable carbon filtration before the RO device usually helps improve RO filtration efficacy and reduce wear. RO systems present two other caveats: about 3–5 l of water is used to produce 1 l of RO water, and the resulting RO water is deprived of useful chemicals (e.g. hardness, alkalinity). RO water may therefore require various salt additions to condition it for use with aquatic animals. Hence, the resource cost (both financial and environmental) of RO must not be neglected.

The alternative is to rely solely on carbon filtration to remove pollutants by adsorption, e.g. chloramines, benzene derivatives, toluene, trichloroethylene, total trihalomethanes, pesticides and other large organic molecules (e.g. anaesthetics, bisphenol). The efficacy of carbon filters can be assessed by laboratories able to test these pollutants in water; the timing of replacing carbon filters can then be controlled. Figure 7.2 illustrates an example of wear markers and decrease in adsorption efficacy over time. Carbon filters do not, however, prevent microbial introduction: this is the job of UV disinfection step later down the water purification line. Also, carbon filters do not reduce the concentration of such a broad spectrum of molecules as does RO. For example, small molecules such as salt will remain unaffected and incoming hard water will remain hard. In brief, there is no solution to suit all set-ups, and each facility must

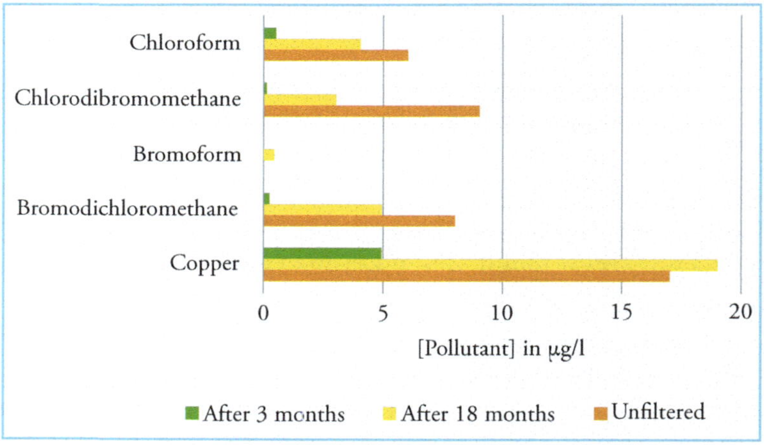

Figure 7.2 Efficacy of carbon filtration after 3 and 18 months of use, compared with unfiltered water. Pollutant concentrations were measured before and after carbon filtration and compared between filters active for 3 and 18 months, respectively. Efficacy was still acceptable after 3 months of use but not after 18 months.

assess the most suitable options for their needs, considering their means and what is available to them. Historically, facilities across the globe have used a variety of water supplies with success; RO is not a must. Some facilities may have access to soft municipal or wild water, others may decide to mix RO and hard municipal water to reach a balanced water quality suitable for zebrafish.

The main aim is that the water supplied to fish systems must not induce disease (see Chapter 4 for more information), and there are a few issues to avoid. First, chlorine must be eliminated from the supplied water. Chlorine derivatives such as chloramines require removal by carbon or RO filtration, whereas chlorine evaporates and is not always fully removed by filtration. Thus it is recommended to provide a step at which water rests before it is introduced to the fish system. In addition to chlorine evaporation, a resting step for water allows stabilization of water temperature. Otherwise, when cold water is introduced to a warm-water system, there is a risk of supersaturation and fish may suffer from gas bubble disease. To facilitate gas evaporation (and chlorine elimination), air bubbles can be added to the resting water reservoir to increase the contact interface between water and air and enhance gas exchange. Because rested water can also become a culture medium for bacteria and algae, to reduce microbial growth the reservoir is usually equipped with UV filtration. Another benefit of having a rested water reservoir is the practical solution in case of contingency, such as a disrupted water supply. Reservoir capacity should be based on an estimation of daily water needs, multiplied by the number of days preceding restoration of normal supply. At a density of up to five zebrafish per litre and in the absence of water renewal at tank level, it is expected that water pollution may not be significantly detrimental for 48 hours if feeding is reduced early on. If water can recirculate, sump water can be added to the total water available to fish and tank density adapted accordingly.

Recirculation of water

Flow-through systems, where water is discarded after it enters fish tanks, are not a common feature of zebrafish laboratories. Indeed, most zebrafish are held in recirculating aquaculture systems (RASs) and water treatment is a simplified version of that used at more intense aquaculture systems. In RAS, water exiting fish tanks is captured in sumps, mechanically filtered, carbon filtered, UV treated, biofiltered, heated or chilled (temperature must be controlled), dosed with salt and pumped back into the fish tanks. Carbon filtration is not always a feature of zebrafish RAS, but it is recommended to compensate for incoming water treatment deficiency and to adsorb organic molecules leaking from pipework and tanks (e.g. polycarbonate tanks and PVC pipes release bisphenol A with an oestrogenic effect), and residual anaesthetics and analgesics (see Chapter 13 for more information). Small quantities of fresh supply water (e.g. RO or dechlorinated water) are directly added to the RAS sump, where it is mixed with recirculating water and chemically balanced. Conductivity probes in the sump are linked to a pump in a salt reservoir (sodium chloride with other minerals in small quantities, depending on salt origin) and recirculating water is dosed in the sump, usually aiming for a conductivity of about 500 µS/cm. This parameter is easily measured and controlled, but it is not the most relevant for zebrafish or system health and it should not focus too much attention because slow variation of a few hundred µS/cm does not seem to affect zebrafish.

pH

pH is a key parameter to control, and pH monitoring should be part of the daily routine tasks, including at weekends. Sump pH is often measured by the system automatically through probe monitoring and feedback. Sodium bicarbonate is added to the sump to increase pH, which would otherwise decrease due to feed degradation. Sump water follows a unidirectional flow, with pH measured at the entrance of the sump and dosing towards the exit. There is no general consensus on the pH range acceptable for zebrafish (see Table 7.1; Aleström et al., 2019; Harper & Lawrence, 2011; Westerfield, 2007), but there is agreement that sudden and extreme variations can induce mortality. Systems have a resistance to pH variation that is measurable, namely alkalinity and thus high alkalinity is an insurance against pH issues. However, high alkalinity

Table 7.1 Published recommended ranges for zebrafish water chemistry. GH, hardness; KH, alkalinity.

Reference	pH	Conductivity (µS/cm)	(mg/l)					
			TAN	NH_3	NO_2^-	NO_3^-	GH	KH
Aleström et al., 2020	6.5–8.0	150–1,700		<0.10	<0.3	<25	50–150	50–150
Harper & Lawrence, 2011	6.8–8.5			<0.02	<1.0	<200	75–200	50–150
Westerfield, 2007	6.8 – 7.5	300–1.000		<0.02	<0.1	<10	50–100	50–100

is often an indicator of high carbon dioxide concentration, and this may be detrimental to fish welfare. Unfortunately, common zebrafish RASs are poorly adapted regarding the addition of calcium or magnesium carbonate salt to increase alkalinity or hardness. The guidelines for zebrafish care do not recommend high hardness or alkalinity (Table 7.1), because fish exposed to high concentrations of calcium or magnesium may develop kidney diseases in the long term (e.g. nephrocalcinosis; see Chapter 4 for more information). As a consequence, zebrafish systems are not exempt from pH issues.

The facility should be organized so that staff are competent in handling pH deviation emergencies and have the necessary equipment ready to do so. In other words, each EU must have a couple of two-point calibrated pH meters. Hand-held pH meters, which are widely available, are easy to use providing the calibration process is performed correctly. pH meters can be one-, two- or three-point calibrated; one-point calibration can still lead to lethal mistakes, and two-point is the minimum requirement. Having one spare pH meter is often insufficient because these devices are fragile but useful in the case of emergency, where another contingency may make the situation uncontrollable. Staff can easily be trained to calibrate, use and maintain pH meters. Nevertheless, pH values should not be over-interpreted; pH meters often have a low resolution, meaning that a pH of 7.5 could be read as 7.4 or 7.6 with the same device. Moreover, comparison of hand-held pH meter values with those measured by the system, despite calibration, can sometimes show constant differences, possibly due to device quality or simply due to pH variation along the RAS. In brief, measuring pH should be done daily to help in the detection of sudden pH variations and mid- to long-term deviation – for example, due to faulty equipment or poor husbandry. Any unexpected pH variation >0.4 should be investigated. In the case of sudden variation the treatment options are to increase water renewal, add acid or alkali (as appropriate) to recirculating water and move fish from the system. To preserve fish welfare when correcting pH, it is recommended not to alter it by more than 0.1–0.2 per hour; ideally, pH should remain between 7.0 and 8.0 at all times. These values are less important for fish welfare because zebrafish have been observed to thrive at both higher and lower pH values. The values presented here are more relevant in regard to maintenance of biofilters and nitrogen pollution.

Ammonia and biofilters

Nitrogen is released in the RAS from feed and animal waste. One nitrogen compound is particularly toxic: ammonia (NH_3; see Chapter 4 for more information). This molecule reacts with the hydrogen in water and is constantly balanced by the concentration of ammonium (NH_4^+), a non-toxicant. Neither ammonia nor ammonium is ever present without the other and, together, they represent total ammonia nitrogen (TAN). Ammonia concentrations are often not measured directly, but instead are extrapolated from TAN according to pH values, as shown in Table 7.2. For a fixed TAN concentration, the higher the pH the more NH_3 toxicant is present. In other words, in the case of ammonia pollution, one option is to reduce pH to tip the balance towards higher NH_4^+ concentrations and reduce NH_3 to residual and safe concentrations. This is simply because more acidic water contains more free/reactive hydrogen ions (H^+) available to bond with NH_3. The correlation between traditional chemistry calculations and evaluation of nitrogen cycle efficacy is blurred by the common employment of so-called nitrogen weights (e.g. NH_3-N, NH_4-N, NO_2-N, NO_3-N) rather than the real weights of these molecules, i.e. inclusion of

Table 7.2 NH_3 concentrations according to pH and TAN. $[NH_3]$ is calculated in ppm for a temperature of 27 °C and conductivity of 500 μS/cm. TAN is measured in NH-N. $[NH_3]$ <0.01 ppm is considered safe and is highlighted in green; $[NH_3]$ >0.02 ppm is considered unsafe and is highlighted in orange; $[NH_3]$ between 0.01 and 0.02 ppm is considered a risk and is highlighted in yellow.

pH	8.50	0	0.054	0.107	0.161	0.214	0.268	0.536
	8.00	0	0.019	0.039	0.058	0.077	0.096	0.193
	7.50	0	0.006	0.013	0.019	0.026	0.032	0.064
	7.25	0	0.004	0.007	0.011	0.015	0.018	0.036
	7.00	0	0.002	0.004	0.006	0.008	0.010	0.020
		0	0.25	0.50	0.75	1.00	1.25	... 2.5
					TAN (NH-N)			

hydrogen or oxygen atoms. A constant can be applied to convert one to the other: $[NH_3] = 1.22\ [NH_3\text{-}N]$; $[NH_4^+] = 1.29\ [NH_4\text{-}N]$; $[NO_2^-] = 3.28\ [NO_2\text{-}N]$; $[NO_3^-] = 4.43\ [NO_3\text{-}N]$. The advantage is that a defined concentration of NO_2-N becomes the same concentration of NO_3-N, for example. If oxygen atoms are included, molecular weights and concentrations will increase. Free calculators are available online and are most useful for such calculations, and others (Arain, 2010).

In practice, a RAS with pH >8.0 is exposed to NH_3 toxic concentrations whereas such intoxication is extremely unlikely with pH <7.0. However, maintaining pH <7.0 is detrimental to certain biomedia. To recycle TAN, RASs are equipped with bacteria called biomedia and set on a biofilter (a bacterium-friendly support with increased surface area and suitable aeration). Biomedia first transform TAN to nitrite (NO_2^-) and then nitrite to nitrate (NO_3^-). Nitrite remains an acute toxicant (with lesser potency than NH_3; see Chapter 4 for more information) whereas nitrate is much better tolerated and eliminated only by water exchange (Table 7.1). Invasion of the biofilter by prolific, active and sufficient biomedia is called 'seeding of the system'. This is a step of utmost importance when setting up a new system, introducing a new biofilter or when biomedia are recovering from an insult (e.g. disinfectant treatment or algal invasion not prevented by regular rinsing of biomedia beads).

Seeding of the system can be enhanced by the addition of biomedia bought from commercial sources, cultivated in-house or by transferring biofilters from another RAS, providing that microbiological statuses are compatible. Ammonium salt can then be added to feed the bacteria and assess their ability to degrade TAN prior to fish introduction. Suitable conditions must be set for the biomedia to thrive: sufficient but not excessive aeration, sufficient TAN provision and pH >7.0. Considering zebrafish RAS temperature, TAN levels in a RAS with a new biofilter will decrease in a few days or weeks, followed by a nitrite increase and then decrease as nitrate concentration increases. Daily measurements are recommended to monitor these curves and confirm that the RAS is seeding as expected. If fish are introduced in a virgin system without the addition of biomedia, natural seeding will take place at a slower pace. In this case, adults or larvae can be introduced at a rate of about 10% of total adult capacity every three weeks, ensuring that fish are added only once TAN and nitrite levels have returned to undetectable. Once seeded, TAN and nitrite concentrations are measured manually at least weekly by either drop tests or electronic devices. This is part of the technical staff's routine duties. Other chemical and physical parameters should be monitored at regular intervals: total dissolved gas or oxygen saturation, alkalinity, hardness and nitrate (with an electronic device to increase nitrate measurement accuracy). By collating these data, trends can be monitored and managed. To ease data comparison and to mitigate variation due to daily events, the best time to sample the RAS is before the first feed. Confirmation of nil scores of TAN and nitrite can then be obtained prior to feeding fish.

Diet distribution

Feed waste not only releases nitrogen into the system but also promotes biofilm growth, which can increase infection pressure to the point of inducing disease outbreak (Mocho, 2016). Abrupt change of diet may induce acute enteritis with gas lodged in the intestine acting against the swim bladder and fish showing transitory buoyancy difficulty in reaching the tank bottom. Thus diet distribution is a real husbandry challenge, with both short- and long-term effects. Further issues with diet quality have already been addressed (see Chapters 3 and 4 for more information).

When 50 larvae (i.e. one Petri dish) are first transferred to a tank and fed at 5 days post fertilization (dpf), diet distribution is based on overfeeding to ensure that larvae can find food without wasting energy and having to swim against the flow. The induced water pollution is tolerated to some extinct by larvae until they reach a developmental stage where they require more oxygen and normal water conditions. Water flow should be introduced at this stage (e.g. by 10 dpf) to ensure that chemical waste is evacuated. The 50 fishes are now in about 3 l of water. However, biofilm build-up is increasing in the tank, thriving on feed waste. After three to four weeks, biofilm at the bottom of the tank is syphoned out to avoid bacterial bloom and mortality. During this first month of life, and for several more weeks until fish are fed as adults, larvae and juveniles receive a range of dry and live diet increasing in size to make use of their growing mouth aperture (Aleström et al., 2019). Organizing the distribution of these diets presents several challenges: feeding the right quantity of diet, ensuring that the technician knows which diet to feed and scheduling diet transition to avoid acute enteritis. To standardize the quantity fed to the fish, feeding guns are available for adults but are more difficult to find for finer diets or live feed. Spatulae or spoons can be labelled to help technicians. Nevertheless, the risk of significant variation remains between individuals and must always be factored into delivery methods. Aquarium staff should rotate duties, at least ensuring competency for weekend and emergency cover. This is the opportunity for the team to share experiences and to discuss how to standardize their practices to avoid abrupt variations in distributed quantities.

Table 7.3 shows an example of a schedule for diet distribution, indicating colour-coded spoons for different diets. Such planning is necessary to clearly communicate which diet to feed and to set transition periods. The schedule does not start with days of the week but rather dpf, because larvae may reach the system on any day of the week. For the first few weeks, tanks should be labelled

Table 7.3 Diet distribution schedule. Zebrafish are fed from 5 dpf, when 50 larvae are set in a tank in a low volume of water. After four weeks of smooth diet transition, the tank of zebrafish is usually sufficiently developed to be separated in smaller groups or to be transferred to a larger tank. Although one simple meal is proposed for a standard tank, the quantity of feed per meal and the number of meals can be adapted to fish housing and activity. For example, large tanks holding high fish numbers can be fed more per meal and even receive an extra meal if intensive breeding is sought. By contrast, standard tanks holding low fish numbers should be fed less. This table indicates ways of adapting accordingly. Colour-coded spoons are dedicated to each diet, but this could be replaced by dedicated feeding guns or other automatic devices that ensure quantity standardization. N/A, not applicable because fish are not fed at that time point.

Week 1	5 dpf	6 dpf	7 dpf	8 dpf	9 dpf	10 dpf	11 dpf
Morning	Red spoon FEED 100	Red spoon FEED 100	Red spoon FEED 100	Red spoon FEED 100	Red spoon FEED 100	Red spoon FEED 100	Red spoon FEED 100
Lunch or Afternoon	*Paramecium* spp.	*Paramecium* spp.	*Paramecium* spp.	*Paramecium* spp.	*Paramecium* spp.	*Paramecium* spp.	*Paramecium* spp.
Week 2	**12 dpf**	**13 dpf**	**14 dpf**	**15 dpf**	**16 dpf**	**17 dpf**	**18 dpf**
Morning	Red spoon FEED 100	Red spoon FEED 100	Red spoon FEED 100	Blue spoon FEED 200	Blue spoon FEED 200	Blue spoon FEED 200	Blue spoon FEED 200
Lunch or afternoon	Brine shrimp	Brine shrimp	Brine shrimp	Brine shrimp	Brine shrimp	Brine shrimp	Brine shrimp
Week 3	**19 dpf**	**20 dpf**	**21 dpf**	**22 dpf**	**23 dpf**	**24 dpf**	**25 dpf**
Morning	Blue spoon FEED 200	Blue spoon FEED 200	Blue spoon FEED 200	Blue spoon FEED 200	Blue spoon FEED 200	Blue spoon FEED 200	Blue spoon FEED 200
Lunch	Brine shrimp	Brine shrimp	Brine shrimp	Brine shrimp	Brine shrimp	Brine shrimp	Brine shrimp
Afternoon (not weekend)	Brine shrimp	Brine shrimp	Brine shrimp	Brine shrimp	Brine shrimp	Brine shrimp	Brine shrimp
Siphon tank if a thick biofilm forms at the bottom before the fish are sufficiently large to be split.							

Weeks 4–6, until split	**Monday**	**Tuesday**	**Wednesday**	**Thursday**	**Friday**	**Saturday**	**Sunday**
Morning	Blue spoon FEED 200	Blue spoon FEED 200	Green spoon FEED 300	Blue spoon FEED 200	Green spoon FEED 300	Green spoon FEED 300	Green spoon FEED 300
Lunch	Blue spoon FEED 200	Blue spoon FEED 200	Blue spoon FEED 200	Blue spoon FEED 200	Blue spoon FEED 200	Blue spoon FEED 200	Green spoon FEED 300
Afternoon	Brine shrimp	Brine shrimp	N/A	Brine shrimp	Brine shrimp	Brine shrimp	N/A
At 4–6 weeks of age (maximum), fish are either separated into standard 3 l tanks at a density of five fish per litre or set as 50 fish in a large 8 l tank							
Fish after split	**Monday**	**Tuesday**	**Wednesday**	**Thursday**	**Friday**	**Saturday**	**Sunday**
Morning	Green spoon FEED 300	Green spoon FEED 300	Green spoon FEED 300	Green spoon FEED 300	Green spoon FEED 300	Green spoon FEED 300	Green spoon FEED 300
Afternoon	Brine shrimp	Brine shrimp	N/A	Brine shrimp	Brine shrimp	Brine shrimp	N/A
The number of green spoons of FEED 300 per meal can be adapted to the number of fish in each 8 l tank; 8 l tanks with >30 breeding zebrafish can be fed extra green spoon(s) of FEED 300 at lunchtime.							
Labelled tanks with <4 fish	**Monday**	**Tuesday**	**Wednesday**	**Thursday**	**Friday**	**Saturday**	**Sunday**
Morning	N/A	N/A	Green spoon FEED 300	N/A	N/A	Green spoon FEED 300	Green spoon FEED 300
Afternoon	Brine shrimp	Brine shrimp	N/A	Brine shrimp	Brine shrimp	N/A	N/A

to indicate which feed they should receive, with the schedule indicating the number and quantity of feeds. Once fish have grown to a sufficient size, distribution can be linked to the day of the week and taking into account weekends when staff are usually unable to feed as many meals, and to the necessity to avoid overfeeding less populated tanks. After four weeks, fish have normally grown sufficiently to be separated into tanks at five per litre. Fish with slower growth may benefit from being fed smaller-sized diets for longer. Tanks with many breeders may benefit from a supplementary meal. Indeed, when willing to 'feed more', we must bear in mind that animals do not have an expandable capacity to ingest food, and feeding a larger diet quantity per meal is only likely to increase dietary waste. To optimize diet assimilation, the preferred option is to wait for the anterior gut segment to have processed the previous meal (3–4 hours for zebrafish) so that there is room for an extra meal. Whatever the developmental stage and the challenges regarding diet provision, a smooth transition must be scheduled when diet distribution, quantity or quality is amended. Table 7.3 proposes examples of transition with only one diet being changed every 3–4 days, making use of weekends and mid-week to reduce the number of distributed meals and to introduce or increase the distribution of new diets. Dry diets often provide more nutritional value than live diets, the latter being provided mainly to create environmental enrichment (see Chapter 2 for more information). To preserve some novelty and avoid potential boredom due to a routine provision, and to optimize nutrient distribution, *Artemia* spp. may not be distributed daily, but perhaps a few times per week.

Health monitoring

Daily care

Daily tasks for the zebrafish technician are very much organized around testing water quality, cleaning equipment and feeding fish, all of which are related to fish health. Notably, while feeding fish the technician can detect abnormal behaviour. The technician should be feeding the fish and not the tank. Thus feeding should not be a race against time, but rather a dedicated time for fish observation. Indeed, health monitoring relies first on daily detection and recording of clinical cases. On a regular basis, technicians can also score body condition, measure length and width or weigh fish, all without sedation. This is achieved with the help of a photograph in a low volume of water (see Chapter 6 for more information) or simply by transferring fish to a weighed receptacle of water. This is really useful in monitoring husbandry output, by detecting clutch heterogeneity (Mocho et al., 2022b). However, care must be taken not to stress fish unnecessarily when measuring growth performance, particularly with genetically altered lines prone to distress.

Mortality

Where fish are affected by pathogens or by non-infectious diseases (see Chapters 4 and 5 for more information), a database needs to be set up to record and calculate both morbidity (i.e. the number of fish affected by a clinical sign or lesion) and mortality (i.e. the number of fish found dead or euthanized). This database facilitates monitoring the detection of disease by comparing the number of fish found dead with that of fish euthanized due to disease, adverse effect or phenotype. Moreover, the database should support data analysis according to age, system, genotype (for comparison with wild-type colonies) etc. This allows the detection of a problem in a specific part of the facility and differentiation between disease and adverse effect or phenotype. The latter will inform severity assessments (see Chapters 9 and 14 for more information). Mortality rates in the EU must be assessed at least on a monthly basis, and a threshold mortality rate can be defined. Veterinary investigation is triggered only when mortality increases beyond this threshold (Mocho, 2016). Sampling fish with clinical signs for further diagnosis can help establish the health status of the EU. However, these are not randomly selected fish, because they have a higher likelihood of being contaminated by a pathogen (Mocho et al., 2022b).

Sentinels

This detail is relevant only when applying epidemiological rules that consider the number of fish to sample according to pathogen prevalence, based on the hypothesis that all fish in the EU have the same risk of being infected (Mocho et al., 2022b). This rule does not apply when setting sentinel fish and sampling the system's environment (Mocho et al., 2017). Sentinels are exposed to water exiting fish tanks (i.e. sump water, pre-filtration) for at least three months before sampling; this water is more likely to carry pathogens and sentinels are therefore more likely to be tested positive. Sentinels can be set in a tank receiving sump water continuously through a dedicated water pump set in the sump, or on a shelf with replacement of sump water three times per week. Sentinels should represent the population, and hence wild types from the most common background are selected. Some pathogens may affect one sex or age group more than another. One option is to set a pool of 25 males and 25 females aged six months and to expose them as sentinel fish for a year (with continuous sump water supply to avoid water chemical pollution). Ten sentinel fish are sampled every quarter. Colony fish are also sampled to complete numbers, ensuring that at least five fish are selected, including some below six months of age. Although PCR is a popular tool used to screen for the most common or dangerous pathogens, histopathology is the only technique that assesses the severity of disease, including non-infectious lesions. Only freshly euthanized fish can be fixed for histopathology analysis (Mocho et al., 2022b).

Environmental samples

The fish environment is also sampled quarterly. Not all sample types need to be taken at the same time: for instance, a different sample type can be sent monthly. Some environmental samples consist of sampling the surface of both sump and live feed culture for *Mycobacterium* spp. identification by PCR. A similar assay is performed with sump sludge, which is also used for the detection of

P. tomentosa by microscopy or PCR. Environmental samples increase the detection rate of certain microbes compared with fish screening (Mocho et al., 2017), but some mycobacteria may be commensal in the environment and do not induce disease. Morbidity and mortality data help interpret such results. The detection of certain pathogens in the environment remains difficult, e.g. *P. neurophilia* (Mocho et al., 2022b).

Screening pattern

As discussed in Chapter 9, the number of samples dictates the capacity of the screening pattern to detect pathogens according to their prevalence (i.e. the proportion of fish infected by the pathogen) (Mocho et al., 2022b; Sergeant, 2018). The limitation to that rule is that it is not possible to demonstrate that an EU is specific-pathogen free (SPF). We can only hypothesize that if a defined pathogen was present, it would be at such a high prevalence that it would be detected. The number of samples sets the prevalence threshold against which we screen, meaning that statistically we should be able to detect those pathogens that are more prevalent than the prevalence threshold; pathogens present at lower prevalence will remain undetected (Mocho & Pereira, 2022). Hence, 'below detection threshold' (BDT) would represent better screening results than SPF. Nevertheless, considering that sentinels and sick fish are screened and that environmental samples can significantly improve our ability to detect certain pathogens, the screening pattern can be organized at different depths depending on the pathogens of concern and the microbiological status that is to be achieved (detailed in Table 7.4). Typically, samples are taken on a quarterly basis for the Screen More Often Pathogens (SMOP) and twice a year for the Screen Less Often Microorganisms (SLOM). SMOP includes the most relevant pathogens, in consideration of their prevalence and virulence (i.e. their potential impact on fish or human health and on models). *Mycobacterium* spp., *P. tomentosa* and *P. neurophilia* are SMOP. The list included in SLOM depends on the facility's biosecurity and context (Mocho et al., 2022a,b). There is little knowledge available on SLOM environmental detectability and, in consideration of the lower frequency of SLOM screening, it is not advisable to claim any SPF or BDT status for SLOM.

For *Mycobacterium* spp., only *M. haemophilum* and *M. marinum* are considered major threats. The presence of other mycobacteria is expected in water systems, with or without fish and, providing that morbidity and mortality are under control, they are not a concern. *M. haemophilum* and *M. marinum* are detectable in environmental samples: surface swab, sludge and feed culture; *P. tomentosa* is also detectable in sludge. We hypothesize that multiplying environmental samples quarterly is therefore a means of reducing the number of screened fish without diminishing our ability to detect these SMOP. Hence, for these three pathogens, higher prevalence thresholds can be set to calculate the required number of fish sampled to demonstrate BDT status (see Table 7.4). Such a status can be claimed only following six quarterly screens that are consistently negative, and when EU biosecurity is optimal in the prevention of contamination (Mocho & Pereira, 2022). The detection of *P. neurophilia* remains challenging in system water, and very high fish numbers would have to be culled and tested to demonstrate BDT for a low prevalence threshold (see Table 7.4). This comes at a very significant cost (both animals and money) and is not suitable for most facilities, even more so considering that most facilities are contaminated by this parasite. Besides monitoring of fish health, screening data are most useful in supporting the safe exchange of colonies between laboratories. Reports can be produced to collate records and husbandry information so that collaborating facilities can be informed and adapt their processes accordingly. Templates for health reports and a description of EU husbandry and biosecurity are available, and facilities are strongly encouraged to share these reports transparently (Mocho et al., 2022a,b). Indeed, preserving fish health is a key step towards maintaining sustainable zebrafish colonies.

Environmental impact

Whenever possible, the aquatic facility should be built with sustainability in mind and managed to minimize its environmental

Table 7.4 Screening patterns for routine and BDT zebrafish health monitoring. A screening pattern concerns only one EU. Samples required to achieve BDT status are in addition to those for routine health monitoring. Fifteen fish are tested quarterly by either PCR or histopathology for routine health monitoring. For the BDT screening pattern for each pathogen, additional sample sizes were calculated using the online software epitools.ausvet.com.au (Sergeant, 2018) for a population of >1,000 fish, with 95% confidence and sensitivity of 95–100%; specificity was considered 100% by the software. Prevalence thresholds were arbitrarily estimated at 10% for *P. tomentosa*, 7% for *M. marinum* and *M. haemophilum* and 3% for *P. neurophilia*. The efficacy and numbers of environmental screening techniques were also considered. Fish found dead are screened by PCR only. Note that quarantine should not be part of the main EU; this is indicated here purely for convenience. In the absence of imports or animals available for screening, the quarantine is sampled quarterly as appropriate (e.g. swabbing of sludge and sump surface: one sample type per quarter and per system).

Screening objective	Origin of sample	Month 1	Month 2	Month 3	Month 4	Month 5	Month 6	Histopathology and/or PCR
Routine Health Monitoring	**Colony fish**	Collect fish with clinical signs		Screen 5 fish <6 months old	Collect fish with clinical signs		Screen 5 fish <6 months old	Month 3: SMOP Month 6: SMOP + SLOM
	Pre-filtration sentinels	Set up sentinel tank		Sample 10 fish			Sample 10 fish	
	Sludge analysis		1 sample for PCR			1 sample for PCR		*Mycobacterium* spp., *P. tomentosa*
	Sump surface swab	1 sample for PCR			1 sample for PCR			*Mycobacterium* spp.
	Feed	1 sample for PCR			1 sample for PCR			*Mycobacterium* spp.
	Quarantine	Screen imports (fish, sludge and/or clutch)						SMOP + SLOM
	Morbidity and mortality	Monitor and increase number of quarterly fish samplings if mortality increases						Histopathology and/or PCR

Table 7.4 (continued)

Screening objective	Origin of sample	Month 1	Month 2	Month 3	Month 4	Month 5	Month 6	Histopathology and/or PCR
Additional samples for BDT screening pattern								
Additional samples for *P. neurophilia*	**Colony fish**			PCR or histopathology on 85 fish			PCR or histopathology on 85 fish	*P. neurophilia*
	Morbidity – Mortality	Monitor and increase number of quarterly fish samplings if mortality increases						Histopathology and/or PCR
Additional samples for *P. tomentosa*	**Colony fish**			PCR or histopathology on 15 fish			PCR or histopathology on 15 fish	*P. tomentosa*
	Sludge analysis		2 samples for PCR			2 samples for PCR		
	Morbidity and mortality	Monitor and increase number of quarterly fish samplings if mortality increases						Histopathology and/or PCR
Additional samples for *Mycobacterium marinum* and/or *Mycobacterium haemophilum*	**Colony fish**			PCR or histopathology on 30 fish			PCR or histopathology on 30 fish	*M. marinum*, *M. haemophilum*
	Sludge analysis		1 sample for PCR			1 sample for PCR		
	Sump surface swab	1 sample for PCR			1 sample for PCR			
	Feed	1 sample for PCR			1 sample for PCR			
	Morbidity and mortality	Monitor and increase number of quarterly fish samplings if mortality increases						Histopathology and/or PCR

impact. Environmental impact arises from, for instance, wasteful activities, unnecessary replacement/refurbishment or incorrect and overzealous protocols. Examples of structural initiatives to reduce environmental impact include the insulation and protection of walls and doors to improve temperature control and prevent early decay due to humidity or trolley banging. Heating, ventilation and air conditioning can be designed to resist extreme weather and/or run on low energy consumption. The use of light-emitting diodes suitable for zebrafish holding rooms can reduce electricity consumption significantly. Another option is to guide sunlight into the fish room, both to reduce energy consumption (lighting and heating) and reproduce a more natural environment. There are numerous initiatives, websites and schemes available for anyone interested in environmental impact, and particularly in laboratory animal science (FELASA, 2019). More specifically, many organizational aspects of the zebrafish facility can significantly impact its environmental impact, whether through stock management, biosecurity or energy and resource consumption. To explore this question further it is necessary to incorporate basic carbon footprint calculations, and we propose some practical applications to illustrate this.

Ordering consumables

The carbon footprint of consumables is always highly correlated to transport (i.e. from raw material to factory, from factory to distributor, from distributor to consumer). Therefore, an easily achievable reduction in the carbon footprint consists of selecting suppliers that use less packaging, and of working with the waste removal company to adapt triage and bin capacities to space rubbish collections. A refined storage management can help too, by limiting the consumable range and the utilization of single-use equipment, ordering larger volumes or more concentrated solutions to minimize packaging and number of deliveries, and regrouping orders to decrease the number of deliveries. These approaches are not challenged by excessive costs or scientific requirements: they depend only on organizational skills and good communication. Similarly, the use of consumables can sometimes be reduced, e.g. Petri dishes can be alcohol disinfected and reused. PPE may not be necessary for some procedures, or it may be reusable or suitable for dedicated recycling routes (James & Sedman, 2023).

The origin of the diet, and its environmental impact from raw material harvest to manufacturing, distributor's storage and finally transit to the aquatic facility, can be challenging but relevant to assess. This is certainly a concern that the aquaculture industry has identified (Alltech Coppens, 2023). Amongst points to be raised, the origin of the proteins in the feed is of particular interest. Fish meal and a shrimp-based diet would likely have a higher carbon footprint than insect- or crop-based diets. However, crops may be genetically modified and more heavily contaminated with pesticides, or they may contain mycotoxins etc. Understanding the source of the diet and its benefits, together with holding a conversation to assess environmental improvements by the manufacturer, could alleviate impacts. Alternatively, developing innovative in-house cultures (e.g. *Daphnia* spp.) could be explored further.

Biosecurity

Setting up a biosecurity programme with sustainability in mind means considering those processes that can damage equipment and the external environment. For example, corrosive chemicals should not be used on sensitive equipment. To prevent zebrafish contamination, detergents are not used in the holding rooms, water and alcohol or hydrogen peroxide being sufficient for routine room cleaning. Waste water can contaminate sewage and rivers with microbes (*Mycobacterium* spp., chytrid from *Xenopus* spp.) and chemicals (e.g. anaesthetics). Before release to the wild, water can be treated with heat to eliminate microbes and/or carbon filtered to adsorb chemicals, depending on risk assessment and local legislation.

Another topic for consideration is tank washing and disinfection. Some facilities autoclave their fish tanks routinely, even when used only in the main holding rooms. Autoclaving requires the production of steam to reach temperatures >121 °C. It is a sterilization process with a particularly high carbon footprint, at each cycle and on building installations. The autoclaved tanks will be contaminated by microbes in the main systems as soon as they are set back on the system. The only perceived advantage is biofilm control, which can be eliminated by a thorough washing system followed by a drying period between uses. Moreover, temperatures in the range of about 70 °C for a few minutes are sufficient to eradicate common zebrafish pathogens (e.g. *Mycobacterium* spp., *P. tomentosa*; Kent et al., 2019). An alternative to autoclaving is to use a washer-disinfector that uses a single cycle to wash and disinfect tanks by thermal disinfection at >80 °C for several minutes, sufficient to eliminate viruses and bacteria, including spores (Mocho et al., 2021). When such eco-friendly cleaning and disinfection equipment is not available, a carbon footprint evaluation can help identify work patterns to reduce carbon footprint. For example, to avoid energy waste, the autoclave should be used only when full (but remaining efficient). The work pattern of tank change can be adapted to mitigate the risk of running a half-empty autoclave. All too often the green-minded and purse-conscious agree when energy consumption is involved. To gain support from financial powers in the establishment, a carbon footprint analysis can help to justify some investments by demonstrating how energy consumption (and therefore running costs) are reduced.

Resource consumption

Regarding water consumption, the biggest question is whether RO water is necessary to supply holding systems. Daily water exchange is often set at 10%, but numerous systems actually run with ≤5% daily water renewal without adverse effects. This lower exchange rate is achieved by not feeding zebrafish excessively, i.e. reducing feed waste, and controlling nitrate levels. This halving of daily water renewal significantly reduces the facility's environmental impact. In a 25,000 l fish-holding system, daily 5% water renewal facilitates the saving of 1,250 l of RO water daily compared with 10% daily water renewal. In the worst-case scenario, where 5 l of municipal water are used to produce 1 l of RO water, the volume of municipal water saved daily by halving water renewal is 6,250 l. Thus, 25,000 l of municipal water (i.e. [25,000/1,000] × 0.378 = 9.450 kg carbon dioxide equivalent (CO_2e) in the UK; see Table 7.5) are saved every

Table 7.5 Conversion factors for water and electricity in the UK and France. The conversion factor for water (in m^3) includes those for water supply and treatment. The given conversion factor for electricity includes transmission and distribution waste (ADEME, 2014; Department for Energy Security and Net Zero, 2023; Executive Office of Energy and Environmental Affairs, 2023).

Country	Water treatment	Electricity
	Kg CO_2e/m^3	**Kg CO_2e/kWh**
France	0.394	0.057
UK	0.378	0.225

four days. In other words, in the case of a system filled with RO water, the volume equivalent in municipal water of the fish holding system is consumed in two days when daily water renewal is set to 10%, and in four days when set to 5%. To this should be added the impact of the production, delivery, maintenance (transport of engineer and membrane) and end of life of the RO unit. Similarly, salt production, delivery and consumption required to condition the system water should be considered. The green alternative seems obvious: reduce the use of RO whenever possible and consider soft dechlorinated municipal water if suitable.

While water consumption represents an environmental impact in itself, clean water production and dirty water recycling at a national or regional level are energy-consuming processes with a dedicated carbon footprint. This is measured in kilograms CO_2e, the universal unit of measurement indicating the global warming potential of one unit of CO_2. To calculate the energy consumption in equivalent greenhouse gas emissions (CO_2e), three scopes of emission-releasing activities are first identified. Scope 1 considers direct emissions from activities owned or controlled by an organization; scopes 2 and 3 focus on indirect emissions due to the energy being produced or consumed by an external and uncontrolled party. Scope 2 is specific to purchased energy resources (i.e. electricity, heat, steam or cooling); scope 3 encompasses the other upstream and downstream consequences of an activity (e.g. travel, waste disposal, purchased materials). Boundaries between scopes very much depend on an organization's activities and controls. For each single emission-releasing activity, an emission factor is defined at national or regional level. To calculate the carbon footprint in CO_2e for an activity in the aquatic facility, energy (e.g. electricity in kWh) and resource consumption (e.g. m^3 of water) under the three scopes are added and multiplied by their respective conversion factor (see Table 7.5; ADEME, 2014; Department for Energy Security and Net Zero, 2023; Executive Office of Energy and Environmental Affairs, 2023).

Carbon footprint comparison

The information required to calculate the carbon footprint of an item of equipment should be provided by the supplier, such as the energy in kWh (and not power in kW) consumed by a tank wash cycle (see Tables 7.6 and 7.7 for examples). The energy value is then multiplied by a conversion factor to obtain the carbon footprint of the wash cycle in kg CO_2e. The conversion factor depends on how the electricity used during the wash cycle is produced. The conversion factor varies with both space and time: countries with nuclear power have lower conversion factors than those relying on fossil fuels; the conversion factor reduces with daylight when solar panels are involved (Executive Office of Energy and Environmental Affairs, 2023). Similar considerations are valuable for municipal water treatment plants. In consequence, each facility must investigate and understand their local context to evaluate their environmental impact and adapt their practices accordingly. Figure 7.3 compares the carbon footprint of equipment use according to country: France versus the UK. It is notable that water consumption represents only a small portion of these carbon footprints. Due to the UK's reliance on fossil fuels, its conversion factor for electricity is about four times that of France (see Table 7.5). As a consequence, the carbon footprint for the use of electric equipment is much higher in the UK than in France. This is particularly obvious for the cycle of a large autoclave (capacity of 180 tanks). Bressler (2021) suggests that the emission of 4,434 t CO_2e causes one excess death globally; this large autoclave would have to run >240,000 cycles in the UK to reach that carbon footprint. Although this threshold seems very unlikely to be reached

Table 7.6 Carbon footprint (kg CO_2e) of equipment in France. Carbon footprint is first calculated for a washer capable of thermal disinfection (TD) (for the start-up phase, a wash-only cycle and a wash plus TD cycle) and for a small and a large autoclave. Carbon footprint is then calculated for various combinations of cycles required to process 180 tanks in a single day. Wash & TD, washer cycle including a wash and thermal disinfection; autoclave 50, small autoclave of 50-tank capacity; autoclave 180, large autoclave of 180-tank capacity; 5 wash & TD + Q autoclave, only tanks coming out of quarantine are autoclaved in a single cycle of the small autoclave and all 180 tanks are washed and thermally disinfected.

France		Consumption per use			
Equipment use	Capacity (tanks)	Electricity (kWh)	Water (l)	Carbon footprint calculation	kg CO_2e
Washer start-up	36	12	200	12 × 0.057 + 0.200 × 0.394	0.76
Wash only		4	50	4 × 0.057 + 0.050 × 0.394	0.25
Wash & TD		8	50	8 × 0.057 + 0.050 × 0.394	0.48
Autoclave 50	50	10	200	10 × 0.057 + 0.200 × 0.394	0.65
Autoclave 180	180	80	900	80 × 0.057 + 0.900 × 0.394	4.91
5 wash only	180	32	450	0.76 + 5 × 0.25	2.00
5 wash & TD		52	450	0.76 + 5 × 0.48	3.14
5 wash & TD + Q autoclave		62	650	3.14 + 0.65	3.79
5 wash and 4 autoclave 50		72	1,250	2.00 + 4 × 0.65	4.60
5 wash and 1 autoclave 180		112	1,350	2.00 + 4.91	6.92

Table 7.7 Carbon footprint (kg CO_2e) of equipment in the UK. The carbon footprint is first calculated for a washer capable of TD (for the start-up phase, a wash-only cycle and a wash plus TD cycle) and for a small and a large autoclave. Carbon footprint is then calculated for various combinations of cycles to process 180 tanks in a single day.

UK		Consumption per use			
Equipment use	Capacity (tanks)	Electricity (kWh)	Water (l)	Carbon footprint calculation	kg CO_2e
Washer start-up	36	12	200	12 × 0.225 + 0.200 × 0.378	2.78
Wash only		4	50	4 × 0.225 + 0.050 × 0.378	0.92
Wash & TD		8	50	8 × 0.225 + 0.050 × 0.378	1.82
Autoclave 50	50	10	200	10 × 0.225 + 0.200 × 0.378	2.33
Autoclave 180	180	80	900	80 × 0.225 + 0.900 × 0.378	18.34
5 wash only	180	32	450	2.78 + 5 × 0.92	7.37
5 wash & TD		52	450	2.78 + 5 × 1.82	11.87
5 wash & TD + Q autoclave		62	650	11.87 + 2.33	14.20
5 wash and 4 autoclave 50		72	1,250	7.37 + 4 × 2.33	16.67
5 wash and 1 autoclave 180		112	1,350	7.37 + 18.34	25.71

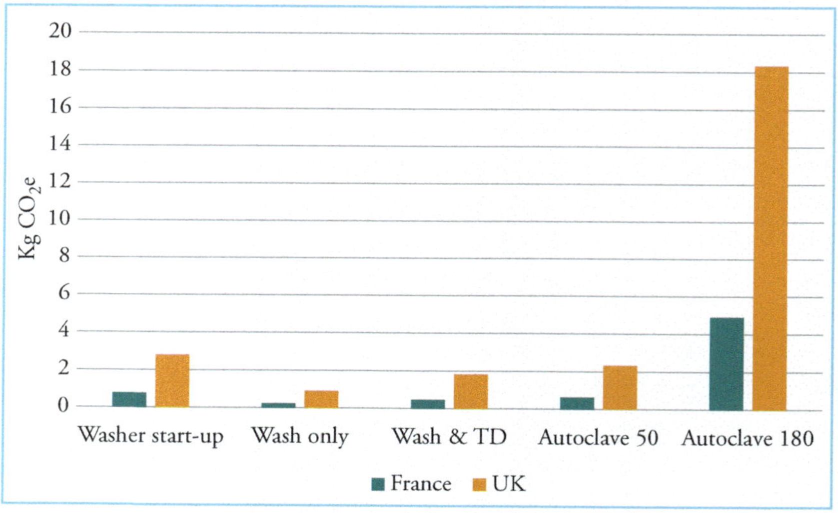

Figure 7.3 Carbon footprint (kg CO_2e) per cycle in France versus the UK.

in many research facilities, an exhaustive carbon footprint analysis would not focus solely on use of the autoclave: it would include its manufacturing and the sourcing of raw materials, as well as other elements of the equipment's life cycle such as installation, removal and fate after disposal. For example, the impact of installation and use of large autoclaves on a building's structure can be significant. This evaluation could not readily be performed by the manager of a zebrafish facility because it requires the help of dedicated professionals. Nevertheless, carbon footprint comparison can be used to organize work in the zebrafish unit.

Table 7.6 and Figure 7.4 show resource consumption (water and electricity) converted to carbon footprint (CO_2e) for various equipment used in France (see Table 7.5 for the conversion factors used). Table 7.7 and Figure 7.5 present the same data calculated using the UK conversion factors. These fictional examples describe the use of a washer-disinfector that can be used for a wash cycle only or for a wash and thermal disinfection cycle. It has a capacity of 36 tanks. Before the first use of the day its water reservoirs must be filled and warmed to the necessary temperature. The start-up phase consumes resources and cannot be ignored. In this example we do not consider the carbon footprint due to chemical use or RO water production. In this scenario, management is considering the need for an autoclave and evaluates various options to process 180 (i.e. 5 × 36) tanks per day. First, the carbon footprint involved in washing 180 tanks is calculated (i.e. start-up plus five wash-only cycles). Then we assess a first option, consisting of the washing and thermal disinfection of all cages with no autoclave use. The second option builds on the first, with the addition of the autoclaving of all tanks coming out of quarantine before they enter the washer. A small autoclave is envisaged for that purpose, with a capacity of 50 tanks. The third and fourth options do not use thermal disinfection: tanks are only washed and then autoclaved. For the third option, the 50-tank-capacity autoclave is used four times to process the 180 tanks. In the final option, a large autoclave with a capacity of 180 tanks is used only once per day to sterilize all the tanks in a single cycle. It should be noted that carbon footprint scales vary between Figures 7.4 and 7.5, but the general comparison draws similar conclusions. The large autoclave does not appear to be a sound investment: indeed, four small autoclave cycles produce one-third less CO_2e than a single large autoclave cycle. The constraints on the building and for installation are also

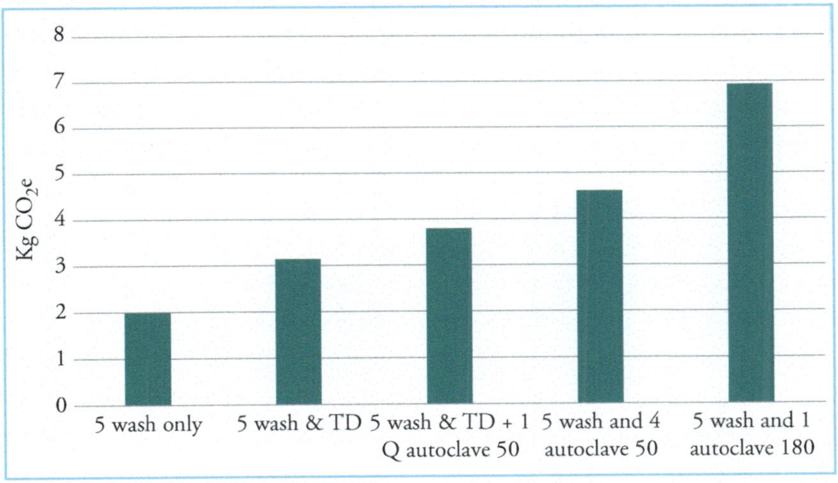

Figure 7.4 Carbon footprint (kg CO_2e) generated by processing 180 tanks in a single day in France.

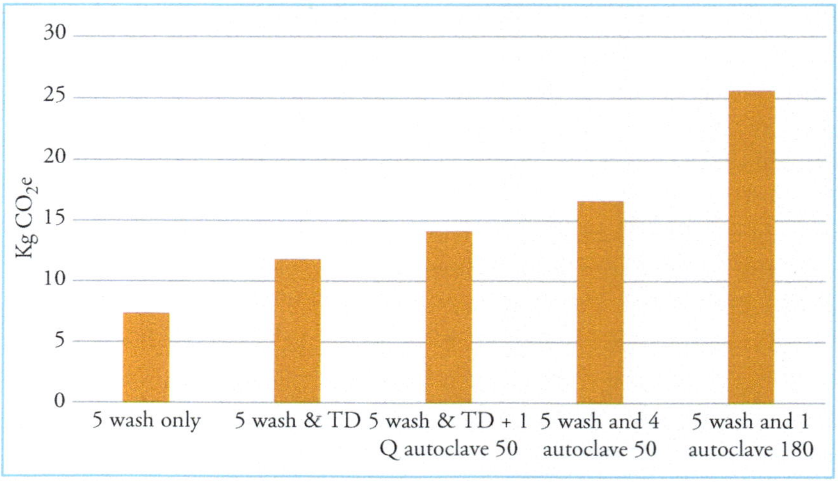

Figure 7.5 Carbon footprint (kg CO_2e) generated by processing 180 tanks in a single day in the UK.

likely to be more sustainable with the small autoclave. However, staff will have to store and process four autoclave runs, and that impact must be considered in regard to the scale of the facility's organization. We could envisage dividing the workload on several days of the week, but the washer's start-up is so resource intensive that this is unlikely to be an environmentally friendly option. The wisest options in this scenario are to thermally disinfect the tanks. There is a reduction of roughly 30% in CO_2e when all tanks are thermally disinfected and none autoclaved. The addition of autoclaving only those tanks coming out of quarantine adds about 17% CO_2e. The necessity to follow that option must be discussed further, in regard to details of installation of the small autoclave. The investment must be justified in regard to the import policy and potential quarantine hazards, bearing in mind the sustainability of the programme.

Conclusion

Times have changed since the first pioneers of zebrafish as a model organism set up their first facilities and primarily concerned themselves with mastering water quality. The modern facility manager must now understand a range of contributing factors and learn a variety of skills. Biosecurity and health monitoring require some microbiological knowledge and, preferentially, some database ability. Refining zebrafish care requires some understanding of therapeutics for anaesthetic, analgesic and euthanasia protocols. Managers are also recruited for their communication and organizational skills to ensure that the facility team works together and complies with the defined protocols and local legislation. Staff should liaise with external colleagues to exchange ideas, be aware of innovations and network. This is a key promoter for the reduction in laboratory animal science's environmental impact. Ideas must be shared, and collective pressure on stakeholders is a significant lever to raise awareness of concerns. The environmental impact of animal research is also an ethical issue. Scientific experiments are approved only when expected benefits are deemed to outweigh the harm induced to animals. However, if the environmental impact is not assessed and considered, it could be argued that a key factor of the equation is missing. This could deter public support and funders. Research establishments ought to tackle this challenge proactively.

References

ADEME. (2014). *Base carbone. Documentation des facteurs d'émissions de la Base Carbone ®*. Version 11.0.0. Agence de l'Environnement et de la Maitrise de l'Energie.

Aleström, P., D'Angelo, L., Midtlyng, P. J., Schorderet, D. F., Schulte-Merker, S., Sohm, F., & Warner, S. (2019). Zebrafish: Housing and husbandry recommendations. *Laboratory Animals*, *54*(3), 213–224. https://doi.org/10.1177/0023677219869037

Alltech Coppens. (2023) Moving beyond Fish in: Fish out ratio measures in sustainable aquaculture feeds. https://www.alltechcoppens.com/en/news/quantifying-feed-sustainability Retrieved August 12, 2023

Arain, H. M. (2010). Aquarium calculators. https://www.hamzasreef.com/Contents/Calculators/ Retrieved August 12, 2023

Bressler, R. D. (2021). The mortality cost of carbon. *Nature Communications, 12*(1), 4467. https://doi.org/10.1038/s41467-021-24487-w

Chang, C. T., Colicino, E. G., DiPaola, E. J., Al-Hasnawi, H. J., & Whipps, C. M. (2015). Evaluating the effectiveness of common disinfectants at preventing the propagation of Mycobacterium spp. isolated from zebrafish. *Comparative Biochemistry and Physiology. Toxicology and Pharmacology*, 178, 45–50. https://doi.org/10.1016/j.cbpc.2015.09.008

Crim, M. J., & Lawrence, C. (2021). A fish is not a mouse: Understanding differences in background genetics is critical for reproducibility. *Lab Animal, 50*(1), 19–25. https://doi.org/10.1038/s41684-020-00683-x

Department for Energy Security and Net Zero. (2023). Greenhouse gas reporting: Conversion factors 2023. https://www.gov.uk/government/publications/greenhouse-gas-reporting-conversion-factors-2023 Retrieved August 12, 2023

Executive Office of Energy and Environmental Affairs. (2023). Greenhouse gas emission intensity of electricity generation in Europe. https://www.eea.europa.eu/ims/greenhouse-gas-emission-intensity-of-1 Retrieved August 12, 2023

FDA. (2014). Reverse osmosis. https://www.fda.gov/inspections-compliance-enforcement-and-criminal-investigations/inspection-technical-guides/reverse-osmosis Retrieved August 12, 2023

FELASA. (2019). 3Rs – Reduce reuse recycle. https://felasa.eu/about-us/library/3rs-reduce-reuse-recycle Retrieved August 12, 2023.

Harper, C., & Lawrence, C. (2011). *The laboratory zebrafish* (1st ed). CRC Press.

James, H., & Sedman, B. (2023). Recycling the "unrecyclable". *In Practice, 45*(5), 292–296. https://doi.org/10.1002/inpr.333

Kent, M. L., Watral, V., Villegas, E. N., & Gaulke, C. A. (2019). Viability of Pseudocapillaria tomentosa Eggs exposed to heat, ultraviolet light, chlorine, iodine, and desiccation. *Zebrafish, 16*(5), 460–468. https://doi.org/10.1089/zeb.2019.1736

Mason, T., Snell, K., Mittge, E., Melancon, E., Montgomery, R., McFadden, M., Camoriano, J., Kent, M. L., Whipps, C. M., & Peirce, J. (2016). Strategies to mitigate a Mycobacterium marinum outbreak in a zebrafish research facility. *Zebrafish, 13*, Suppl. 1, S77–S87. https://doi.org/10.1089/zeb.2015.1218

Mendoza, G., Jr., Afolalu, A. A., Lertpiriyapong, K., Lipman, N. S., & Lieggi, C. (2022). Evaluation of media conductivity and a combination of iodine and sodium hypochlorite surface disinfection on zebrafish (Danio rerio) embryo viability and morphology. *Zebrafish*, 19(5), 190–199. https://doi.org/10.1089/zeb.2022.0028

Mocho, J. P. (2016). Three-dimensional screen: A comprehensive approach to the health monitoring of zebrafish. *Zebrafish*, *13*, Suppl. 1, S132–S137. https://doi.org/10.1089/zeb.2015.1200

Mocho, J. P., & Pereira, N. (2022) Chapter 4. Health monitoring, disease, and clinical pathology. In L. D'Angelo & P. de Girolamo (Eds.), *Laboratory fish in biomedical research* (pp. 81–100). Academic Press.

Mocho, J. P., & von Krogh, K. (2022). A FELASA working group survey on fish species used for research, methods of euthanasia, health monitoring, and biosecurity in Europe, North America, and Oceania. *Biology*, *11*(9), 1259. https://doi.org/10.3390/biology11091259

Mocho, J. P., Collymore, C., Farmer, S. C., Leguay, E., Murray, K. N., & Pereira, N. (2022a). FELASA-AALAS recommendations for monitoring and reporting of laboratory fish diseases and health status, with an emphasis on zebrafish (Danio rerio). *Comparative Medicine*, *72*(3), 127–148. https://doi.org/10.30802/AALAS-CM-22-000034

Mocho, J. P., Collymore, C., Farmer, S. C., Leguay, E., Murray, K. N., & Pereira, N. (2022b). FELASA-AALAS recommendations for biosecurity in an aquatic facility, including prevention of zoonosis, introduction of new fish colonies, and quarantine. *Comparative Medicine*, *72*(3), 149–168. https://doi.org/10.30802/AALAS-CM-22-000042

Mocho, J. P., Coutot, R., Douglas, M., Szpiro, L., Bouchami, D., Durimel, L., Moulès, V., & Hardy, P. (2021). Assessment of microbial reduction by cage washing and thermal disinfection using quantitative biologic indicators for spores, viruses and vegetative bacteria. *Journal of the American Association for Laboratory Animal Science*, *60*(5), 529–538. https://doi.org/10.30802/AALAS-JAALAS-21-000026

Mocho, J. P., Martin, D. J., Millington, M. E., & Saavedra Torres, Y. (2017). Environmental screening of Aeromonas hydrophila, Mycobacterium spp., and Pseudocapillaria tomentosa in zebrafish systems. *Journal of Visualized Experiments: JoVE*, (130), 55306. https://doi.org/10.3791/55306

Norris, L. J., Watral, V., & Kent, M. L. (2018). Survival of bacterial and parasitic pathogens from zebrafish (Danio rerio) after cryopreservation and thawing. *Zebrafish*, *15*(2), 188–201. https://doi.org/10.1089/zeb.2017.1528

Peneyra, S. M., Lerpiriyapong, K., Riedel, E. R., Lipman, N. S., & Lieggi, C. (2020). Impact of pronase, sodium thiosulfate, and methylene blue combinations on development and survival of sodium hypochlorite surface-disinfected zebrafish (Danio rerio) embryos. *Zebrafish, 17*(5), 342–353. https://doi.org/10.1089/zeb.2020.1917

Sergeant, E. S. G. (2018). *EpiTools Epidemiological Calculators. Ausvet.* http://epitools.ausvet.com.au Retrieved August 17, 2023

Westerfield, M. (2007). *The zebrafish book. A guide for the laboratory use of zebrafish (Danio rerio)* (1st ed). University of Oregon Press.

ZIRC. (2019). Embryo surface sanitation. https://zebrafish.org/wiki/_media/protocols/ess/customer_egg_bleaching.pdf (retrieved August 12, 2023).

Part II

Practical research methodology

Chapter 8

The 3Rs in zebrafish research

Kamar E. Ameen-Ali and Claire Allen

Abstract

The 3Rs (replacement, reduction and refinement) are an ethical framework for carrying out scientific procedures involving animals and are embedded in international law. With the significant increase in the use of zebrafish in the biosciences, it has never been more necessary to emphasize the importance of applying the 3Rs to studies involving zebrafish. This can include finding opportunities for replacement, such as using mathematical or computer modelling in place of zebrafish, or embryonic or foetal forms of vertebrates. Any refinements must be robustly validated to ensure that they improve welfare, reduce potential harm and suffering and are designed to encourage zebrafish to engage in species-specific behaviours. Researchers and facility staff should be adequately supported to implement refinements when there is strong evidence that they will enhance the lives of laboratory zebrafish. There are resources and guidelines available to assist in the design and reporting of zebrafish experiments, and animal research more broadly, with the aims of improving the reproducibility of these studies and minimizing the number of animals required to obtain statistically meaningful results. Embedding the 3Rs in zebrafish research ensures that the highest standards of welfare are applied and has the potential to advance scientific progress through better reporting and more reproducible studies, and the development of new methods and technologies.

Claire Allen and Jean-Philippe Mocho (eds) **Zebrafish**
DOI: 10.1079/9781800629431.0008,

Keywords

replacement, reduction, refinement, zebrafish, enrichment, welfare

Introduction

The 3Rs principles (replacement, reduction and refinement) were first proposed by Russel and Birch (1959) as a guide for the promotion of greater ethical use of animals in scientific research. Over 60 years later the 3Rs act not only as a framework for animal-based research, but they are also embedded within national and international legislation to help regulate the use of animals in scientific procedures.

The aims of this chapter are to highlight the importance of applying the 3Rs to zebrafish research, and to emphasize how these can be used both as an ethical framework for carrying out scientific procedures using animals and to advance scientific progress through the development of new methods and technologies. Here we discuss ethics and how the 3Rs can be applied to the use of zebrafish in research, as well as a range of resources that can be used to improve the design and reporting of experiments involving zebrafish.

The 3Rs

When using animals in research it is a legal requirement in most regions to use the least sentient organism to appropriately address the scientific question. There has been a significant increase in the use of zebrafish in biomedical research, particularly with the application of CRISPR-Cas9 genome editing that allows for the more efficient generation of transgenic lines (Irion et al., 2014). Many research areas across, for example, developmental biology, neuroscience and genetics, have previously involved the use of rodents but can now be carried out with zebrafish with increased knowledge of their biology, and at a fraction of the cost of rodent research. This does not necessarily constitute replacement, even though one species is being used in place of what may be regarded as a more sentient one. Replacement refers to methods or technologies that replace the use of animals in studies where they would have previously been used. Replacement may include the use of mathematical and computer models, or human tissue or cells, which are examples of full replacement. Replacement also includes the use of invertebrates not protected by legislation (such as *Drosophila*), and embryonic and foetal forms of vertebrates, which are examples of partial replacement. Therefore, categorizing the use of zebrafish in place of rodents as replacement would depend on the stage of zebrafish development. Zebrafish are vertebrates and are protected under national and international legislation when used for scientific procedures. The time point in their maturity from which they become protected varies between certain countries, but the UK Animals (Scientific Procedures) Act 1986 and European Directive 2010/63 specify that zebrafish larvae are protected from the point at which they are capable of independent feeding (120 hours post fertilization (hpf) at 28 ± 0.5 °C). Because zebrafish are not fully developed before 120 hpf, they are limited in terms of the external stimuli they can experience, so their use in scientific research before this stage in place of a legally protected species would, in this case, be considered replacement. In fact, research suggests that zebrafish at 120 hpf are capable of responding to noxious stimuli, which can then be ameliorated by analgesics (Lopez-Luna et al., 2017a,b). It is possible that further research investigating behavioural and physiological responses in larval zebrafish could lead to future legislative change if such research concludes that the data are comparable to those observed in adult zebrafish. This would mean that the welfare of larval zebrafish would similarly be affected by environments or procedures with the potential to cause harm or suffering (Sneddon, 2018), and zebrafish at lower developmental stages should become legally protected.

Reduction refers to methods that either minimize the number of animals required to obtain statistically meaningful results or maximize the amount of information obtained per animal. Using more animals than necessary to obtain statistically meaningful results can potentially lead to unnecessary animal suffering; however, using too few animals can result in an experiment that is too underpowered to derive any meaningful conclusions. There are methods that can be used a priori to determine an appropriate sample size sufficient to detect an effect at a specific significance level (Festing & Altman, 2002). It is important to consider whether certain animal models are likely to have an attrition rate, in which case a slightly higher number of animals would need to be factored in; for example, certain transgenic lines might have a high spontaneous early mortality rate that should be calculated for when determining overall group sample sizes (Wilson et al., 2019).

Methods that can be used to promote reduction may include novel imaging technologies that allow for longitudinal imaging of the same animal rather than requiring multiple cohorts of animals, or sharing of data and resources. It is important to address any potential welfare implications from repeat procedures, and to recognize the need for appropriately designed and analysed experiments that maximize the power to obtain a scientifically valid result that is reproducible (see Chapter 9 for more information). In some cases the use of zebrafish can lead to a reduction in the use of mammalian models later down the experimental pipeline if, for example, such studies act to high-throughput pre-screen compounds before investigation in a mammalian model.

Refinement refers to methods that minimize animal suffering and improve welfare. This may include appropriate housing and husbandry that maximizes the ability for species-specific behaviours to be expressed. Refinement also relates to the scientific procedures used with animals and may include, for example, refined handling techniques to reduce stress and the use of appropriate analgesia and anaesthesia protocols to minimize pain. There is a requirement to train the researcher in the husbandry and welfare of the species, their preferences and qualities in order to achieve critically considered refinements. Examples of such refinements will be discussed in more detail in the next section.

The use of zebrafish in research can offer a range of scientific and practical advantages over the use of mammalian models that have commonly been used. This species are omnivorous photoperiodic breeders with regenerative abilities and their transparent embryos can be harvested easily and manipulated from birth. Transparency is perhaps the most crucial of all these qualities because this facilitates the microscopic visualization of cellular processes including blood flow and wound healing. It was chosen as a model organism, after careful consideration, for its robust ability to tolerate a wide range of water quality parameters. Because zebrafish breed readily and produce many eggs, one advantage for scientific research is an increase in the efficiency of studies that may involve rigorous repeats. Genetic modifications in zebrafish are simpler to carry out compared with mammalian models because the embryos are laid and fertilized externally, allowing genetically modified lines to be generated quickly and effectively.

The 3Rs in practice: validating and implementing refinements in zebrafish research

When discussing refinements to zebrafish research it is worth being reminded of what constitutes a refinement. A refinement is often used to refer to an improvement in a method or protocol, providing scientific benefits. However, in a 3Rs context these examples would be regarded as a refinement only if they minimize animal suffering and improve animal welfare.

As research with zebrafish in the biosciences increases, good practice for their husbandry and welfare is critical. There remains clear variation in terms of standards and protocols employed by different laboratories that cannot be attributed solely to legislative or regional reasons. Many facilities have historically developed their own guidelines and practices based on experience and operated within a legislative and ethical framework. A survey by Lidster et al. (2017) reported variation in the husbandry and practices of many aspects, from standards of animal welfare to methods of euthanasia. This survey highlighted many opportunities to implement refinements in this emerging area of research. The key themes of this survey related to anaesthesia and euthanasia, housing and husbandry, breeding and production and refinement opportunities, with responses received from 22 countries.

Researchers and facility staff have become increasingly informed of the behavioural and welfare needs of zebrafish, particularly as studies of their behaviour have become more common (see Chapter 12 for more information). It is necessary to gain an understanding of the natural behaviour of zebrafish and then to put welfare measures in place in the laboratory to try and promote these behaviours. Animal welfare and research outcomes are intrinsically linked (Poole, 1997), so to gain the most robust scientific studies it is essential not only to minimize the potential for animal suffering and mitigate any that may occur because of certain procedures, but also to promote welfare that allows animals to engage in species-specific behaviours. Not doing so can lead to behaviours and physiological responses that can confound the data, give misleading results and negatively contribute to the irreproducibility of studies.

Welfare measures can include direct observation of behaviour (Lawrence, 2011), measuring performance in terms of development and reproduction (Castranova et al., 2011) and physiological and biochemical measures (Filby et al., 2010), such as cortisol as an indicator of stress (Ellis et al., 2012). Cortisol can be measured non-invasively either through the water (Felix et al., 2013), analysis of scale cortisol levels (Aerts et al., 2015) or by swabbing (Jorge et al., 2023). Increased understanding of zebrafish behaviour can inform the development of more sensitive welfare measures and ultimately promote natural behaviours. Examples of refinement through promotion of environmental enrichment, measurement and assessment of pain and the methods of euthanasia to minimize suffering will now be discussed.

Environmental enrichment

Understanding the natural social behaviour of zebrafish would provide a baseline that could be used to develop housing conditions that promote species-specific behaviours, and would also provide a valuable indicator of welfare if it were known what behaviours could be classified as 'normal' (Bracke & Hopster, 2006; Špinka, 2006).

For zebrafish, environmental enrichment can include the type of substrate used in the tank and the inclusion of plants or objects that can be used for hiding (Williams et al., 2009; see Chapter 2 for more information). Providing environmental enrichment for laboratory zebrafish is not standard practice. There is currently a lack of consensus in terms of the best form of environmental enrichment, and the degree of benefit it can provide to the welfare of zebrafish. Moreover, some zebrafish users and facilities have expressed concerns around how enrichment might impact on water quality pathogen load due to increased biofilm within each tank (Stevens et al., 2021), and whether structures may lead to an increase in aggression due to competition over high-value resources. Inclusion of enrichment in the form of either substrates, a hiding place or structures such as plants has been shown to reduce cortisol in response to stress (Giacomini et al., 2016; Marcon et al., 2018). Environmental enrichment in the form of plastic plants has also been shown to facilitate enhanced growth in fish, particularly males, compared with those without enrichment (Estes et al., 2021). Interestingly, research concerning aggression associated with the presence of structures has been very variable, with some studies finding evidence of increased aggression (Bhat et al., 2015; Woodward et al., 2019) and others finding either no difference or reduced levels (Basquill & Grant, 1998; Carfagnini et al., 2009; Hamilton & Dill, 2002; Keck et al., 2015). Reducing stock densities and increasing the number of structures in a tank might mitigate any potential aggression over competition for these high-value resources.

The effect of enrichment on zebrafish behaviour has been assessed using preference tests, which help to identify preferred environmental enrichments (Osborne et al., 2016). For example, Kistler et al. (2011) reported that zebrafish previously housed in barren tanks showed a significant preference for an environment enriched with stones and plants when given the choice. Research by Schroeder et al. (2014) further investigated environmental preference and found that a range of enrichments (sand or gravel images placed underneath transparent tanks, submerged or floating plants) were all preferred over a barren tank. Moreover, the time the zebrafish spent occupying the space with gravel images was comparable to when the actual substrate was used. Such tests not only provide evidence that zebrafish prefer enriched environments over barren environments, but also usefully help researchers and facility staff identify what type of enrichment is preferred and provide options for facilities that may have historically avoided the use of substrates due to hygiene concerns.

A study by Collymore et al. (2015) reported that single-housed zebrafish in a barren environment showed altered exploratory behaviours compared with single-housed zebrafish with enrichment, group-housed zebrafish or group-housed zebrafish with enrichment. The single-housed zebrafish benefitted from enrichment in the form of an artificial plant, but this enrichment did not have the same benefit for the group-housed zebrafish. These findings suggest that, because zebrafish are a shoaling species, group housing is preferred but, if they are required to be individually housed, for genotyping or breeding for example, the addition of enrichment can be beneficial to their welfare.

Enrichment does not have to be structurally based and can be in the form of food. A survey by Lidster et al. (2017) highlighted variation in food quality and quantity, with both dry and frozen food being used as well as live prey. Live foods, generally in the form of *Artemia* spp., *Paramecium* spp. or salt or freshwater *Rotifer* spp., are high in protein and act as a form of enrichment because they promote natural foraging behaviours (Osborne et al., 2016). Unfortunately, because the nutritional content of live foods is variable and can result in contaminated water, they are not ideal in all contexts. Watts et al. (2016) reported a lack of knowledge and standardization relating to the optimum diet for zebrafish, which not only presents a refinement opportunity to improve animal welfare but also to minimize variation across laboratories, which would ultimately improve study reproducibility (Lawrence, 2016; Watts et al., 2016). More information on zebrafish food and diet is included in Chapter 3.

Refining methods of euthanasia

According to EU and UK legislation, from 120 hpf (or ~5 dpf at 28.0 ± 0.5 °C), zebrafish that are no longer required must be humanely euthanized. Three methods are listed according to EU legislation for zebrafish euthanasia: anaesthetic overdose, concussion and electrical stunning. There is currently a lack of consensus regarding which of these is the most humane (Wilson et al., 2009), and some methods such as concussion may be unsuitable if, for example, brain tissue is required for post-mortem analysis (Valentim et al., 2016). Anaesthetic overdose, particularly with tricaine methanesulphonate (e.g. MS-222), is very widely used although it has been reported to be aversive (Readman et al., 2013; Wong et al., 2014).

Research often drives the requirement for a particular method, and publications may influence these decisions further. Care must be taken to align published instructions with local procedures and legislation. Death should occur without any sign of panic, pain or suffering and involve the minimum time until unconsciousness is achieved. Choosing the correct euthanasia method for optimal research outcome is crucial, because some methods do not satisfy the biological readout and histological quality demands of research. The quality of the resulting tissue integrity, both structurally and molecularly, has the potential to degrade if the wrong method is chosen. For instance, confirmation of death by rigor mortis takes time and severely degrades tissue.

Despite the wide use of anaesthetic overdose by MS-222, there is evidence to suggest rapid cooling, by placing adult zebrafish in ice slush (<4 °C) to induce hypothermal shock, may be more humane (Wilson et al., 2009). This method is widely used for killing fish in many regions, with some evidence suggesting that it induces a quicker death with a reduced potential for causing stress (Matthews & Varga, 2012; Strykowski & Schech, 2015; Wallace et al., 2018; Wilson et al., 2009). Rapid cooling to cull is not permitted under EU or UK legislation, and Lidster et al. (2017) reported that there is little evidence regarding how humane the method is. Euthanasia of zebrafish by this method does not have a clearly defined end point for when brain activity becomes sufficiently disrupted. More research is clearly needed to identify the most humane methods of euthanasia, and there is scope for refinements to be applied in these areas (see Chapter 13 for more information).

Measurement and assessment of pain

Sneddon et al. (2014) presented detailed criteria that must be met for an animal to be considered capable of experiencing pain. The criteria fell under the following two key principles: how an animal responds to a potentially painful stimulus compared with an innocuous stimulus; and a behavioural change following a potentially painful event.

Research has shown that fish have nociceptors that mediate the ability to respond to stimuli, facilitating the potential to give rise to the sensation of pain (Ashley et al., 2006, 2007; Mettam et al., 2012; Sneddon 2002, 2003a, 2018). In addition, studies have reported both behavioural and physiological changes in fish in response to exposure to noxious stimuli (Dunlop et al., 2006; Maximino, 2011; Mettam et al., 2011; Reilly et al., 2008; Roques et al., 2010; Sneddon, 2003b; Sneddon et al., 2003a,b; White et al., 2017), all of which can be mitigated with analgesics (Lopez-Luna et al., 2017a,b,c; Mettam et al., 2011; Newby et al., 2009; Nordgreen et al., 2009; Schroeder & Sneddon, 2017; Sneddon, 2003b, 2012; Taylor et al., 2017). To reduce pain, zebrafish cannot self-select food medicated with analgesics because they suspend feeding when exposed to noxious stimuli (Sneddon, 2009); however, they have been shown to endure a cost in order to access pain relief. For example, Sneddon (2012) demonstrated that acid-treated zebrafish opted to spend more time in a non-preferred barren chamber with lidocaine dissolved in the water over an enriched chamber with no analgesic. Controls chose to spend time in the enriched lidocaine-free chamber. This study therefore showed that zebrafish are willing to endure a cost, through spending time in a non-enriched environment over an enriched environment, to receive analgesia following exposure to noxious stimuli.

Because there is very strong evidence that zebrafish can experience pain, scientific procedures should be refined to provide appropriate analgesia following a potentially painful event. For instance, fin clipping of zebrafish is a routine procedure for genotyping involving removal of a small portion of the fin (usually the caudal fin), typically carried out under anaesthetic. Providing post-procedure analgesics is not routine in all regions, but could reduce pain following a fin clip when added to tank water (Schroeder & Sneddon, 2017; Sneddon et al., 2017). Schroeder et al. (2021) have published a useful peri-operative analgesia protocol following fin clipping. Research has shown that fin clipping can be avoided altogether because swabbing the skin of non-anaesthetized zebrafish is a potentially reliable alternative that can yield suitable DNA concentrations and purities without the need for anaesthesia or analgesia (Breacker et al., 2017). Swabbing also reduced cortisol release, suggesting that this refinement has the potential to improve fish welfare (Tilley et al., 2020), because fin clipping has been shown to be associated with a number of negative welfare indicators such as reduced activity (Deakin et al., 2019a,b; Schroeder & Sneddon, 2017) and decreased feeding (De Lombaert et al., 2017). See Chapter 10 for more information and details on genotyping larvae, which markedly reduces the number of fish used above 120 hpf.

Certain procedures may require the use of analgesia to minimize pain, or anaesthesia to reduce stress or potential damage during procedures. The challenge is knowing quickly and easily whether the fish is in pain and to what extent it is distressed. More research is needed to determine the efficacy of analgesics and anaesthetics, including an evaluation of the most appropriate doses and routes of administration, which are important for both scientific reasons and improved animal welfare (see Chapter 13 for more information). Tolerance to procedures can be enhanced by refining the life experience of the zebrafish; preferences such as remaining with, or at least within sight of, tank mates and acclimatization to new environments reduce visible distress.

Assessing zebrafish welfare can be difficult because they are prey animals and changes in behaviour may be subtle, making them difficult to identify visually (Stasiak et al., 2003). Deakin et al. (2019a,b) developed the fish behaviour index (FBI), which uses cameras and tracking software to monitor and quantify activity and distance swum before and after procedures to automatically detect these subtle changes free from error and bias. In addition to the FBI developed for individual fish monitoring, Thomson et al. (2019) developed the chromatic fish analyser (CFA) that uses spatial chromatic analysis applied to images of recordings of groups of zebrafish and can detect abnormal behaviour of fish within a group. These monitoring tools have the potential to advance the assessment of pain in zebrafish, and their welfare more broadly, whether they are individually or group housed. Moreover, these tools may become particularly important in identifying early humane end points in zebrafish. Welfare assessments rely heavily on classifying clinical signs visually and, if this process was, at least in part, automated there would be the potential to reduce suffering through identifying when an early humane end point had been reached (Ellis & Katsiadaki, 2021).

Despite opportunities for refinements in zebrafish research being identified, implementing these may present challenges. Such challenges include time to implement, assessing biosecurity risks, the effect on scientific data and covering associated financial costs. Training stakeholders to an appropriate level of competency is also essential and helps to develop further refinements. Despite these potential barriers, facilities should work towards implementing opportunities to improve the welfare of laboratory zebrafish, even though there is sometimes a lack of evidence or strategy in place to evaluate the degree to which a refinement provides a welfare benefit. Although the evidence for refinements is growing, more needs to be done to assess, validate and ultimately lead to consensus within the field on the most appropriate refinements to both enrich the lives of zebrafish and minimize potential pain and suffering during scientific procedures.

Improving the design and reporting of animal experiments

The significant increase in the use of zebrafish in bioscience research has coincided with efforts to improve the design and reporting of animal research. The lack of reproducibility and translatability of animal research has been widely reported (Begley & Ellis, 2012; Garner, 2014; Howells et al., 2014; van der Worp et al., 2010) and can have significant long-term effects if such experiments lead to clinical studies that are ultimately not effective (Begley & Ellis, 2012;

Begley & Ioannidis, 2015; Scott et al., 2008). These experiments potentially result in a waste of animal lives (Chalmers et al., 2014; Macleod et al., 2014; Munafò et al., 2017), meaning that the lack of reproducibility is at odds with the 3Rs principles in cases where animals are being used in studies of little or no scientific benefit.

A number of guidelines have been produced to improve animal experiments and their subsequent reporting; these are not unique to zebrafish research. Unfortunately, due to the amount of unregulated work that occurs with zebrafish <5 dpf (in the UK), research may be particularly vulnerable to a lack of training in good experimental design and reporting. The two main guidelines that have been widely adopted are PREPARE (planning research and experimental procedures on animals: recommendations for excellence; Smith et al., 2018) and ARRIVE (animal research: reporting of in vivo experiments; Kilkenny et al., 2010).

The PREPARE guidelines are designed to encourage both researchers and animal facility staff to consider certain factors at the planning stage of a study, which should lead to improved scientific validity, reproducibility and animal welfare (Smith et al., 2018). Key topics that the checklist covers include the formulation of a study; creating a dialogue between researchers and animal facility staff; and quality control with the study (such as considering information in the experimental animals, the experimental procedures in which they will be involved and the housing and husbandry). Recommendations are given under each of the 15 main topics of the checklist.

The ARRIVE guidelines followed a comprehensive review of published animal research that identified major omissions in the reporting of animal experiments (Kilkenny et al., 2009). The survey reported a lack of clear statements relating to study objectives and hypotheses and poor reporting of animal numbers and characteristics. In addition, most studies failed to report the use of randomization and blinding, which are important for minimizing systematic bias. The ARRIVE guidelines were developed to improve the reporting of animal experiments, and they include a 20-item checklist of information deemed necessary for scientists to fully understand how and why a study was carried out, assess the reliability and validity of the study and ultimately improve its reproducibility. The checklist items are broadly grouped into the typical sections of a scientific paper (Introduction, Methods, Results, Discussion) and the information covered by these items allows for the internal validity of the study to be assessed, providing sufficient detail for the study to be accurately reproduced. The items also provide the context and relevance of why the animal experiment was carried out in the first place.

The ARRIVE guidelines are very widely endorsed by scientific journals in the biomedical sciences (Percie du Sert et al., 2017a), funding bodies and universities, with some taking the initiative to ask authors to complete the checklist with a submitted manuscript specifying exactly where in the manuscript the information has been included. However, when the impact of the ARRIVE guidelines on the quality of reporting of in vivo studies was measured, adherence to the guidelines was not improved solely by mandating an ARRIVE checklist with a submitted manuscript (Hair et al., 2019). This led to a revision of the guidelines into a tiered system that improves clarity and prioritizes a subset of items (ARRIVE 2.0; Percie du Sert et al., 2020). 'ARRIVE Essential 10' describes the minimum information required for inclusion in a manuscript and includes information on the experimental design, sample size, inclusion/exclusion criteria, how bias was reduced (blinding and randomization), clearly stated outcome measures, statistical analysis methods, animal species, experimental procedures and the results. The 'Recommended Set' classifies the remaining items that provide context to the study and include information on ethics, declaration of interest, housing, husbandry, protocol registration and data access, some of which are new additions to the guidelines. Through ensuring compliance with ARRIVE Essential 10, journal editors should be able to assess whether these core items have been reported in submitted manuscripts, with a move towards eventual inclusion of all items from the Recommended Set, which will subsequently improve the reporting of in vivo studies (Percie du Sert et al., 2020). The aspiration is that the zebrafish research community will start to publish more experimental details and move away from simply stating that the zebrafish were maintained according to standard practice. As reported by

Lidster et al. (2017), there is little standardization between zebrafish facilities, so this variability should be reflected in the Methods section of published papers. An excellent way to allow publications to include all the environmental and housing conditions experienced within a facility is for the facility to publish annual data sets (Lawrence et al., 2020).

Experimental design is considered an essential step in research that yields both qualitative and quantitative variables. The correct controls and measures should be anticipated at this stage, prior to data collection, which in turn should lead to the ARRIVE guidelines being satisfied with the appropriate information reported in manuscripts. To assist researchers in the design and analysis of in vivo experiments, the NC3Rs (National Centre for the Replacement, Refinement, and Reduction of Animals in Research) have developed a free web-based tool named the Experimental Design Assistant (EDA; Percie du Sert et al., 2017b). When building an EDA diagram of a typical experiment, the system provides feedback highlighting potential inconsistencies and missing information in the design. Moreover, the system has an inbuilt power analysis tool to help address the issue of underpowered studies that use too few animals for the results to be considered reliable. Support for carrying out blinding and randomization is provided to minimize bias that may inadvertently arise from, for example, how the animals are allocated to groups, how the researcher interprets behaviour and how the data are handled. More details of the work that the NC3Rs are doing with zebrafish can be found at https://www.nc3rs.org.uk/zebrafish-welfare.

Researchers who have access to statistical support can use the EDA to facilitate the process of identifying the crucial aspects of an experiment required by a statistician in order to provide helpful advice. Similarly, the EDA can be used within laboratory groups to facilitate discussion of experiments before they are carried out. The EDA diagram and accompanying report (containing all the information embedded within the diagram in a standardized box format) can be made available to funding bodies, ethical review committees and in other scenarios where peer review of in vivo research is required (Percie du Sert et al., 2017b). Moreover, studies could be pre-registered to demonstrate that they were subsequently carried out and reported as originally intended, with no unplanned changes to outcome measures and analyses. Encouraging animal researchers to spend more time thinking about these crucial aspects of experimental design should yield more robust, reliable and, importantly, reproducible animal studies.

Summary

This chapter has outlined the importance of applying the 3Rs to zebrafish research, which is not only a legal and ethical requirement but can also advance scientific progress. The use of non-protected species and foetal or embryonic life forms in scientific research is advancing, and offers clear opportunities to replace some mammalian studies or, at the very least, reduce the number of animals that may be required further down the experimental pipeline. Ultimately, careful consideration is needed when selecting a model of choice in line with a study's objectives, and models and methods that are relied on purely for historical reasons need to be reduced to allow alternatives to be explored.

Emerging refinements should be robustly evaluated. For example, environmental enrichment can act to enhance the quality of life of laboratory zebrafish but consideration is needed regarding the type of enrichment to provide, how frequently it should be cleaned and by what method.

Evidence is accumulating on the most appropriate form of euthanasia, and on the degree to which some anaesthetics are aversive. Assessments of overall suffering are essential when balancing the potential harm of using certain sedatives or anaesthetic agents, with the benefit of reducing overall harm or suffering that an animal might experience because of the procedures to which it will be subjected within its lifetime (Hawkins et al., 2011). Repeated procedures may not on their own be associated with a degree of harm but do have the potential to contribute to cumulative suffering if an animal is involved in multiple protocols and might not have sufficient recovery time in between.

There are various resources that can be used to improve the monitoring and assessment of zebrafish welfare, which is an important advancement due to the overwhelming evidence that zebrafish experience pain (Deakin et al., 2019a,b; Thomson et al., 2019). A good example of research aiming to validate a refinement is the investigation into swabbing by Tilley et al. (2020), who set out to determine whether swabbing zebrafish adults was indeed a refinement. They investigated cortisol readouts and behavioural and molecular responses. This collection of tests could be used to assess other procedures that are potential refinements. All protocols used to monitor welfare, whether behavioural or a molecular assay, should be properly validated. It is proposed that, by using sensitive monitoring tools, analgesia can be appropriately provided when previously it might have been difficult to assess whether, and to what degree, a fish might be suffering following a potentially painful event.

This chapter has also highlighted the resources that can be used to improve the design and reporting of experiments involving zebrafish, which is essential due to the increase in the use of this species in biomedical research as an alternative to the use of rodents. Unintended behavioural and physiological responses can occur when the potential for animal suffering is not minimized or mitigated, and welfare that allows animals to engage in species-specific behaviours must be promoted. Not doing so can confound data, give misleading results and negatively contribute to the irreproducibility of studies.

Everyone who works with zebrafish should have the 3Rs embedded in all the work they are involved in, as these not only act as an ethical and legislative framework that regulates how zebrafish are used in scientific research but have the potential to advance scientific progress through ensuring the highest standards of welfare, and the development of new methods and technologies.

References

Aerts, J., Metz, J. R., Ampe, B., Decostere, A., Flik, G., & De Saeger, S. (2015). Scales tell a story on the stress history of fish. *PLOS ONE*, *10*(4), e0123411. https://doi.org/10.1371/journal.pone.0123411

Ashley, P. J., Sneddon, L. U., & McCrohan, C. R. (2006). Properties of corneal receptors in teleost fish. *Neuroscience Letters*, *410*(3), 165–168. https://doi.org/10.1016/j.neulet.2006.08.047

Ashley, P. J., Sneddon, L. U., & McCrohan, C. R. (2007). Nociception in fish: Stimulus-response properties of receptors on the head of trout *Oncorhynchus mykiss*. *Brain Research*, *1166*, 47–54. https://doi.org/10.1016/j.brainres.2007.07.011

Basquill, S. P., & Grant, J. W. (1998). An increase in habitat complexity reduces aggression and monopolization of food by zebrafish (*Danio rerio*). *Canadian Journal of Zoology*, *76*(4), 770–772. https://doi.org/10.1139/z97-232

Begley, C. G., & Ellis, L. M. (2012). Drug development: Raise standards for preclinical cancer research. *Nature*, *483*(7391), 531–533. https://doi.org/10.1038/483531a

Begley, C. G., & Ioannidis, J. P. (2015). Reproducibility in science: Improving the standard for basic and preclinical research. *Circulation Research*, *116*(1), 116–126. https://doi.org/10.1161/CIRCRESAHA.114.303819

Bhat, A., Greulich, M. M., & Martins, E. P. (2015). Behavioral plasticity in response to environmental manipulation among zebrafish (*Danio rerio*) populations. *PLOS ONE*, *10*(4), e0125097. https://doi.org/10.1371/journal.pone.0125097

Bracke, M. B. M., & Hopster, H. (2006). Assessing the importance of natural behavior for animal welfare. *Journal of Agricultural and Environmental Ethics*, *19*(1), 77–89. https://doi.org/10.1007/s10806-005-4493-7

Breacker, C., Barber, I., Norton, W. H. J., McDearmid, J. R., & Tilley, C. A. (2017). A low-cost method of skin swabbing for the collection of DNA samples from small laboratory fish. *Zebrafish*, *14*(1), 35–41. https://doi.org/10.1089/zeb.2016.1348

Carfagnini, A. G., Rodd, F. H., Jeffers, K. B., & Bruce, A. E. E. (2009). The effects of habitat complexity on aggression and fecundity in zebrafish (*Danio rerio*). *Environmental Biology of Fishes*, *86*(3), 403–409. https://doi.org/10.1007/s10641-009-9539-7

Castranova, D., Lawton, A., Lawrence, C., Baumann, D. P., Best, J., Coscolla, J., Doherty, A., Ramos, J., Hakkesteeg, J., Wang, C., Wilson, C., Malley, J., & Weinstein, B. M. (2011). The effect of stocking densities on reproductive performance in laboratory zebrafish (*Danio rerio*). *Zebrafish*, *8*(3), 141–146. https://doi.org/10.1089/zeb.2011.0688

Chalmers, I., Bracken, M. B., Djulbegovic, B., Garattini, S., Grant, J., Gülmezoglu, A. M., Howells, D. W., Ioannidis, J. P., & Oliver, S. (2014). How to increase value and reduce waste when research priorities are set. *Lancet, 383*(9912), 156–165. https://doi.org/10.1016/S0140-6736(13)62229-1

Collymore, C., Tolwani, R. J., & Rasmussen, S. (2015). The behavioral effects of single housing and environmental enrichment on adult zebrafish (*Danio rerio*). *Journal of the American Association for Laboratory Animal Science, 54*(3), 280–285.

De Lombaert, M. C., Rick, E. L., Krugner-Higby, L. A., & Wolman, M. A. (2017). Behavioral characteristics of adult zebrafish (*Danio rerio*) after MS222 anesthesia for fin excision. *Journal of the American Association for Laboratory Animal Science, 56*(4), 377–381.

Deakin, A. G., Buckley, J., AlZu'bi, H. S., Cossins, A. R., Spencer, J. W., Al'Nuaimy, W., Young, I. S., Thomson, J. S., & Sneddon, L. U. (2019a). Automated monitoring of behaviour in zebrafish after invasive procedures. *Scientific Reports, 9*(1), 9042. https://doi.org/10.1038/s41598-019-45464-w

Deakin, A. G., Spencer, J. W., Cossins, A. R., Young, I. S., & Sneddon, L. U. (2019b). Welfare challenges influence the complexity of movement: Fractal analysis of behaviour in zebrafish. *Fishes, 4*(1), 8. https://doi.org/10.3390/fishes4010008

Dunlop, R., Millsopp, S., & Laming, P. (2006). Avoidance learning in goldfish (*Carassius auratus*) and trout (*Oncorhynchus mykiss*) and implications for pain perception. *Applied Animal Behaviour Science, 97*(2–4), 255–271. https://doi.org/10.1016/j.applanim.2005.06.018

Ellis, T., & Katsiadaki, I. (2021). Clarification of early end-points for refinement of animal experiments, with specific reference to fish. *Laboratory Animals, 55*(3), 244–253. https://doi.org/10.1177/0023677220971002

Ellis, T., Yildiz, H. Y., López-Olmeda, J., Spedicato, M. T., Tort, L., Øverli, Ø., & Martins, C. I. M. (2012). Cortisol and finfish welfare. *Fish Physiology and Biochemistry, 38*(1), 163–188. https://doi.org/10.1007/s10695-011-9568-y

Estes, J. M., Altemara, M. L., Crim, M. J., Fletcher, C. A., & Whitaker, J. W. (2021). Behavioral and Reproductive Effects of Environmental Enrichment and *Pseudoloma neurophilia* infection on Adult zebrafish (*Danio rerio*). *Journal of the American Association for Laboratory Animal Science, 60*(3), 249–258. https://doi.org/10.30802/AALAS-JAALAS-20-000113

Félix, A. S., Faustino, A. I., Cabral, E. M., & Oliveira, R. F. (2013). Noninvasive measurement of steroid hormones in zebrafish holding-water. *Zebrafish, 10*(1), 110–115. https://doi.org/10.1089/zeb.2012.0792

Festing, M. F. W., & Altman, D. G. (2002). Guidelines for the design and statistical analysis of experiments using laboratory animals. *ILAR Journal, 43*(4), 244–258. https://doi.org/10.1093/ilar.43.4.244

Filby, A. L., Paull, G. C., Bartlett, E. J., Van Look, K. J., & Tyler, C. R. (2010). Physiological and health consequences of social status in zebrafish (*Danio rerio*). *Physiology and Behavior, 101*(5), 576–587. https://doi.org/10.1016/j.physbeh.2010.09.004

Garner, J. P. (2014). The significance of meaning: Why do over 90% of behavioral neuroscience results fail to translate to humans, and what can we do fix it? *Institute for Laboratory Animal Research Journal, 55*, 438–456. https://doi.org/10.1093/ilar/ilu047

Giacomini, A. C. V. V., Abreu, M. S., Zanandrea, R., Saibt, N., Friedrich, M. T., Koakoski, G., Gusso, D., Piato, A. L., & Barcellos, L. J. G. (2016). Environmental and pharmacological manipulations blunt the stress response of zebrafish in a similar manner. *Scientific Reports, 6*, 28986. https://doi.org/10.1038/srep28986

Hair, K., Macleod, M. R., Sena, E. S., & IICARus Collaboration. (2019). A randomised controlled trial of an intervention to improve compliance with the ARRIVE guidelines (IICARus). *Research Integrity and Peer Review, 4*, 12. https://doi.org/10.1186/s41073-019-0069-3

Hamilton, I. M., & Dill, L. M. (2002). Monopolization of food by zebrafish (Danio rerio) increases in risky habitats. *Canadian Journal of Zoology, 80*(12), 2164–2169. https://doi.org/10.1139/z02-199

Hawkins, P., Dennison, N., Goodman, G., Hetherington, S., Llywelyn-Jones, S., Ryder, K., & Smith, A. J. (2011). Guidance on the severity classification of scientific procedures involving fish: Report of a Working Group appointed by the Norwegian Consensus-Platform for the Replacement, Reduction and Refinement of animal experiments (Norecopa). *Laboratory Animals*, 1–13.

Howells, D. W., Sena, E. S., & Macleod, M. R. (2014). Bringing rigour to translational medicine. *Nature Reviews. Neurology, 10*(1), 37–43. https://doi.org/10.1038/nrneurol.2013.232

Irion, U., Krauss, J., & Nüsslein-Volhard, C. (2014). Precise and efficient genome editing in zebrafish using the CRISPR/Cas9 system. *Development, 141*(24), 4827–4830. https://doi.org/10.1242/dev.115584

Jorge, S., Félix, L., Costas, B., & Valentim, A. M. (2023) Housing conditions affect adult zebrafish (*Danio rerio*) behavior but not their physiological status. *Animals (Basel)*, *13*(6):1120.

Keck, V. A., Edgerton, D. S., Hajizadeh, S., Swift, L. L., Dupont, W. D., Lawrence, C., & Boyd, K. L. (2015). Effects of habitat complexity on pair-housed zebrafish. *Journal of the American Association for Laboratory Animal Science*, *54*(4), 378–383.

Kilkenny, C., Browne, W. J., Cuthill, I. C., Emerson, M., & Altman, D. G. (2010). Improving bioscience research reporting: The ARRIVE Guidelines for reporting animal research. *PLOS Biology*, *8*(6), e1000412. https://doi.org/10.1371/journal.pbio.1000412

Kilkenny, C., Parsons, N., Kadyszewski, E., Festing, M. F. W., Cuthill, I. C., Fry, D., Hutton, J., & Altman, D. G. (2009). Survey of the quality of experimental design, statistical analysis and reporting of research using animals. *PLOS ONE*, *4*(11), e7824. https://doi.org/10.1371/journal.pone.0007824

Kistler, C., Hegglin, D., Würbel, H., & König, B. (2011). Preference for structured environment in zebrafish (*Danio rerio*) and checker barbs (*Puntius oligolepsis*). *Applied Animal Behaviour Science*, *135*(4), 318–327. https://doi.org/10.1016/j.applanim.2011.10.014

Lawrence, C. (2011). Advances in zebrafish husbandry and management. *Methods in Cell Biology*, *104*, 429–451. https://doi.org/10.1016/B978-0-12-374814-0.00023-9

Lawrence, C. (2016). New frontiers for zebrafish management. *Methods in Cell Biology*, *135*, 483–508. https://doi.org/10.1016/bs.mcb.2016.04.015

Lawrence, C., Best, J., James, A., Hurley, S., Shia, M., Urh, M., Hirshfeld, B., Bakker, N., Perdomo, H., & Krueger, A. (2020). Zebrafish (*Danio rerio*) environmental summary, aquatic resources program, Boston Children's Hospital *v.7 Protocols.io*.

Lidster, K., Readman, G. D., Prescott, M. J., & Owen, S. F. (2017). International survey on the use and welfare of zebrafish *Danio rerio* in research. *Journal of Fish Biology*, *90*(5), 1891–1905. https://doi.org/10.1111/jfb.13278

Lopez-Luna, J., Al-Jubouri, Q., Al-Nuaimy, W., & Sneddon, L. U. (2017a). Impact of analgesic drugs on the behavioural responses of larval zebrafish to potentially noxious temperatures. *Applied Animal Behaviour Science*, *188*, 97–105. https://doi.org/10.1016/j.applanim.2017.01.002

Lopez-Luna, J., Al-Jubouri, Q., Al-Nuaimy, W., & Sneddon, L. U. (2017b). Activity reduced by noxious chemical stimulation is ameliorated by immersion in analgesic drugs in zebrafish. *Journal of Experimental Biology*, *220*(8), 1451–1458. https://doi.org/10.1242/jeb.146969

Lopez-Luna, J., Al-Jubouri, Q., Al-Nuaimy, W., & Sneddon, L. U. (2017c). Impact of stress, fear and anxiety on the nociceptive responses of larval zebrafish. *PLOS ONE*, *12*(8), e0181010. https://doi.org/10.1371/journal.pone.0181010

Macleod, M., Michie, S., Roberts, I., Dirnagl, I., Chalmers, J., Ioannidis, J. P. A., Al-Shahi Salman, R., Chen, A.-W., & Glasziou, P. (2014). Biomedical research: increasing value, reducing waste. *The Lancet*, 382, 101–104.

Marcon, M., Mocelin, R., Benvenutti, R., Costa, T., Herrmann, A. P., De Oliveira, D. L., Koakoski, G., Barcellos, L. J. G., & Piato, A. (2018). Environmental enrichment modulates the response to chronic stress in zebrafish. *Journal of Experimental Biology*, *221*(4), jeb176735. https://doi.org/10.1242/jeb.176735

Matthews, M., & Varga, Z. M. (2012). Anesthesia and euthanasia in zebrafish. *ILAR Journal*, *53*(2), 192–204. https://doi.org/10.1093/ilar.53.2.192

Maximino, C. (2011). Modulation of nociceptive-like behavior in zebrafish (*Danio rerio*) by environmental stressors. *Psychology and Neuroscience*, *4*(1), 149–155. https://doi.org/10.3922/j.psns.2011.1.017

Mettam, J. J., McCrohan, C. R., & Sneddon, L. U. (2012). Characterisation of chemosensory trigeminal receptors in the rainbow trout (*Oncorhynchus mykiss*): Responses to irritants and carbon dioxide. *Journal of Experimental Biology*, *215*(4), 685–693. https://doi.org/10.1242/jeb.060350

Mettam, J. J., Oulton, L. J., McCrohan, C. R., & Sneddon, L. U. (2011). The efficacy of three types of analgesic drugs in reducing pain in the rainbow trout, Oncorhynchus mykiss. Applied Animal Behaviour Science, 133(3–4), 265–274. https://doi.org/10.1016/j.applanim.2011.06.009

Munafò, M. R., Nosek, B. A., Bishop, D. V. M., Button, K. S., Chambers, C. D., Percie du Sert, N., Simonsohn, U., Wagenmakers, E.-J., Ware, J. J., & Ioannidis, J. P. A. (2017). A manifesto for reproducible science. *Nature Human Behaviour*, *1*(1). https://doi.org/10.1038/s41562-016-0021

Newby, N. C., Wilkie, M. P., & Stevens, E. D. (2009). Morphine uptake, disposition and analgesic efficacy in the common goldfish (*Carassius auratus*). *Canadian Journal of Zoology*, *87*(5), 388–399. https://doi.org/10.1139/Z09-023

Nordgreen, J., Garner, J. P., Janczak, A. M., Ranheim, B., Muir, W. M., & Horsberg, T. E. (2009). Thermonociception in fish: Effects of two different doses of morphine on thermal threshold and post-test behaviour in goldfish (*Carassius auratus*). *Applied Animal Behaviour Science, 119*(1–2), 101–107. https://doi.org/10.1016/j.applanim.2009.03.015

Osborne, N., Paull, G., Grierson, A., Dunford, K., Busch-Nentwich, E. M., Sneddon, L. U., Wren, N., Higgins, J., & Hawkins, P. (2016). Report of a Meeting on Contemporary Topics in Zebrafish Husbandry and Care. *Zebrafish, 13*(6), 584–589. https://doi.org/10.1089/zeb.2016.1324

Percie du Sert, N., Bamsey, I., Bate, S. T., Berdoy, M., Clark, R. A., Cuthill, I., Fry, D., Karp, N. A., Macleod, M., Moon, L., Stanford, S. C., & Lings, B. (2017b). The experimental design assistant. *PLOS Biology, 15*(9), e2003779. https://doi.org/10.1371/journal.pbio.2003779

Percie du Sert, N., Hurst, V., Ahluwalia, A., Alam, S., Avey, M. T., Baker, M., Browne, W. J., Clark, A., Cuthill, I. C., Dirnagl, U., Emerson, M., Garner, P., Holgate, S. T., Howells, D. W., Karp, N. A., Lazic, S. E., Lidster, K., MacCallum, C. J., Macleod, M., . . . Würbel, H. (2020). The ARRIVE guidelines 2.0: Updated guidelines for reporting animal research. *PLOS Biology, 18*(7), e3000410. https://doi.org/10.1371/journal.pbio.3000410

Poole, T. (1997). Happy animals make good science. *Laboratory Animals, 31*(2), 116–124. https://doi.org/10.1258/002367797780600198

Readman, G. D., Owen, S. F., Murrell, J. C., & Knowles, T. G. (2013). Do fish perceive anaesthetics as aversive? *PLOS ONE, 8*(9), e73773. https://doi.org/10.1371/journal.pone.0073773

Reilly, S. C., Quinn, J. P., Cossins, A. R., & Sneddon, L. U. (2008). Behavioural analysis of a nociceptor event in fish: Comparisons between three species demonstrate specific responses. *Applied Animal Behaviour Science, 114*(1–2), 248–259. https://doi.org/10.1016/j.applanim.2008.01.016

Roques, J. A. C., Abbink, W., Geurds, F., van de Vis, H., & Flik, G. (2010). Tailfin clipping, a painful procedure: Studies on Nile tilapia and common carp. *Physiology and Behavior, 101*(4), 533–540. https://doi.org/10.1016/j.physbeh.2010.08.001

Russel, W. M. S., & Birch, R. L. (1959). *The principles of humane experimental technique*. Universities Federation for Animal Welfare.

Schroeder, P., Jones, S., Young, I. S., & Sneddon, L. U. (2014). What do zebrafish want? Impact of social grouping, dominance and gender on preference for enrichment. *Laboratory Animals, 48*(4), 328–337. https://doi.org/10.1177/0023677214538239

Schroeder, P., Lloyd, R., McKimm, R., Metselaar, M., Navarro, J., O'Farrell, M., Readman, G. D., Speilberg, L., & Mocho, J. P. (2021). Anaesthesia of laboratory, aquaculture and ornamental fish: Proceedings of the first LASA-FVS Symposium. *Laboratory Animals, 55*(4), 317–328. https://doi.org/10.1177/0023677221998403

Schroeder, P. G., & Sneddon, L. U. (2017). Exploring the efficacy of immersion analgesics in zebrafish using an integrative approach. *Applied Animal Behaviour Science, 187*, 93–102. https://doi.org/10.1016/j.applanim.2016.12.003

Scott, S., Kranz, J. E., Cole, J., Lincecum, J. M., Thompson, K., Kelly, N., Bostrom, A., Theodoss, J., Al-Nakhala, B. M., Vieira, F. G., Ramasubbu, J., & Heywood, J. A. (2008). Design, power, and interpretation of studies in the standard murine model of ALS. *Amyotrophic Lateral Sclerosis, 9*(1), 4–15. https://doi.org/10.1080/17482960701856300

Smith, A. J., Clutton, R. E., Lilley, E., Hansen, K. E. A., & Brattelid, T. (2018). PREPARE: Guidelines for planning animal research and testing. *Laboratory Animals, 52*(2), 135–141. https://doi.org/10.1177/0023677217724823

Sneddon, L. U. (2002). Anatomical and electrophysiological analysis of the trigeminal nerve in a teleost fish, *Oncorhynchus mykiss. Neuroscience Letters, 319*(3), 167–171. https://doi.org/10.1016/s0304-3940(01)02584-8

Sneddon, L. U. (2003a). The evidence for pain in fish: The use of morphine as an analgesic. *Applied Animal Behaviour Science, 83*(2), 153–162. https://doi.org/10.1016/S0168-1591(03)00113-8

Sneddon, L. U. (2003b). Trigeminal somatosensory innervation of the head of a teleost fish with particular reference to nociception. *Brain Research, 972*(1–2), 44–52. https://doi.org/10.1016/s0006-8993(03)02483-1

Sneddon, L. U. (2009). Pain perception in fish indicators and end points. *ILAR Journal, 50*(4), 338–342. https://doi.org/10.1093/ilar.50.4.338

Sneddon, L. U. (2012). Clinical anaesthesia and analgesia in fish. *Journal of Exotic Pet Medicine, 21*(1), 32–43. https://doi.org/10.1053/j.jepm.2011.11.009

Sneddon, L. U. (2018). Comparative physiology of nociception and pain. *Physiology, 33*(1), 63–73. https://doi.org/10.1152/physiol.00022.2017

Sneddon, L. U., Braithwaite, V. A., & Gentle, M. J. (2003a). Do fishes have nociceptors? Evidence for the evolution of a vertebrate sensory system. *Proceedings. Biological Sciences*, *270*(1520), 1115–1121. https://doi.org/10.1098/rspb.2003.2349

Sneddon, L. U., Braithwaite, V. A., & Gentle, M. J. (2003b). Novel object test: Examining nociception and fear in the rainbow trout. *Journal of Pain*, *4*(8), 431–440. https://doi.org/10.1067/s1526-5900(03)00717-x

Sneddon, L. U., Elwood, R. W., Adamo, S. A., & Leach, M. C. (2014). Defining and assessing pain. *Animal Behaviour*, *97*, 201–212. https://doi.org/10.1016/j.anbehav.2014.09.007

Sneddon, L. U., Halsey, L. G., & Bury, N. R. (2017). Considering aspects of the 3Rs principles within experimental animal biology. *Journal of Experimental Biology*, *220*(17), 3007–3016. https://doi.org/10.1242/jeb.147058

Špinka, M. (2006). How important is natural behaviour in animal farming systems? *Applied Animal Behaviour Science*, *100*(1–2), 117–128. https://doi.org/10.1016/j.applanim.2006.04.006

Stasiak, K. L., Maul, D., French, E., Hellyer, P. W., & Vandewoude, S. (2003). Species-specific assessment of pain in laboratory animals. *Contemporary Topics in Laboratory Animal Science*, *42*(4), 13–20.

Stevens, C. H., Reed, B. T., & Hawkins, P. (2021). Enrichment for laboratory zebrafish – A review of the evidence and the challenges. *Animals: An Open Access Journal from MDPI*, *11*(3), 698. https://doi.org/10.3390/ani11030698

Strykowski, J. L., & Schech, J. M. (2015). Effectiveness of recommended euthanasia methods in larval zebrafish (*Danio rerio*). *Journal of the American Association for Laboratory Animal Science*, *54*(1), 81–84.

Taylor, J. C., Dewberry, L. S., Totsch, S. K., Yessick, L. R., DeBerry, J. J., Watts, S. A., & Sorge, R. E. (2017). A novel zebrafish-based model of nociception. *Physiology and Behavior*, *174*, 83–88. https://doi.org/10.1016/j.physbeh.2017.03.009

Thomson, J. S., Al-Temeemy, A. A., Isted, H., Spencer, J. W. & Sneddon, L. U. (2019). Assessment of behaviour in groups of zebrafish (Danio rerio) using an intelligent software monitoring tool, the chromatic fish analyser. *Journal of Neuroscience Methods*, *1*, 328.

Tilley, C. A., Carreno, H., Gutierrez, Sebire, M., Obasaju, O., Reichmann, F., Katsiadaki, I., Barber, I., & Norton, W. H. J. (2020). Skin swabbing is a refined technique to collect DNA from model fish species. *Scientific Reports*. 10, 18212.

Valentim, A. M., van Eeden, F. J., Strähle, U., & Olsson, I. A. S. (2016). Euthanizing zebrafish legally in Europe: Are the approved methods of euthanizing zebrafish appropriate to research reality and animal welfare? *EMBO Reports*, *17*(12), 1688–1689. https://doi.org/10.15252/embr.201643153

van der Worp, H. B., Howells, D. W., Sena, E. S., Porritt, M. J., Rewell, S., O'Collins, V., & Macleod, M. R. (2010). Can animal models of disease reliably inform human studies? *PLOS Medicine*, *7*(3), e1000245. https://doi.org/10.1371/journal.pmed.1000245

Wallace, C. K., Bright, L. A., Marx, J. O., Andersen, R. P., Mullins, M. C., & Carty, A. J. (2018, January 1). Effectiveness of rapid cooling as a method of euthanasia for Young zebrafish (*Danio rerio*). *Journal of the American Association for Laboratory Animal Science*, *57*(1), 58–63.

Watts, S. A., Lawrence, C., Powell, M., & D'Abramo, L. R. (2016). The vital relationship between nutrition and health in zebrafish. *Zebrafish*, *13* (Suppl. 1), S72–S76. https://doi.org/10.1089/zeb.2016.1299

White, L. J., Thomson, J. S., Pounder, K. C., Coleman, R. C., & Sneddon, L. U. (2017). The impact of social context on behaviour and the recovery from welfare challenges in zebrafish, *Danio rerio*. *Animal Behaviour*, *132*, 189–199. https://doi.org/10.1016/j.anbehav.2017.08.017

Williams, T. D., Readman, G. D., & Owen, S. F. (2009). Key issues concerning environmental enrichment for laboratory-held fish species. *Laboratory Animals*, *43*(2), 107–120. https://doi.org/10.1258/la.2007.007023

Wilson, J. M., Bunte, R. M., & Carty, A. J. (2009). Evaluation of rapid cooling and tricaine methane sulfonate (MS-222) as methods of euthanasia in zebrafish (*Danio rerio*). *Journal of the American Association for Laboratory Animal Science*, *48*(6), 785–789.

Wilson, R. P., Holton, M., Wilson, V. L., Gunner, R., Tysse, B., Wilson, G. I., Quintana, F., Duarte, C., & Scantlebury, D. M. (2019). Towards informed metrics for examining the role of human-induced animal responses in tag studies on wild animals. *Integrative Zoology*, *14*(1), 17–29. https://doi.org/10.1111/1749-4877.12328

Wong, D., von Keyserlingk, M. A., Richards, J. G., & Weary, D. M. (2014). Conditioned place avoidance of zebrafish (*Danio rerio*) to three chemicals used for euthanasia and anaesthesia. *PLOS ONE*, *9*(2), e88030. https://doi.org/10.1371/journal.pone.0088030

Woodward, M. A., Winder, L. A., & Watt, P. J. (2019). Enrichment increases aggression in zebrafish. *Fishes*, *4*(1), 22. https://doi.org/10.3390/fishes4010022

Chapter 9

Statistical considerations for the breeding of zebrafish and their use in experiments

Amy Spencer, Peter Laud, Claire Allen, James Bull, Eleanor Stillman and Sara Hilditch

Abstract

There are important statistical considerations required for research endeavours using the zebrafish as a model organism. Zebrafish are a specialist species with some unique qualities that must be taken into account in data acquisition. This popular and versatile fish requires statistical knowledge for stock management, health screening and in order to produce robust experimental data. We demonstrate how to use known or inferred information about stock to choose how to manage aspects of a zebrafish colony and provide information on the most common methods used for statistical testing in experiments. Whilst only giving a brief introduction to most topics relating to experimental data, we aim to expand on areas where we believe difficulties commonly arise for the zebrafish researcher. These are illustrated with topical examples in the main body of the text, whilst underlying statistical formulae are given in an Appendix to the chapter. We want to motivate enquiry and highlight requirements and options to empower the researcher to take these ideas and expand upon their principles. In today's research network, support is easy to access, even in areas outside the core discipline; for example, many universities have statistics advisory services. Higher expectations exist to justify animal research and there

Claire Allen and Jean-Philippe Mocho (eds) **Zebrafish**
DOI: 10.1079/9781800629431.0009, © CAB International, 2024

are increasingly obligatory minimum guidelines to follow within reporting. Planning is essential because it is now widely acknowledged that publishing research is futile if data are deficient, and peer reviewers will pick up on this.

Keywords

zebrafish, statistical methods, stock management, health screening, sample size calculation, hypothesis testing

Introduction

The use of animal models for scientific endeavour requires careful justification. The ethics and principles of the 3Rs (reduction, refinement and replacement) in such research are clear (Graham & Prescott; 2015; Sneddon et al., 2017), and the correct use of statistics can be particularly impactful upon reduction. It is widely understood that sample size calculations should be carried out for scientific experiments, and these calculations produce the minimum number of animals to achieve a properly powered result. The scientific validity of experimental results will also depend on minimization of bias in the resulting data by using some of the methods outlined in this chapter and the appropriate analysis of those data. We will also cover something that is perhaps less well documented, but also highly important: the use of statistics to calculate the appropriate numbers of fish kept as breeding stock and used in routines such as health screens.

Statistics is at the core of any experiment that produces quantitative data. Those data need to be analysed to determine the strength of the evidence regarding a pre-specified hypothesis. Preparation is crucial from a statistical point of view. Without adequate experimental design there is a risk that irrelevant or inadequate data collection will ensue, failing to fully address the original question. Poor data collection is not only a waste of time and resources but, more importantly, a waste of animal life, which is ethically unacceptable

(Nuffield Council on Bioethics, 2005). In order to mitigate such issues it should be clear what data will be collected and how they will be analysed, justifying this by using, for example, pilot data, previously published work or related research that gives a sound basis for a new avenue of investigation.

Published guidelines for researchers in in vivo research fields highlight the importance of statistics (Gosselin, 2019). Grant-funding bodies and journals are now encouraging researchers and insisting that they follow such guidelines (for instance, this is included in the funding policy for animal research by BBSRC, MRC and Wellcome in the UK). The Planning Research and Experimental Procedures on Animals: Recommendations for Excellence (PREPARE) guidelines (Smith et al., 2017) include 'Experimental design and statistical analysis' as one of their 15 topics, placing it under the general area 'Formulation of the study'. The Animal Research: Reporting of In Vivo Experiments (ARRIVE) guidelines (Percie du Sert et al., 2020), which are now considered the gold standard for reporting, form an invaluable checklist that has far-reaching statistical implications. However, although the original version of these guidelines was published in 2010, independent reports several years later suggest that they are not being followed, even in journals that support their use (Hair et al., 2019; Leung et al., 2018). Another checklist that researchers may find helpful is the Reducing IrreProducibility in labOratory STudiEs (RIPOSTE) framework (Masca et al., 2015). This is general guidance for all biomedical laboratory-based studies and so may be familiar to, or more readily understood by, a wider audience.

A practical tool, often endorsed, is the Experimental Design Assistant (EDA) (Percie du Sert et al., 2017). Designed with the assistance of statisticians, this interactive graphical tool allows the user to add steps to their experimental diagram then prompts them to enter details relating to that step, for instance *n* (number of experimental units, for example animals) or types of outcome measures. It may help researchers to choose an appropriate statistical test and, in certain circumstances, to carry out a sample size calculation. Aspects of good study design such as blinding and randomization are also highlighted, with tools included to assist in the application of these processes. It also highlights designs that will not produce data that can be analysed using simple and common tests, or when it is not obvious which test to choose at all. These are examples of situations in which the advice of an expert statistician should be sought. Both the EDA and the ARRIVE guidelines were developed by the UK's National Centre for the Replacement, Refinement and Reduction of Animals in Research (NC3Rs).

Statistical considerations for zebrafish research is a broad topic that all too often becomes overshadowed by considerations of higher vertebrates. We aim to cover three areas that require statistical consideration within zebrafish research. First we provide calculations related to the keeping of zebrafish stocks and offer several look-up tables detailing example situations. This is followed up with some introductory, and principally statistical, ideas pertaining to health-screening considerations (Collymore et al., 2016). Finally we turn our attention to the use of statistical tests in scientific experiments, with examples of common issues. This taster should instil an appreciation for the most fundamental considerations required and motivate the zebrafish researcher to plan and prepare in a more considered way.

Breeding and stock management

Breeding zebrafish is more involved than indicated in the published data (Nasiadka & Clark, 2012), hence the need for the ARRIVE guidelines (Percie du Sert et al., 2020). Added to this, inbreeding impacts both fecundity and survival (Mrakovčič & Haley, 1979). Researchers must be able to justify the numbers of protected animals required for a study in order to comply with ethical review processes and demonstrate reduction, as part of the 3Rs, while still having sufficient animals to investigate the hypothesized scientific endpoints appropriately.

The core issue is that, while sample size calculations may be carried out for the experimental animals, they are rarely considered for the breeding stock. In fact, many fish experiments use few, or even no, fish of protected age and the legally protected animals are the parent shoal. This shoal size is typically dictated by cost and not statistically sound principles that ensure security, genetic integrity or deliverability. Statistical methods can be used to calculate the ideal number of fish required for breeding purposes, and so support optimal line management.

Required number of offspring in line management

Having an understanding of how many offspring are required is sometimes straightforward but can be complicated by factors such as genetics (particularly involving multiple alleles) and in-house survival rates. Generally equal numbers of each sex will be desired, although it is common for there to be an uneven ratio of males to females in zebrafish populations (Krug, 2019). The exact ratio varies depending on the environmental conditions, breeding regimes of the facility and origins/characteristics of the particular line. If the approximate ratio for a facility or line is known, this can be used to determine the probability of acquiring a certain number of fish of either sex. Since not all offspring will survive to adulthood, each facility should know its normal background morbidity and mortality rate in order to monitor when levels rise significantly above this. Assuming that all living offspring can be used (there is no restriction related to a particular genotype, for example), Table 9.1 shows the total numbers of offspring that would need to be produced to have 90 or 95% confidence in ensuring that there were at least 50 living females. These values are given for a range of background mortality rates and male:female ratios. The calculation for these values is given in the Appendix to this chapter.

Many zebrafish experiments are carried out on genetically modified lines where only the fish with a particular genotype will be of interest. Not all the fish that are bred, therefore, will be useful. Table 9.2 shows the approximate numbers of fish with genotypes of various frequencies that would be expected to survive from a pool of 200, a typical number from a single mating, given different background mortality rates. Figure 9.1 shows the proportions that would be expected to survive from any size pool of offspring.

Table 9.1 Number of total offspring that would be needed to be produced to have 90 and 95% probability of at least 50 female fish reaching experimental age.

Male:female ratio	Background mortality rate (%)						
	1	2	5	10	15	20	25
90% probability							
50:50	114	116	119	126	134	143	153
55:45	128	129	133	141	150	159	170
60:40	144	146	151	159	169	180	192
65:35	166	168	173	183	194	207	221
95% probability							
50:50	119	120	124	131	139	149	159
55:45	133	134	139	147	156	166	178
60:40	150	152	157	166	176	188	201
65:35	173	175	180	191	202	215	230

Table 9.2 Number of fish of the required genotype expected to survive from a pool of 200 offspring, given various genotype frequencies and background mortality rates.

Genotype frequency	Background mortality rate (%)						
	1	2	5	10	15	20	25
1 (all fish)	198	196	190	180	170	160	150
2/3	132	131	127	120	113	107	100
1/2	99	98	95	90	85	80	75
1/3	66	66	64	60	57	53	50
1/4	50	49	48	45	43	40	38
1/8	25	25	24	23	21	20	19
1/16	13	13	12	12	11	10	9

The formula used to calculate the values in Table 9.2 can be found in the Appendix to this chapter.

Fish can be considered as being in one of two categories: the correct genotype or not, and so if the overall probability of a fish having the correct genotype is known, the binomial distribution can be used to determine the probability of obtaining sufficient numbers of these fish from a particular sample. For example, with 24 fish in total and a genotype frequency of 1/4, even though the

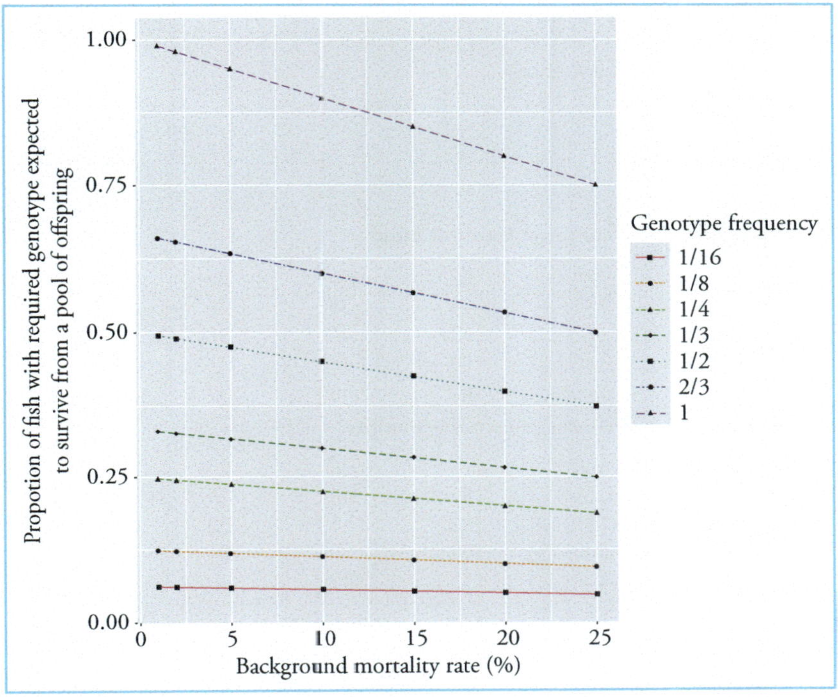

Figure 9.1 Proportion of fish with the correct genotype (various frequencies given) and that are expected to survive depending on background mortality rate.

most likely number of fish with this genotype is 24/4 = 6, there is only an 18.5% chance of obtaining exactly six and a 57.8% chance of six or more. In the Appendix to this chapter the binomial probability mass function and cumulative distribution function (CDF) are given, and it is demonstrated how these are used to calculate these values.

The CDF can be solved in order to determine the minimum number of fish required for there to be a certain probability of obtaining the required fish. So, applying this to the above example where the genotype frequency was 1/4, in order to be 95% sure of obtaining at least six fish of this genotype one would need 40 fish in total. Table 9.3 gives the total numbers of fish that would be required to be 90 or 95% sure of getting at least five to at least nine

Table 9.3 Total pool of fish required to obtain various minimum numbers of a specific genotype with 90 and 95% probability. Numbers are given for a variety of genotype frequencies.

Number of fish with the required genotype	Genotype frequency					
	1/16	1/8	1/4	1/3	1/2	2/3
90% probability of obtaining at least this number						
5	126	62	30	22	14	10
6	147	73	35	26	17	12
7	167	82	40	30	19	14
8	186	92	45	33	21	15
9	206	102	50	37	24	17
95% probability of obtaining at least this number						
5	144	71	34	25	16	11
6	166	82	40	29	18	13
7	187	92	45	33	21	15
8	208	103	50	37	23	16
9	228	113	55	40	26	18

fish for various genotype frequencies. The table does not account for background mortality.

Example 1. Estimating the necessary number of matings to produce the required offspring

Tables 9.1–9.3 offer guidance as to the target number of initial offspring that should be produced. For example, a pilot experiment is planned with eight healthy mature fish of a genotype that occurs only in one-quarter of fish. In the aquarium in which the experiment will take place there is a background mortality rate of 20%. Using Table 9.2 we can see that, for all genotypes (genotype frequency = 1) and with this mortality rate, from a single mating producing 200 offspring, 160 would be expected to survive. Looking at Table 9.3, we can see that 50 fish, including siblings that do not have the correct genotype, would be required to be 95% sure of obtaining eight with the correct genotype. Therefore, only a single pair-mating would be required for this pilot experiment. For other experiments, such as large screens, many more will be needed. Depending on the experimental design, Table 9.1 or 9.2 may be helpful to estimate how many matings will be needed.

Genetic management of zebrafish lines can be problematic because inbred strains that parallel the breeding regimes of the rodent world are simply not viable (Crim & Lawrence, 2021). Instead, a drive to maintain heterozygosity for good reproductive performance, which minimizes inbreeding depression, is generally preferred. In order to retain some genetic variation, usually at least 20 pairs of wildtype fish and a minimum of four pairs of fish with each specific genetic modification are kept for breeding. Outcrossing to prevent inbreeding is an important tool in zebrafish line management, whereby genetic lineage is managed by identification of phenotypes or genotypes specific to the line (Allen et al., 2021). These factors also need to be accounted for when calculating the number of fish to use for breeding and managing different generations. Optimal spawning capabilities, fertility and fecundity are dependent on husbandry practices (Spence et al., 2008), but we can expect pairs to produce weekly clutches that exceed 100 embryos from the age of 6–18 months.

Survival of new genotypes

When raising fish with a new genetic modification there is a risk that the mutation could be unexpectedly fatal. It is reasonable to expect slightly higher mortality in a new line, by chance alone. Based on the background mortality rate for the wild-type colony with the same background as the new transgenic line, we explain (in the Appendix to this chapter) a method to calculate how many fish may be observed to die before there is evidence that the mutation is increasing mortality. This information is given for some example background mortality rates and numbers of fish of the new genotype in Table 9.4.

Table 9.4 Number of dead fish that would provide 95, 90 and 80% confidence that the mortality rate has increased above the background known for the wild-type colony with the same background as the new transgenic line (various given) for different-sized pools of fish.

Number of fish of new genotype	Background mortality rate (%)						
	1	2	5	10	15	20	25
95% confidence							
50	3	4	6	10	13	16	19
100	4	6	10	16	22	28	33
150	5	7	13	22	31	39	47
200	6	8	16	28	39	50	61
90% confidence							
50	2	3	6	9	12	15	17
100	3	5	9	15	21	26	32
150	4	6	12	21	29	37	45
200	5	8	15	27	38	48	59
80% confidence							
50	2	3	5	8	11	13	16
100	3	4	8	13	19	24	30
150	3	5	11	19	27	35	43
200	4	7	14	25	35	46	56

Statistics in health screening

How many fish should be screened?

In aquaria it will be necessary to carry out regular health screens to detect infections in the fish. Assuming that an infection is equally prevalent at a given level throughout the whole aquarium and that fish are sampled randomly, we can calculate how many fish would need to be sampled in order to have a particular probability of including fish with the infection in that sample.

For various sample sizes between 10 and 70, Figure 9.2 and Table 9.5 give some examples of the probabilities that a fish with the infection would be included in a sample where the overall number of fish infected is between 1% and 30%. For instance, with a sample of 30 fish there is a probability of 0.958 of sampling a fish with the

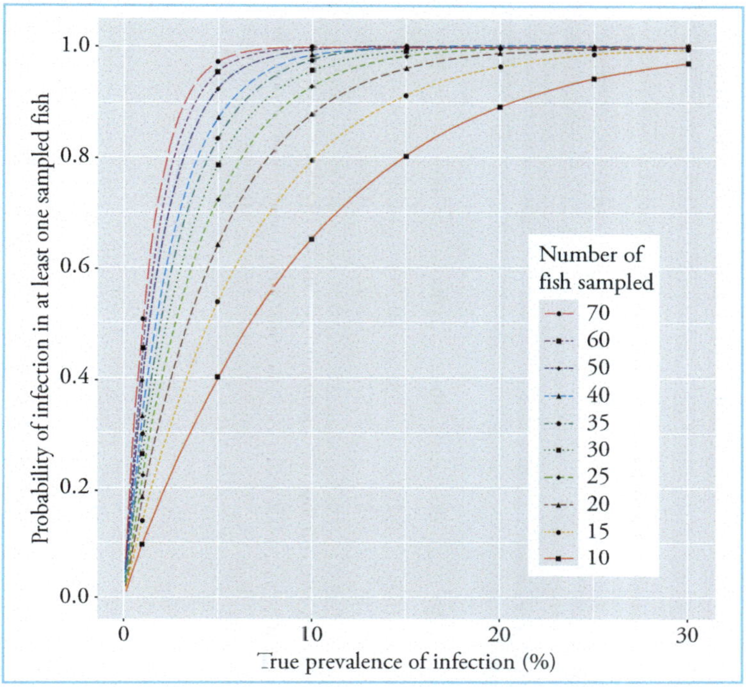

Figure 9.2 Probability of sampling at least one infected fish dependent on the true prevalence of the infection and number of fish sampled.

infection, where it is present in 10% of fish overall, but a probability of only 0.785 if the true prevalence is 5%. The person overseeing the screen would have to determine the most suitable sample size by judging, for example, whether the increase in probability when 30 fish are sampled over 25 is worth the extra five fish. If it were considered necessary to have at least an 80% chance (probability of 0.8) of obtaining an infected fish for infections prevalent in at least 5% of fish, a sample of at least 35 fish would be required.

The lower the prevalence the more unlikely it is that a fish with the infection would be included in a sample, and an example prevalence of 1% is given in Table 9.5. In this case there is only around a 50% chance of an infected fish being sampled, even with a sample of 70 fish. Further details on the probability of sampling an infected fish in less widespread infections are given in Table 9.6.

Table 9.5 Probability of sampling at least one infected fish for different levels of true prevalence and numbers of fish sampled.

True prevalence of infection		Number of fish sampled									
%	**Propn**	**10**	**15**	**20**	**25**	**30**	**35**	**40**	**50**	**60**	**70**
1	**0.01**	0.096	0.140	0.182	0.222	0.260	0.297	0.331	0.395	0.453	0.505
5	**0.05**	0.401	0.537	0.642	0.723	0.785	0.834	0.871	0.923	0.954	0.972
10	**0.10**	0.651	0.794	0.878	0.928	0.958	0.975	0.985	0.995	>0.999	>0.999
15	**0.15**	0.803	0.913	0.961	0.983	0.992	>0.999	>0.999	>0.999	>0.999	>0.999
20	**0.20**	0.893	0.965	0.988	>0.999	>0.999	>0.999	>0.999	>0.999	>0.999	>0.999
30	**0.30**	0.972	>0.999	>0.999	>0.999	>0.999	>0.999	>0.999	>0.999	>0.999	>0.999

Table 9.6 Values of true prevalence that yield the specified probability of sampling at least one infected fish for different sample sizes.

Probability of sampling ≥1 infected fish	Number of fish sampled									
	10	15	20	25	30	35	40	50	60	70
0.50	6.7%	4.5%	3.4%	2.7%	2.3%	2.0%	1.7%	1.4%	1.1%	1.0%
0.60	8.8%	5.9%	4.5%	3.6%	3.0%	2.6%	2.3%	1.8%	1.5%	1.3%
0.70	11.3%	7.7%	5.8%	4.7%	3.9%	3.4%	3.0%	2.4%	2.0%	1.7%
0.80	14.9%	10.2%	7.7%	6.2%	5.2%	4.5%	3.9%	3.2%	2.6%	2.3%

Again considering a sample of 30 fish, there is an 80% chance of one of these being infected if the prevalence were 5.2%, but only a 50% chance if the true prevalence were 2.3%. This can also be read from Figure 9.2. The formulae for these tables are given in the chapter Appendix.

What is the prevalence of infection?

It is important to note that the values given in Tables 9.5 and 9.6 and Figure 9.2 relate to the probability of sampling infected fish. However, no screening tests will have 100% sensitivity (always yield positive results when a fish is infected) and 100% specificity (always yield negative results when a fish is not infected). When gold standard tests with high expected sensitivity and specificity are used, the probability of a positive test result in a sample will be very close to the probability of an infected fish being included in that sample. However, if it is known that the sensitivity and/or specificity of the test is unlikely to be very high, this should always be accounted for. Tests with low sensitivity will require more fish to be sampled before obtaining a satisfactory probability of detecting infection in a fish, and the probability of obtaining a positive test result is likely to be lower than the values given in Table 9.5. On the other hand, those with low specificity will have higher probabilities of positive test results and are likely to give an over-inflated estimate of the prevalence of the infection. Results from a test with both low sensitivity and specificity would be extremely difficult to interpret.

Once the screening has been carried out, we want to estimate the true prevalence of the infection. Here we do that assuming that test results are 100% sensitive and specific, but how appropriate that is should be judged based on the known properties of the screen being used. A confidence interval (CI) should be given to describe the uncertainty around the estimate, and a recommended method for calculating this is the skewness-corrected asymptotic score (SCAS; Laud, 2017, 2018). In this scenario it is the maximum level of infection that is of most interest, so it is appropriate to use a one-sided CI. Table 9.7 shows what the estimated prevalence and upper limit of a one-sided SCAS 95% CI would be when between zero and six fish were found to be infected in samples of varying size. If no infected fish is found in the sample that does not, of course, mean that the infection is definitely not present. The upper limit of the CIs in the row of Table 9.7 for 'zero fish found' gives a value for which we can be 95% confident that the true prevalence is lower. The true prevalence of infection across the whole aquarium should be above this upper limit in only 5% of cases. For example, the true

Table 9.7 Estimated prevalence of infection (and one-sided 95% CIs calculated using the SCAS method) based on the number of infected fish in different sample sizes.

Number of infected fish in sample	Number of fish sampled				
	20	25	30	35	40
0	0% (<9.8%)	0% (<7.9%)	0% (<6.7%)	0% (<5.8%)	0% (<5.1%)
1	5% (<18.3%)	4% (<14.9%)	3.3% (<12.6%)	2.9% (<10.9%)	2.5% (<9.6%)
2	10% (<25.3%)	8% (<20.6%)	6.7% (<17.4%)	5.7% (<15.1%)	5% (<13.3%)
3	15% (<31.6%)	12% (<25.9%)	10% (<21.9%)	8.6% (<19.0%)	7.5% (<16.7%)
4	20% (<37.4%)	16% (<30.7%)	13.3% (<26.1%)	11.4% (<22.6%)	10% (<20.0%)
5	25% (<43.0%)	20% (<35.4%)	16.7% (<30.1%)	14.3% (<26.1%)	12.5% (<23.1%)
6	30% (<48.3%)	24% (<39.9%)	20% (<33.9%)	17.1% (<29.5%)	15% (<26.1%)

prevalence should only be >6.7% in 5% of screens where 0 out of 30 fish tested were found to have the infection. Note how wide the interval can be, especially when using small samples; for example, if we screen 20 fish and find that none is infected, we can really be fairly confident only that the prevalence is below about 10% (9.8%) rather than that it is truly absent.

Statistics in experimental design

Statistical methodologies for experiments carried out with zebrafish are generally the same as for any other type of experiment and are widely available in many publications. Two excellent resources are *The Analysis of Biological Data*, 3rd edn (Whitlock & Schluter, 2020) and *Design and Analysis of Experiments*, 2nd edn (Dean et al., 2017); we give further recommendations in the section below on regression models. In the remainder of this chapter we will cover some of the basics and try to clarify some points that often cause confusion.

Importantly, each experiment within a project should be designed individually, and a good place to begin is by making sure you have clear answers to the following questions:

- What is the key objective of the experiment, the question that you are trying to answer?
- What are you choosing to vary between groups of animals: for example the genotype or a treatment?
- What is the main outcome that you will measure?

It is important to try and choose a single primary outcome, which may depend on things including the key objective, how easy measurements are to obtain with accuracy and how much prior information (published, or from pilot studies) you have on different measurements.

Once these things have been decided, the design of the experiment can begin. The design involves, in addition to all biological and practical aspects, choosing the most appropriate statistical analysis method and calculating a suitable sample size to use. These factors are all closely linked and, importantly, must be planned before any data are collected. The following sections cover these topics.

An important point to make regarding the statistical analysis of experimental data is that although much of the focus tends to be the 'success' of an experiment, based on whether a test statistic meets the threshold for statistical significance, this is not the whole story. Even if the outcome is non-significant (there is not enough evidence from the data to reject the null hypothesis) it is important to report all experiments carried out, along with an estimated size of any measured outcome and a CI. This can be extremely valuable in knowledge sharing among the research community and reducing unnecessary repetition of similar experiments.

Choosing a method of analysis

Although the statistical testing on the experimental data will not happen until after all have been collected, the type of analysis should be chosen at the planning stage. This is because the required sample size will need to be calculated and this calculation will be specific to the analysis method that is going to be used.

The most commonly used methods can be thought of as belonging to two main groups: hypothesis testing and regression modelling. It should be noted that these groups are not exclusive. For example, several of the hypothesis-testing techniques discussed below are derived from simple regression models. Thinking back to the first of the three questions above – 'What is the key objective of the experiment?' – is it simply to determine whether there is a difference between groups or does it also encompass estimating the size of that difference and potentially also predicting the outcome in future observations? The type of answer you want will determine the type of analytical method you should choose.

Before discussing individual methods, it might be helpful to define a few statistical terms in order to help clarify this section.

- variable – a measurable or classifiable feature that can take a range of values
- continuous or numerical variable – a variable that takes a numerical value
- categorical variable or factor – a variable that has a finite number of values (classes/categories)
- ordinal variable – a categorical variable for which the classes have a clear order
- independent, or explanatory, variable – variables for which the values are set at the beginning of the experiment and which, theoretically, should not depend on the values of any other variable
- dependent or response variable, outcome measure – the variable that is measured in an experiment and is expected to vary according to the values or classes of the independent variables
- covariate – an independent variable that is included in the analysis but is not the one of primary interest
- model – in this chapter 'model' refers to a statistical model that is fitted to the data and describes the inferred relationship between the independent and dependent variables. It is important to distinguish this from the 'animal model' discussed elsewhere in this book
- residual variation, residuals – when the statistical model is fitted, it assigns some of the variation observed in the outcome variable as having been caused by the independent variable(s). The residual variation is any additional variation that cannot be accounted for by these variables. The 'residuals' are the difference between the actual observed values and the expected or predicted values based on the fitted model.

Hypothesis testing

Hypothesis testing involves formulating a 'null' hypothesis of no effect (of the conditions that are different in the experiment), collecting data and comparing them with a statistical distribution in order to weigh the evidence against the null hypothesis. All hypothesis tests require the data to meet certain assumptions, so it is important to choose the correct test for the correct situation and to check that its assumptions are met. In addition, when carrying out a hypothesis test on data from a small sample of a much larger population, there is no way of knowing exactly how well that sample reflects the whole population. Therefore there is always a chance of obtaining a false positive (or type I error) in a hypothesis test, and we choose to test at a particular significance level (α), typically 0.05 (5%). If the data do meet the assumptions of the chosen test, this is the probability that the null hypothesis would be rejected if it were actually true (a false positive result). Similarly, there is a chance that the null hypothesis is actually not correct but will not be rejected. This is a false negative result, also called type II error, and these occur at a rate signified by β. Often we talk of the statistical power of a test, which is $1 - \beta$, the true positive probability, and when we carry out sample size calculations we choose a target power (for the assumed distribution of the data), typically 0.8 (80%).

Figure 9.3 shows some of the most commonly used hypothesis tests and may be helpful in choosing which one could be appropriate for your experiment.

The type of outcome data is key to the choice of test. If the outcome can be assigned to a group with a label, for example 'survived to three months' or 'did not survive to three months' then it is categorical. Numerical outcomes are usually continuous, but sometimes a categorical outcome may be measured on a scale with numeric labels. If there are more than two categories, it should also be taken into consideration whether there is an obvious order to the categories. An example of ordered categories would be 'no impairment', 'partial impairment', 'severe impairment'.

For grouped categorical data, if the data can be summarized, or cross-classified, in a two-way table of counts (e.g. number of fish surviving/not surviving versus treatment/no treatment), then a χ^2 test (chi-squared test; Pearson, 1900) or a Fisher's exact test (Fisher, 1922; Fisher, 1954) may be used. Both methods test the null hypothesis that the two factors are independent (e.g. whether a fish survives does not depend on whether it was treated or not). With larger sample sizes the χ^2 test is usually preferable whereas with

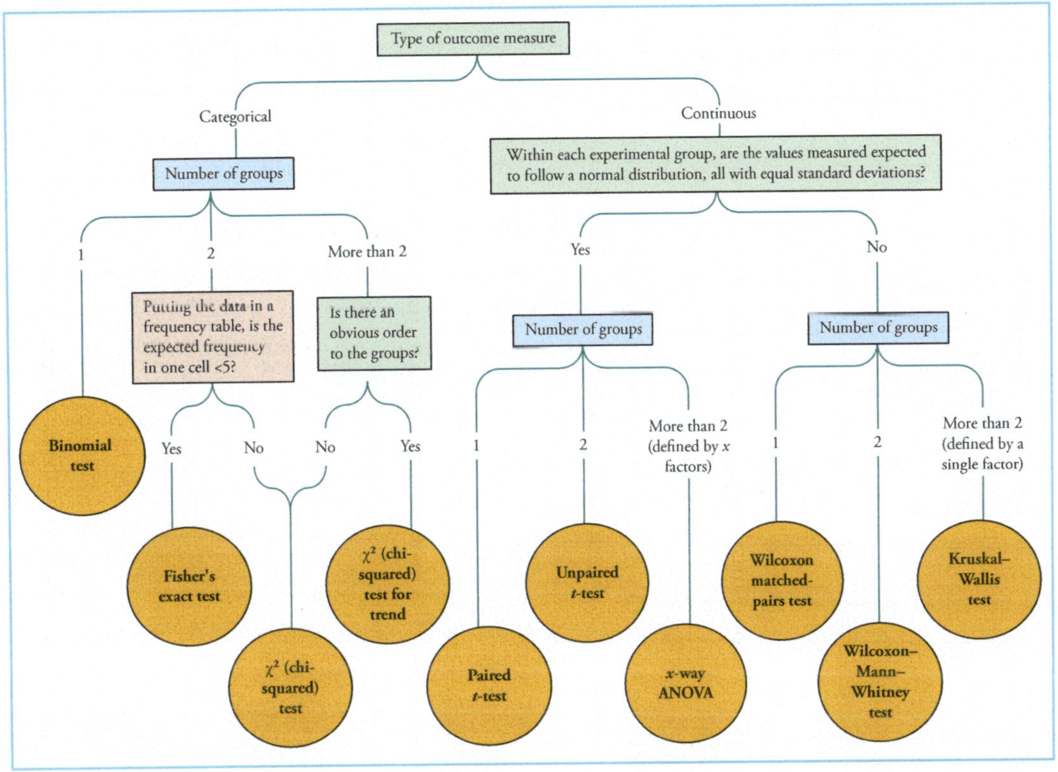

Figure 9.3 Some common hypothesis testing methods for comparison of measured outcomes under different experimental conditions.

small or unevenly distributed samples a Fisher's exact test should be used. A common rule of thumb is to choose a Fisher's exact test if the expected number (given what was seen overall for each factor) in >20% of cells is <5.

x-Way ANOVA (analysis of variance) refers to a simultaneous analysis of *x* factors, comparing the means of the different levels of each factor, and also (in the case where $x > 1$) whether there is any additional effect of interaction between the factors. This type of test may be suitable when the dependent variable is continuous, but all independent variables are categorical. Where the residuals are non-normally distributed, a Kruskal–Wallis test can be used as a non-parametric alternative to a one-way ANOVA.

An ANOVA is a simple type of linear regression model but, in the framework of hypothesis testing methods, can also be extended to allow for more flexibility in the types of variable collected. One such extension is ANCOVA (analysis of covariance), which enables the incorporation of continuous covariates that also may have an important effect on the outcome (e.g. time that the animals have been kept under a particular condition, weight of the animals).

If the outcome is continuous it is important to consider whether the residuals (see the definition in the previous section) are likely to follow the well-known normal distribution (sometimes called Gaussian distribution). This describes data with a central average (can be calculated as the mean, the mode or the median) and decreasing occurrences of values further from this average, measured by the standard deviation (SD or σ) and gives a bell-shaped curve when plotted.

Once the data have been collected they should be checked (including using plots) to ensure that they satisfy the assumptions of the planned analysis. Summary checks of continuous data will often involve calculating measures such as:

Mean, median or mode – three different ways to summarize the 'average' value of the data. The mean is the sum of all values divided by the total number of observations and is often used when the data are approximately symmetrically distributed (for example, normal distribution); the median is the value in the exact middle when all values are ordered and is often used when the data are skewed or non-symmetrical; the mode is the value that occurs most often in the observations.

- Minimum and maximum values – always worth checking to ensure that there isn't something outside of the range of logical values.
- Lower and upper values of the interquartile range (1st and 3rd quartiles) – the median is the 2nd quartile. If you take the lower (or upper) half of the ordered data and find the midpoint of that, you obtain the 1st (or 3rd) quartile. As with using the median, this is often helpful when investigating whether data have a symmetrical distribution (you would expect the median to fall half-way between them if it does) or reporting on a known non-symmetrical dataset.
- Variance – a measure of spread of the data. It is based on the squared distances of each individual observed value from the mean of those values. Therefore it (and other measures of spread derived from it) is helpful when the data are symmetrically distributed.
- Standard deviation (SD) – calculated as the square root of the variance.
- Standard error of the mean (SEM) – this is calculated as SD divided by the square root of the sample size and is used as a measure of confidence in the estimate of the mean.

If the test you want to use can compare distinct groups, you can plot the raw data for each of these groups separately, making sure to take into account all permutations of the different factors, to determine whether they are all normally distributed with a similar standard deviation (see example 3). If you have continuous covariates you will need to analyse your data first then check whether the residuals are normally distributed. If the residuals of continuous data do not satisfy the assumptions of a normal distribution with constant variance, applying a transformation (such as a log-transformation) may improve the model fit. When planning an experiment, some idea of whether residuals are likely to be normally distributed may be obtained by inspecting data from the control arm of a previous similar experiment, or from a pilot study.

Example 2. Why one-way ANOVA is only a starting point when comparing multiple groups

Consider an experiment with a zebrafish model of a particular disease. There will be four groups of fish, one of which is a control group, and each of the others is given a different treatment. After a certain number of days, a measurement of the outcome will be taken. Previous experiments have been done using untreated fish of this model, so the researcher has a good idea of the mean outcome in the control group and the SD and, from plotting those data, it does appear to be normally distributed. In this model it is reasonable to assume that the SD will be the same in the other groups and, because the treatment will be the only thing varied between fish, this suggests that residuals will be normally distributed and the assumptions of a one-way ANOVA are valid, although we should check the residuals once we have fitted the model.

What would the results of the ANOVA test actually produce? The null hypothesis is that all the groups have the same mean outcome. If the test results in a P value of <0.05, this null hypothesis can be rejected at the 5% significance level. A P value of ≥ 0.05 means that the null hypothesis cannot be rejected, which means there is no evidence to suggest there is any difference between the groups. However, even in the case of $P <0.05$, all that would be demonstrated is a difference between at least two of the groups, and it is unlikely that this would be a particularly helpful result as it does not definitively answer which of the four groups are different to each other.

If an ANOVA produces a significant P value, it is usually followed up with 'post hoc' t-tests between pairs of treatment groups, to determine which groups differ from each other. Whether or not the investigator wishes to initially carry out an ANOVA, they should plan the analysis and calculate the sample size taking into consideration the number of t-tests that may need to be carried out in order to ensure that the overall significance level is retained whilst accounting for multiple testing. This is because the more tests that are carried out, the higher the probability of false positive results occurring if no such multiplicity adjustment is made. Two options to consider are:

A. Compare all possible pairs (six *t*-tests, sometimes called Tukey's method):

Control versus treatment 1	Treatment 1 versus treatment 2
Control versus treatment 2	Treatment 1 versus treatment 3
Control versus treatment 3	Treatment 2 versus treatment 3

B. Compare all active treatments with control (three *t*-tests, sometimes called Dunnett's method):

Control versus treatment 1
Control versus treatment 2
Control versus treatment 3

A new (adjusted) significance level should be calculated based on the number of *t*-tests that will be performed. Usually when there are only a small number this can be done simply and adequately using the Bonferroni adjustment where the original significance level is divided by the number of tests. In the case of option A, the new significance level will be 5%/6 = 0.83%. If the only item of interest is whether the treatments are different to the control, but not whether they are different from each other, the appropriate choice is option B and a significance level of 5%/3 = 1.67%. This should be the level used in the sample size calculation, and therefore the number of relevant post hoc tests needs to be considered at the planning stage. When multiple *t*-tests are carried out manually, these are the boundaries below which the researcher should consider a P value to be significant (e.g. $P < 0.0083$ or < 0.0167 rather than the usual < 0.05). However, some statistical software automatically adjusts for multiple testing in post hoc tests (for example, GraphPad Prism; GraphPad Software, 2021), meaning that an adjusted P value will be reported that should be compared with the overall significance threshold (usually 0.05). It is generally preferable to carry out as few tests as possible while still being able to meet the objective of the experiment, thereby minimizing the necessary significance level adjustment.

A special case is where the active treatment groups are actually three doses of the same treatment:

C. Compare all doses with control (up to three *t*-tests):

Control versus dose 1
Control versus dose 2
Control versus dose 3

It is safe to assume that if there is a significant difference between the control and any particular dose of treatment, there would also be a significant difference between the control and a higher dose of this treatment. However, the aim is to find the lowest dose for which the difference is significant. In this case it is possible to carry out each test at the 5% level by using a hierarchical testing strategy. The differences in response between each dose and the control are tested in order, beginning with the highest dose and, if this difference is found to be significant, it is acceptable to also test the next highest dose at the 5% level. Once a test is non-significant at this level, no further testing is carried out and the final dose found to produce a significantly different response to the control is the 'optimal' dose.

Statistical methods for dealing with repeats

Some types of experiment are commonly split into blocks, and these blocks are carried out at different times or in different locations. This may be because it is unfeasible to carry out an experiment of the necessary size all in one go, or because there is believed to be a need to capture some of the effects of uncontrollable environmental variation that can influence biological systems. Whilst the overall treatment effect will remain constant, the exact outcome values may vary between blocks. Pooling the data and analysing them together will enable the treatment effect to be estimated and the statistical test will have more power than separate tests on the data from the individual blocks. It may be useful to compare separate plots of the blocks to determine whether they appear to be noticeably different. A blocked experimental design can be taken into account in a single analysis, for example using a blocked ANOVA, or a Friedman test as a non-parametric alternative.

These are types of 'mixed-effects' analyses, and these allow for the inclusion of effects that cannot be controlled in an experiment. They include the usual 'fixed-effects' factors that are being compared in the experiment (for example, treatment group and sex), but also the nuisance factor(s) that you do not want to test for differences between, yet still need to be considered (like experimental block). These are included in the model as 'random effects'. Although no test results will be produced comparing block 1 with block 2, and so on, the results for treatment group and sex will have been adjusted allowing for the variation in measured outcome between experimental blocks.

A different type of repeat in an experiment is when multiple measures of the outcome are taken from each animal. If there is only one time point at which these are collected, a single average value can be taken, allowing for a relatively simple analysis. This is suitable providing that using an average response is considered to be an appropriate outcome measure, without the amount of variability in the individual measurements being accounted for. However, if this isn't the case, or if the same animals are measured at different times in the experiment, the analysis will need to be different. Whilst changes may be expected over time, it is logical to expect that the measures taken from one single animal are likely to be more similar than those taken from different animals. In experiments of this kind, the appropriate type of mixed-effects model to use may be a repeated-measures ANOVA or Friedman test.

Although these types of method should be used when the same animal is measured multiple times, with zebrafish it can be very difficult to track the same individual over time. Methods to overcome this problem such as visible implant elastomer (VIE) tagging are currently under development (Rácz et al., 2021). There are also small differences in the stripe marking of individual zebrafish and methods have been developed recently to help distinguish fish based on computer images (Al-Jubouri et al., 2018). Depending on the experiment, it may be possible to house small sets of fish with different visible phenotypes such as pigmentation and tail length differences, or it may even be considered necessary to sacrifice the normal shoaling environment by housing fish in solitary confinement. If none of these methods is employed and a standard fixed-effects analysis has to be carried out, it is important to bear in mind (and report with any results) that, by not taking into account the fact that multiple measures have been taken from the same fish, the power will be artificially inflated, thereby increasing the chances of false positive results.

It is recommended that a professional statistician is consulted to assist with these types of analysis.

Regression models

Hypothesis testing may be sufficient for many experiments carried out on zebrafish due to their usual context in the early, and even exploratory, pre-clinical phase. However, fitting regression models allows for estimation of the magnitude of effect that each variable has in addition to testing hypotheses, and covers a wider range of scenarios. Once the magnitudes of these effects have been estimated it is possible to predict expected values, including for combinations of variables that may not have been observed in the experiment. More complex experiments can also be analysed using regression models. Such modelling can be used for many types of data and is particularly useful when there are several, or many, independent variables (either categorical factors or continuous) that may have an effect on the outcome.

Whilst linear models are used for continuous outcome data, generalized linear models can often be used where the outcome takes a different form, such as a binary outcome (logistic regression) or a count outcome (Poisson regression). Regression models can be extended to allow for the inclusion of random effects, such as repeated measures.

It is important to remember that all types of regression have assumptions related to the data, similar to those discussed previously for hypothesis tests. Software may produce output for a particular analysis of a dataset when asked to, but the results will not be meaningful unless the data satisfy the relevant assumptions. In particular, most methods of statistical analysis require that all observations are independent and that the covariates are not highly correlated (as little as possible collinearity). Of course the outcome variable also needs to be the appropriate type, and care needs to be taken particularly around categorical data that may look like continuous data due to having numerical labels (e.g. stages of severity). For linear models, the residuals must be normally distributed.

Survival analysis, as the name suggests, is commonly used when analysing how long something survives, usually with an element of censoring. However, survival analysis techniques can also be used for other outcomes of the same type (time until the occurrence of an event). In the case of modelling actual survival, for example in a clinical trial comparing treatments for a late-stage cancer, death is the event of interest and censoring would occur because the trial period came to an end, or subjects left the study, before all times until death were known. It is illegal to take animals to an end point of death in experiments in the UK, but the same statistical methods might be used when the experiment involves measuring time to reaching a certain stage of disease severity, for example, or time to observation of a particular phenotype. Kaplan–Meier plots show the estimated survival probability taking into account how many animals have been followed up as time progresses, and log-rank tests are used to compare the results for different groups.

Types of regression such as the Cox proportional hazards model (Cox, 1972) allow the changing hazard of the outcome occurring over time to be described based on multiple variables. 'Proportional hazards' refers to the fact that an additional assumption of this type of model is that each independent variable has the same effect on the hazard over time. There is a baseline risk of reaching the end point that applies to all patients in a clinical trial, or fish in a laboratory experiment, for example, and this changes over time. However, in this experiment a drug may halve the hazard of fish developing a specific clinical sign and proportional hazards means that a model of this type will assume that halving of the hazard is true one week into the experiment, two weeks in or at any other time. Similarly, type ABC fish may have a 15% higher hazard of developing the clinical sign throughout the length of the experiment compared with type DDD1 fish. More complex types of survival model allow this assumption to be relaxed, but the results are much more difficult to interpret.

Many statistical texts give much more detailed information on these and other regression techniques. A detailed source is Harrell (2015), which also provides information to get the user started using R statistical analysis software (R core team, 2017). Armitage et al. (2002) give clear guidance on the use of many methods in the context of clinical and epidemiological research, but much of this can easily be transferred to laboratory research. Many different types of experiment and the analysis of their data are explored in Dean et al. (2017), again providing advice on how to carry these out practically, in this case in both R and SAS (SAS Institute Inc., 2004). Zuur et al. (2009) provide a guide through various types of regression methods with a focus on applications in ecology and covering many practical issues that occur in real data, including collinearity, zero inflation and non-independence. This book also uses R software in its examples.

Calculation of sample size

It is vital to carry out an appropriate sample size calculation at the planning stage. This can be done either using power calculations for standard hypothesis tests or simulations where a more complex or unusual analysis has been proposed. These methods allow one to enter what is already known or assumed about the type of data one is going to collect, as well as what one wants to be able to detect (e.g. a magnitude of difference between groups), and will output a minimum sample size that would maintain levels of false positive (α) and false negative (β) results under the maximum that would be acceptable under those circumstances. This is particularly important in studies using protected species from an ethical standpoint, so that as few animals as possible are used in each experiment. As well as this, it is critical that sufficient numbers of animals should be used to have the appropriate power to test the hypothesis of interest. A poorly planned experiment with too few animals is also very unethical.

Power calculators are widely available, for example G*power (Faul et al., 2007), although care does need to be taken because they offer so many options and the correct sample size will be calculated only if all the correct options are selected. The power calculation is specific to the type of analysis and, if calculation software does not have an option for a particular kind of analysis, a statistician should be contacted. For most common tests, power calculations require the following parameters:

- Expected outcome (e.g. mean, m_1) in at least one group, usually the control group, and generally obtained from previous published or pilot experiments.
- A minimum biologically relevant difference between this group and any experimental groups ($m_2 - m_1$). A statistician will not be able to determine this, and neither should it should be determined from previous experimental data (it is not an expected difference). Consider whether you would actually be interested in seeing an increase or decrease of 1, 5, 10, 20, etc. Ask yourself, at what point does the difference become interesting or meaningful, or what would be the smallest difference that would be considered worth publishing?
- The variability expected within the groups usually measured by SD. This should be estimated from pilot experiments or previous published data. It is often assumed that the SD is the same in each group, but if that is not expected then it might be more appropriate to choose a test where this is not necessary (see Figure 9.3). If the SD is not included in a previously published paper but the 95% CI is, it can be derived using SD = confidence interval width/3.92. It is important to note that the SD is not the same thing as SEM. SEM = SD/$\sqrt{n}$ and, as such, will vary depending on the sample size in the

study. SD does not depend on sample size (although a larger sample size should give a better estimate of the population SD).

- Significance level (α) the proportion of times that the test will give a significant result by chance when there is actually no difference (a false positive). Usually it is set at 0.05 (5%), but a different value can be chosen with justification and, if you are planning on doing lots of tests using data from the same animals, you should make a multiplicity adjustment to account for the fact that doing more tests will result in a higher probability of obtaining a false positive result (see Example 2).
- Power ($1 - \beta$) the proportion of times that the test will give a significant result when the observed effect size ($(m_2 - m_1)/\mathrm{SD}$) is actually as, or more, extreme than the given value from the calculation (a true positive). Usually it is set at 0.8 or 0.9 (80 or 90%), but a different value can be chosen with justification.
- A one- or two-sided test whether you would be interested in a significant change in just one direction (e.g. $m_2 > m_1$) or in either direction ($m_2 \neq m_1$). Usually, a two-sided test should be used as it is also important to know whether the effect opposite to that expected occurs. In fact, journals will often not accept use of one-sided tests because of their inability to detect unexpected results. There are, however, exceptions, such as non-inferiority tests. Usually, tests such as *t*-tests are used to try to demonstrate that one group performs 'better' than the other by disproving the hypothesis that the means in two groups are equal. However, it may actually be sufficient, or of more interest (e.g. when comparing two active treatments), to demonstrate that the response in the one group is 'at least as good as' (non-inferior to) that in the other. This needs to be done by disproving a hypothesis that, for example, the mean in the comparator group is greater than that in the experimental treatment group by some specified amount (the non-inferiority margin). Several of the tests previously discussed can be set up as non-inferiority tests, by

determining whether a confidence interval crosses the non-inferiority margin.

When using power calculation software, be aware of the default settings because they may not be appropriate for your problem. In particular, check whether it assumes that the test is one- or two-sided and that the power is set at the level you require.

It is also important to consider whether you will be applying a transformation to your data (for example, to ensure validity of the normality assumption one might look at log data). If this is likely to be the case, the sample size should be calculated using the m_1, m_2 and SD values of the previously collected and hypothesized log-transformed data. A sample size calculated in this way will give you the required power to carry out the analysis that is planned.

Often you are expected to have an idea of your sample size before you have the necessary information to calculate it. In particular, if you have not yet collected any data under your specific experimental conditions, you may not be able to estimate the variability. A good way to deal with this is to run an internal pilot within your main experiment. This involves measuring the outcome in a smaller number of the experimental animals (for example, half the original estimate of the overall sample size), using these values to estimate the SD within each of the groups and utilizing this to recalculate the overall sample size. It is statistically valid to include these internal pilot animals in the final analysis, because the data for this subset are used to calculate only the new estimate of the SD; no statistical test between the groups is carried out at this stage, so the new sample size estimate should not be subject to any related bias. There are practical reasons why this may not always be possible: for example, if all animals have to enter the experiment concurrently.

The bodies that fund and license animal research need to understand the scale of the experiments and will expect a justification of expected sample size, even if pilot data are yet to be obtained. An internal pilot is not always possible and, even when this method is used, an initial estimate of the final sample size is still required. In these cases, whatever information is available must be used. Whatever previous experiments have been done (by yourself or published by others) that have led you to progress to this one will give you some insight. This may be in vitro work, experiments using compounds of a similar type or work on wild-type fish, for example.

The smaller the difference between the two group means (m_1 and m_2) that is required, and the larger the SD, the larger the sample size will need to be. To give an idea of how the necessary sample size changes depending on these parameters, Table 9.8 and Figure 9.4 show the required numbers of animals for some combinations in a *t*-test assuming equal SD values in the two groups. In order to make these more general, the size of m_2 proportional to m_1 is presented rather than the difference between m_1 and m_2 (where $m_2 = A^*m_1$; $m_2 - m_1 = (A - 1)^*\ m_1$). Figure 9.5 illustrates what a bar chart of the theorized data would look like for four of the scenarios in Table 9.8 (shaded cells). Note that the error bars, illustrating variability around the means, give SDs in this figure rather than SEMs or CIs. It is more difficult to detect true differences between groups when the variability around their means overlaps by more. It can be seen that the sample size needs to be larger when the lower extreme of the SD bar for group 2 is below the upper extreme of the SD bar for group 1 than when the SDs are small in comparison with the difference between the means.

Table 9.8 Sample sizes (per group) required to achieve 80% power in a Student's *t*-test at a 5% significance level. Various possible sample sizes are given based on a range of differences in means (m_1 and m_2) and SDs of data in the two groups.

m_2	SD						
	$0.10 \times m_1$	$0.25 \times m_1$	$0.50 \times m_1$	$0.75 \times m_1$	$1.00 \times m_1$	$1.25 \times m_1$	$1.50 \times m_1$
1.10 × m_1	17	100	394	884	1,571	2,454	3,533
1.25 × m_1	4	17	64	143	253	394	567
1.50 × m_1	3	6	17	37	64	100	143
1.75 × m_1	2	4	9	17	29	45	64
2.00 × m_1	2	3	6	10	17	26	37
2.50 × m_1	2	2	4	6	9	12	17
3.00 × m_1	2	2	3	4	6	8	10

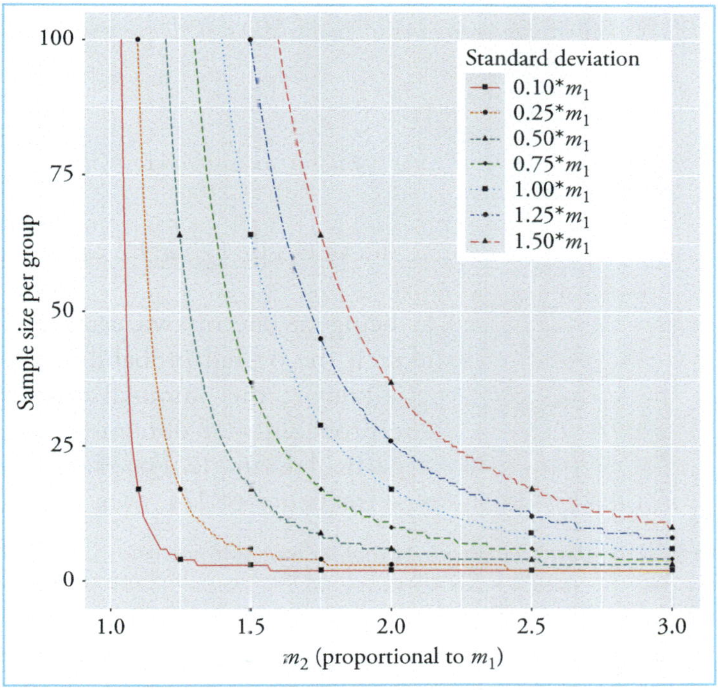

Figure 9.4 Curves showing the sample size required per group to achieve 80% power in a Student's *t*-test at a 5% significance level. These are shown for experiments where the mean in the experimental group (m_2) is up to three times larger than that in the control group (m_1), and the SD of both groups is between 0.1 and 1.5 times m_1. Sample sizes corresponding to values in Table 9.8 are highlighted by symbols.

Carrying out the experiment

At this stage you have calculated how many experimental units you will need in each of your groups. With more mature fish, the experimental units are likely to be the fish themselves but, especially when using larvae, these could be a tank or a Petri dish. For example, you might have an experiment where the experimental unit is a Petri dish containing larval fish and the sample size is six per group, meaning there are six dishes under each different experimental condition.

Every experiment will be different, but the procedures outlined below should be followed wherever possible. If these things are

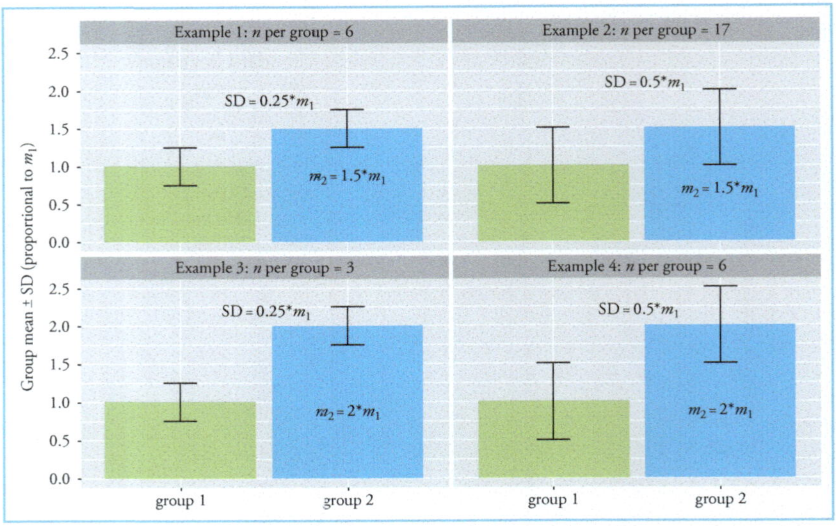

Figure 9.5 Bar charts illustrating four of the example scenarios from Table 9.8, with error bars showing one SD either side of the mean. In these scenarios, m_2 = 1.5 or 2.0 times m_1 and the SD of both groups is 0.25 or 0.50 times m_1.

not taken into account, the results of any statistical analyses may not be valid.

- Experimental units should be allocated to groups randomly.
- If possible, the person collecting the data should be blinded to (unaware of) which groups the units belong to. This may not be possible if the group is genotype based and the different genotypes produce obvious phenotypes. The EDA (Percie du Sert et al., 2017) tool has a feature that generates random allocation of units to groups, and it will send the resulting instructions to the email address of another person so that the EDA user themselves may be blinded to the allocation strategy.
- Other than the condition(s) being investigated, all other treatment of the animals should be the same across groups: for example, diet, water temperature and lighting.
- If possible, control groups should be subjected to a type of intervention similar to that of the experimental group(s): for

example, a similar substance with known effect, a placebo or vehicle.

- In experiments where the same animals receive two or more treatments sequentially, the order in which they are given should be randomized. This will reduce bias from carry-over (effects of an earlier treatment that carry on after a later treatment has started), environmental and age-related effects of the treatments. This is known as a cross-over design.
- With larger sample sizes, blocking should be used. For example, where the experimental unit is mature fish, blocks could be tanks, assessment days or cohorts containing equal numbers of fish from each experimental group:
  - If four tanks of fish are required, is it possible to have equal numbers from each experimental group in each tank?
  - If only 12 fish can be measured on each day, can these be taken equally from each group?
  - If the experiment is repeated on multiple cohorts to achieve the required sample size, the data can all be analysed together if each cohort is treated as a block.

Welfare and husbandry can also have a large bearing on the measured outcome. Fish living in poor conditions and already struggling may not respond to treatments in the way that well-maintained fish would. Refinement of animal welfare is not only important from an ethical point of view, but also in order to produce the most scientifically sound results possible.

Exploratory data analysis

Once the data have been collected, it is valuable initially to examine them using appropriate plots. You may also wish to produce summary statistics such as calculating the means and SDs of the different groups (see the section above on hypothesis testing for further helpful summary statistics). Hopefully this initial data exploration will give you a very good idea of whether or not your groups are in fact producing different outcomes. It may also help you to identify something that has gone wrong or not turned out as expected.

If you have multiple continuous independent variables, plot them against each other to check for correlation. If the association between them is high, it is usually not a good idea to include them all in the main analysis. A common example of this from studies with human subjects is that height and weight are often measured but are likely to be highly correlated. Depending on what is considered most appropriate for the analysis, it may be decided to choose just one of these, or to use instead body mass index (BMI; a composite value of both).

Refer back to the planned method of analysis. All statistical analyses require the data to meet certain assumptions: for example, that the residuals are normally distributed or that SDs in the groups are the same. There is much more information on this in the section on choosing a method of analysis. It must be checked whether the assumptions of the chosen test are met (at least approximately). If this is not the case, carrying out data transformation (such as taking the log-values of the outcomes) may produce data that will then meet the assumptions. If this is still not the case, an alternative method should be sought that would test the relevant hypothesis and for which the assumptions are met. Remember that SD is not the same as SEM, which is a measure of uncertainty in the calculated mean and depends on the sample size (for example, the number of animals) in each group. Therefore, if you have different numbers you cannot compare SEMs to determine whether the variability in groups is equal. You need to compare the SDs, which are not dependent on sample size.

Example 3. Do my data meet the assumptions of the analysis?

Imagine an experiment with two factors of interest: for example, one featuring both diet and an experimental treatment in a zebrafish model of a particular disease. Half of the animals in the experiment are fed the standard facility diet (diet 1) and the remainder are fed an enhanced diet (diet 2). Each of these is then split into two further groups, one of which is given the treatment and the other is used as a control. These will be the only two factors taken into account in the analysis. This leads to a total of four groups, each with (say) 18 fish. After a reasonable period of time a measure of disease severity is taken. The planned primary analysis may be a two-way ANOVA to test the effect of treatment and that of diet.

Figure 9.6 shows box plots (also called box-and-whisker diagrams) of data for an experiment of this type. These kinds of plot are very useful for assessing the approximate distribution of the data, being constructed around the quartiles of the data. The median is shown on the diagram as the thick line in the middle of a box, and the lower and upper quartiles are displayed as the lower and upper extremities of a box. The 'whiskers' extending from the box may reach out to the maximum and minimum observed values, but we can see here that the default setting in the software used to create this plot has considered points in three of the groups to be outliers and plotted them individually. This plot was produced using R statistical software (R Core Team, 2017), and the default is to use

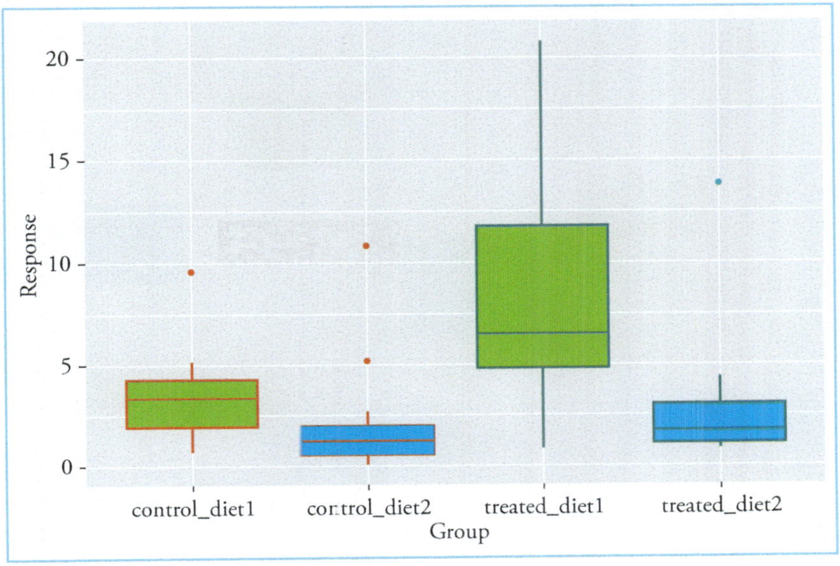

Figure 9.6 Box plots of response by group.

Tukey whiskers (1.5 times the interquartile range from the edge of the box) and define anything more extreme as outliers, which are shown individually. Different software can have different criteria for an outlier in a box plot.

In the case of these particular data we can see that groups having a higher median value also tend to have a wider spread of values, indicated by the height of the box (interquartile range) and the height of the whiskers and outliers (full range). They are also positively skewed, meaning that the response values are not symmetrical around the median and they tend to extend out further from the median to higher values than to lower values. Both of these observations are common features where the response variable is constrained to be above zero, and indicate that the residuals of the analysis may well not be normally distributed, but log-normally distributed. Transforming the raw data by taking the $\log_{10}$ values of these numbers can be used to check this. Plotting the log-transformed values in the same way gives the box plots shown in Figure 9.7. It would be equally valid (and yield a plot that looked visually the same) to take a different log, for example $\log_2$ or $\log_e$ (ln) and, if you are

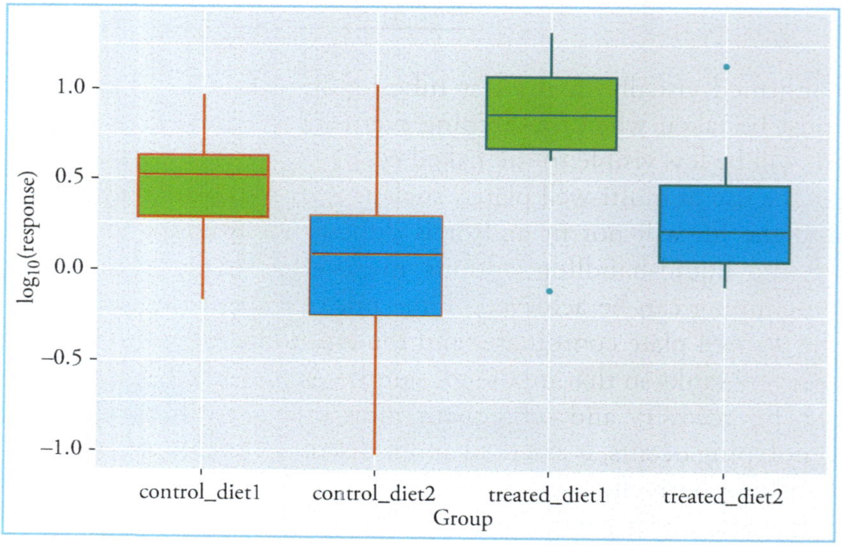

Figure 9.7 Box plots of log-transformed response by group.

more familiar with working with any particular base, it is probably most appropriate to use that. Transforming the data in this way has resulted in much more symmetrical boxes and whiskers, with similar variability in all of the groups. These look much more like normally distributed data.

In this example it looks like a two-way ANOVA may be a suitable analysis to use, but it should be performed on the log-transformed data rather than the raw data. If transforming the data had not produced approximately normally distributed data, non-parametric tests would have been required. These are the types of statistical analysis that do not assume normal distribution when comparing similarity between groups. There is no universal non-parametric equivalent of the two-way ANOVA, but several Mann–Whitney–Wilcoxon tests could be used to compare individual pairs of groups, using a multiple testing adjustment to take account of how many were used. Unfortunately, this would not be able to quantify any interaction between the two factors.

Analysing the data and interpreting the results

If all the planning has been done carefully, the analysis of the data may well be the easiest part! Choose the correct test, and the correct piece of software to do that test, carefully enter the raw data and with a few clicks of a button the answer should be revealed!

Of course, that is slightly oversimplifying things. First, it really does have to be a correct test because simply carrying out an analysis that produces some output does not necessarily mean it is a suitable analysis. We emphasized, in the section on choosing a method of analysis, that the choice should be made when the experiment is first being planned, and certainly before any data are collected. It should be a method that tests a relevant hypothesis to answer a key question of interest. The specific test might sometimes need to be adjusted, for example if distributional assumptions about the data proved not to be correct (see Example 3); nevertheless it needs to test a suitable hypothesis using the data available.

The software then needs to be chosen and used appropriately. GraphPad Prism (GraphPad Software, 2021) and SPSS (IBM Corp., 2020) are popular choices for biologists due to their user-friendly interfaces, examples and documentation. For more numerically minded researchers, or for those who have the assistance of a statistical expert, other specialist software such as R (R Core Team, 2017) or SAS (SAS Institute Inc., 2004) may allow for more flexibility in the analysis method and the output that is generated. R may also appeal to users who are new to data science, or who don't have the option to pay for one of the alternatives, because it is a freely available software with a large amount of support and documentation available online, as well as being used in example problems in many applied statistics textbooks.

A P value of <0.05 in a single statistical test means that, if the null model being tested were true, the probability of a test statistic as, or more, extreme than that which was collected occurring randomly is <0.05. This is usually considered sufficiently strong evidence to reject the null hypothesis, but there is still a chance of a false positive. A P value >0.05 means that there is not enough evidence to reject the null hypothesis, but it does not provide evidence to accept it, either. It may just be that there were not enough data collected.

Using a P value threshold of 0.05 (alpha of 5%) may not always be appropriate and, in particular, a multiple testing adjustment should be carried out if several hypotheses are being tested in parallel. Examples of this are when multiple groups are compared with each other (see Example 2) or co-primary outcomes are tested in the same experiment. Some statistical software makes it easy for the user to implement appropriate multiple testing adjustments (for example, it is the default in post hoc testing in GraphPad Prism), whereas in other cases it may need to be done manually. A simple and conservative way to do this is to apply a Bonferroni adjustment, 'splitting the alpha' between the different tests, so that the sum of the P value thresholds is 0.05. However, as the number of tests being adjusted for increases, the more the power to detect a true positive is reduced. Alternative adjustments such as the Šidák correction (Šidák, 1967), Holm method (Holm, 1979) or the

Benjamini–Hochberg procedure, which controls the false discovery rate (Benjamini & Hochberg, 1995), help to compensate for this and may be more appropriate when a large number of tests have been carried out.

Whether the null hypothesis is rejected or not, the relevant statistics should be reported along with some quantification of uncertainty. For example, if it is the difference between two groups that is being investigated, the mean and 95% confidence interval of the difference are usually the most appropriate values to report. Whilst SDs or even standard errors are sometimes reported as measures of uncertainty, they can be difficult to interpret. Confidence intervals give limits within which most values of the outcome would be expected to fall, based on what has been observed. When numerical data are not normally distributed, the median and interquartile range should be reported.

It is often much easier to read statistical information from a figure rather than a large chunk of text or a table. When comparing a one-off outcome under different conditions, plotting the data as box plots or bar charts of the means or medians of the groups with CIs or interquartile ranges included as interval lines is simple, but will help to show which groups are similar and which are very different. In particular, if there are multiple factors used to define the groups (e.g. treatment and diet combinations), added design features such as shading of the boxes can help to distinguish between these. However, bear in mind that they may be printed in black and white, or the people reading them may be colour-blind, so greyscale or patterns may be more effective than multiple bright colours.

Measurements taken over time, or at different values of a continuous independent variable (e.g. doses of a drug), can be displayed graphically as a profile plot of time or dose on the x axis and average (mean or median) outcome values on the y axis joined by a line. If there are multiple groups, these can be plotted with different symbols and line-types, such as circles versus squares and solid versus dashed.

Some software will automatically plot the data for you, but check that the plot that is produced is intuitive to read; it can always be adjusted. Try and keep things simple: too much information in one figure can make it difficult to read.

Reporting on the statistical analysis

When reporting on an experimental project involving animals, the ARRIVE guidelines (Percie du Sert et al., 2020) should always be used. With regard to the statistical considerations, it is particularly helpful to report:

- The key objective of each experiment or the hypothesis to be tested.
- The type of analysis for each experiment separately, rather than in a single summary paragraph at the end of the 'Methods' section.
- How the sample size was chosen, including information about the data used to estimate the SD (variability) and what significance level and power were used in the calculation.
- Results! If possible, include these in the text or a table and in a figure. Indicate the point estimate (e.g. mean) and a range (e.g. 95% CI) and the approximate *P* value. Indicate whether a multiplicity adjustment has been made and, if so, which one (e.g. Bonferroni, Šidák, Holm, Benjamini–Hochberg).
- Ensure that it is clear which analysis is the primary one and which are secondary or exploratory. If an experiment has been designed around a primary outcome and analysis, any claims made should focus on this. Secondary and exploratory analyses hold less weight but may still be used to illustrate interesting findings and, importantly, change the focus of future work.
- Include results that were not significant! If a whole experiment did not produce any significant results, at least try to include a mention of it in a paper about other related experiments so that it is published and other researchers are less likely to repeat it.

Summary

This chapter has only touched on the aspects of statistics in zebrafish breeding, health screening and experimentation that seemed to the authors to be either most important, lacking in availability elsewhere or where common mistakes or misunderstandings occur. It by no means covers every statistical scenario that someone working in a zebrafish laboratory might come across, but hopefully will clarify some general points and give guidance as to what type of approach should be used. Much of the information given is also much more widely applicable to many scientific disciplines, especially the parts of the chapter relating to experimental design, analysis and reporting, but the examples given are tailored to zebrafish.

Using more animals than is necessary is unethical, but not having enough within breeding stock, health screens or experimental cohorts to be able to reach valid conclusions means that any fish that are used are all wasted animals as well. A suitable sample size is crucial to deliver significance and ensure that the use of the animal was justified.

Whilst breeding and health screening are two areas that are perhaps often left to the facility management team or more junior members of a laboratory, they are crucial for ARRIVE (Percie du Sert et al., 2020) and planning. Therefore, it is important for these staff to have an understanding of the appropriate number of fish that they should be using and also for researchers to have an appreciation of how these things are managed by their colleagues.

You may wish to consider attending a course in basic statistics to gain a better understanding or organizing for someone to come in and give one to your research group. There are also, of course, many texts that cover topics in this chapter in much more detail. However, you don't have to do all of this yourself. Never forget that there are people out there with specialist training and vast experience with statistics. They might be found at the institution you work for or could be sought at external consultancy companies. It would not be expected that an expert in a particular area of biomedical science, for example, was also a statistical expert.

There are also many different statistical software packages on the market, but not all are able to carry out every kind of analysis and some types of analysis are not straightforward to perform. If you are unsure what type of analysis you should be doing, what package you can use, what options to select to carry it out or how to interpret the results, always seek help from a statistician. Funding bodies generally look kindly upon the suggestion of a portion of funding being put towards statistical support for projects, with some now insisting on experimental design and statistics details in the proposal itself (Gosselin, 2019).

When planning experiments, remember that the appropriate sample size needs to be chosen before they can be carried out. The calculation of that sample size will be specific to the type of analysis that is planned, so that method also needs to be chosen right at the very beginning. All these things need to be recorded and reported to aid other researchers carrying out related work.

Statistics is an important part of much of zebrafish husbandry, welfare and research methodology. This can be daunting to practitioners who don't have advanced mathematical qualifications, but we have tried to provide usable tables and figures relating to common issues and references to appropriate texts that cover specific topics in more depth. In the Appendix we give more detail of the formulae underlying the calculations presented in the tables and figures. These may be adapted by researchers for cases not covered in the examples.

Appendix of formulae for breeding, stock management and health screening

Please refer to Table 9.1. These values were calculated by solving the cumulative binomial distribution function to determine the minimum n such that $P(x \geq 50) \geq 0.90$ or 0.95 as required.

$$P(X \geq 50) = 1 - P(X \leq 49) = 1 - \sum_{i=0}^{49} ({}^{n}/{}_{i})\, p^{i}(1-p)^{n-i},$$

where $({}^{n}/{}_{i})$ is the binomial coefficient, giving the number of possible combinations of i successes and n-i failures in n trials, and p is the proportion of females multiplied by the background mortality rate (as a proportion).

The values in Table 9.2 were calculated using $n = 200 * f * (1 - m)$, where f is the genotype frequency and m is the background mortality rate (as a proportion). Figure 9.1 uses simply $f* (1 - m)$ to give the proportion of any sized pool of fish.

The example in the text (Table 9.3) is that, with 24 fish in total and a genotype frequency of 1/4, even though the most likely number of fish with this genotype is 24/4 = 6, there is only an 18.5% chance of obtaining exactly six and a 57.8% chance of obtaining six or more. The first of these values comes from the binomial probability mass function, where we calculate the probability of observing exactly $i = 6$ of the desired genotype from a pool of $n = 24$ where $p = 0.25$ as

$$\begin{aligned}(^n/i)\,p^i(1-p)^{n-i} &= (24\ 6\,)\,0.25^6(1-0.25)^{24-6}\\ &\approx 134596 * 0.00024 * 0.00564\\ &\approx 0.185.\end{aligned}$$

$(^n/i)$ is the binomial coeﬃcient, or the number of possible combinations of i successes and n-i failures in n trials.

Similarly, the CDF,

$$P(X \geq x) = 1 - P(X \leq x - 1) = 1 - \sum_{i=0}^{x-1} (^n/i)\,p^i(1-p)^{n-i}$$

is used to obtain $P(X \geq 6) \approx 0.578$.

To be Y% sure (say, 90 or 95%, as given in Table 9.3) that the required genotype is observed in at least x of them, we solve the cumulative binomial distribution function to find the minimum n such that $P(X \geq x) \geq 0.90$ or 0.95 as required. So, applying this to the above example, where the genotype frequency was 1/4, in order to be 95% sure of obtaining at least six fish of this genotype, you would need 40 fish in total.

The data in Table 9.4 are based on one-sided Clopper–Pearson confidence limits (Clopper & Pearson, 1934). The value given is the lowest number of dead fish (x) from the total number of fish shown in the left-hand column (n) that gives a lower 95% (or 90 or 80%, as required) confidence limit $>p$, where p is the background mortality rate. This is the confidence limit that is calculated when a binomial test is carried out to test the null and alternative hypotheses:

$$H_0: P(mortality) = p$$
$$H_A: P(mortality) > p$$

and x successes (deaths) are observed in n trials (individual fish), with $1 - \beta = 0.95$ (or 0.9 or 0.8). A lower confidence limit $>p$ indicates that, having observed this many deaths, we can be $100 * (1 - \beta)\%$ confident that the mortality rate has risen above the original background mortality rate of p.

The data in Table 9.5 were calculated using $p = 1 - (1 - z/100)^n$, where z is the percentage of infected fish in the background population and n is the size of the sample. For instance, with a sample of 30 fish there is a probability of 0.958 of sampling an infected fish where the prevalence of infection is 10% ($1 - 0.9^{30} \approx 0.958$), but only a probability of 0.785 if the true prevalence is 5% ($1 - 0.95^{30} \approx 0.785$). The dynamics of this are illustrated in Figure 9.2. Note that these are the probabilities of sampling an infected fish, and do not take into account the sensitivity and specificity of screening tests.

The values in Table 9.6 were derived from rearranging the formula for Table 9.5, to give

$$z = 100*\left(1 - exp\left(\frac{ln(1-p)}{n}\right)\right).$$

The examples given in the text use $n = 30$ with $p = 0.8$ $\left(z = 100*\left(1 - exp\left(\frac{ln(0.2)}{30}\right)\right) = 100*(1 - exp(-0.054)) \approx 5.2\%\right)$ and $p = 0.5$ $\left(z = 100*\left(1 - exp\left(\frac{ln(0.5)}{30}\right)\right) = 100*(1 - exp(-0.023)) \approx 2.3\%\right)$.

Acknowledgements

The authors would like to acknowledge the valuable contributions of Kathleen Baster, Lucy Gelder and Rosie Taylor at the Statistical Services Unit of The University of Sheffield to various discussions around the content of this chapter. We also thank the editors of the book for their helpful comments and additions.

References

Al-Jubouri, Q., Al-Azawi, R. J., Al-Taee, M., & Young, I. (2018). Efficient individual identification of zebrafish using Hue/Saturation/Value color model. *Egyptian Journal of Aquatic Research*, *44*(4), 271–277. https://doi.org/10.1016/j.ejar.2018.11.006

Allen, C., Argenton, F., Van Eeden, F. J., & Tiso, N. (2021). Transgenesis, mutagenesis, knockdown, and genetic colony management. In L. D'Angelo & P. de Girolamo (Eds.), *Laboratory fish in biomedical research* (pp. 139–152). Elsevier.

Armitage, P., Berry, G., & Matthews, J. N. S. (2002). *Statistical methods in medical research* (4th ed). Blackwell Publishing

Benjamini, Y., & Hochberg, Y. (1995). Controlling the false discovery rate: A practical and powerful approach to multiple testing. *Journal of the Royal Statistical Society: Series B*, *57*(1), 289–300. https://doi.org/10.1111/j.2517-6161.1995.tb02031.x

Biotechnology and Biological Sciences Research Council. (2021). BBSRC research grants guide. https://www.ukri.org/publications/bbsrc-research-grants-guide/ Retrieved October 4, 2021.

Clopper, C. J., & Pearson, E. S. (1934). The use of confidence or fiducial limits illustrated in the case of the binomial. *Biometrika*, *26*(4), 404–413. https://doi.org/10.1093/biomet/26.4.404

Collymore, C., Crim, M. J., & Lieggi, C. (2016). Recommendations for health monitoring and reporting for zebrafish research facilities. *Zebrafish*, *13* (Suppl. 1), S138–S148. https://doi.org/10.1089/zeb.2015.1210

Cox, D. R. (1972). Regression models and life tables [With discussion]. *Journal of the Royal Statistical Society: Series B*, *34*(2), 187–202. https://doi.org/10.1111/j.2517-6161.1972.tb00899.x

Crim, M. J., & Lawrence, C. (2021). A fish is not a mouse: Understanding differences in background genetics is critical for reproducibility. *Lab Animal*, *50*(1), 19–25. https://doi.org/10.1038/s41684-020-00683-x

Dean, A., Voss, D., & Dragulić, D. (2017). *Design and analysis of experiments* (2nd ed) Springer Texts in Statistics.

Faul, F., Erdfelder, E., Lang, A. G., & Buchner, A. (2007). G*Power 3: A flexible statistical power analysis program for the social, behavioral, and biomedical sciences. *Behavior Research Methods, 39*(2), 175–191. https://doi.org/10.3758/bf03193146

Fisher, R. A. (1922). On the interpretation of χ^2 from contingency tables, and the calculation of P. *Journal of the Royal Statistical Society, 85*(1), 87–94. https://doi.org/10.2307/2340521

Fisher, R. A. (1954). *Statistical methods for research workers.* Oliver & Boyd.

Gosselin, R. D. (2019). Guidelines on statistics for researchers using laboratory animals: The essentials. *Laboratory Animals, 53*(1), 28–42. https://doi.org/10.1177/0023677218783223

Graham, M. L., & Prescott, M. J. (2015). The multifactorial role of the 3Rs in shifting the harm-benefit analysis in animal models of disease. *European Journal of Pharmacology, 759,* 19–29. https://doi.org/10.1016/j.ejphar.2015.03.040

GraphPad Software (2021). GraphPad Prism version 9.1.2 for Windows. GraphPad Software, San Diego, CA, USA. www.graphpad.com

Hair, K., Macleod, M. R., & Sena, E. S. (2019) A randomised controlled trial of an Intervention to Improve Compliance with the ARRIVE guidelines (IICARus). *Research Integrity and Peer Review, 4*(1), 12. https://doi.org/10.1186/s41073-019-0069-3

Harrell, Jr., F. E. (2015). *Regression modelling strategies* (2nd ed). Springer Series in Statistics.

Holm, S. (1979). A simple sequentially rejective multiple test procedure. *Scandinavian Journal of Statistics, 6*(2), 65–70.

IBM, Corp. (2020). *IBM SPSS Statistics for windows* version 27.0. IBM, Corp.

Krug, L. (2019). *The effect of increased feeding in proportion to density on sex ratios in zebrafish (Danio rerio)* [Master's Thesis]. Harvard Extension School.

Laud, P. J. (2017). Equal-tailed confidence intervals for comparison of rates. *Pharmaceutical Statistics, 16*(5), 334–348. https://doi.org/10.1002/pst.1813

Laud, P. J. (2018). Equal-tailed confidence intervals for comparison of rates. *Pharmaceutical Statistics, 17*(3), 290–293. https://doi.org/10.1002/pst.1855

Leung, V., Rousseau-Blass, F., Beauchamp, G., & Pang, D. S. J. (2018). ARRIVE has not ARRIVEd: Support for the ARRIVE (animal research: reporting of in vivo experiments) guidelines does not improve the reporting quality of papers in animal welfare, analgesia or anesthesia. *PLOS ONE, 13*(5), e0197882. https://doi.org/10.1371/journal.pone.0197882

Masca, N. G., Hensor, E. M., Cornelius, V. R., Buffa, F. M., Marriott, H. M., Eales, J. M., Messenger, M. P., Anderson, A. E., Boot, C., Bunce, C., Goldin, R. D., Harris, J., Hinchliffe, R. F., Junaid, H., Kingston, S., Martin-Ruiz, C., Nelson, C. P., Peacock, J., Seed, P. T., . . . Teare, M. D. (2015). RIPOSTE: A framework for improving the design and analysis of laboratory-based research. *eLife, 4*(4), e05519. https://doi.org/10.7554/eLife.05519

Medical Research Council. (2021). Proposals involving animal use. https://www.ukri.org/councils/mrc/guidance-for-applicants/proposals-involving-animal-use/ Retrieved October 4, 2021.

Mrakovčič, M., & Haley, L. E. (1979). Inbreeding depression in the zebra fish Brachydanio rerio (H. Buchanan). *Journal of Fish Biology, 15*(3), 323–327. https://doi.org/10.1111/j.1095-8649.1979.tb03612.x

Nasiadka, A., & Clark, M. D. (2012). Zebrafish breeding in the laboratory environment. *ILAR Journal, 53*(2), 161–168. https://doi.org/10.1093/ilar.53.2.161

Nuffield Council on Bioethics. (2005). *The ethics of research involving animals.* Nuffield Council on Bioethics.

Pearson, K. (1900). X. On the criterion that a given system of deviations from the probable in the case of a correlated system of variables is such that it can be reasonably supposed to have arisen from random sampling. *London, Edinburgh, and Dublin Philosophical Magazine and Journal of Science, 50*(302), 157–175. https://doi.org/10.1080/14786440009463897

Percie du Sert, N., Bamsey, I., Bate, S. T., Berdoy, M., Clark, R. A., Cuthill, I., Fry, D., Karp, N. A., Macleod, M., Moon, L., Stanford, S. C., & Lings, B. (2017). The experimental design assistant. *PLOS Biology, 15*(9), e2003779. https://doi.org/10.1371/journal.pbio.2003779

Percie du Sert, N., Ahluwalia, A., Alam, S., Avey, M. T., Baker, M., Browne, W. J., Clark, A., Cuthill, I. C., Dirnagl, U., Emerson, M., Garner, P., Holgate, S. T., Howells, D. W., Hurst, V., Karp, N. A., Lazic, S. E., Lidster, K., MacCallum, C. J., . . . Würbel, H. (2020). Reporting animal research: Explanation and elaboration for the ARRIVE Guidelines 2.0. *PLOS Biology, 18*(7), e3000411. https://doi.org/10.1371/journal.pbio.3000411

R Core Team. (2017). *R: A Language and Environment for Statistical Computing*. http://www.r-project.org/. R Foundation for Statistical Computing.

Rácz, A., Allan, B., Dwyer, T., Thambithurai, D., Crespel, A., & Killen, S. S. (2021). Identification of individual zebrafish (Danio rerio): A refined protocol for VIE tagging whilst considering animal welfare and the principles of the 3 Rs. *Animals: An Open Access Journal from MDPI*, *11*(3), 616. https://doi.org/10.3390/ani11030616

SAS Institute, Incorp. (2004). *SAS/STAT.* ® 9.1 *Users Guide*. SAS Institute, Inc.

Šidák, Z. K. (1967). Rectangular confidence regions for the means of multivariate normal distributions. *Journal of the American Statistical Association*, *62*(318), 626–633. https://doi.org/10.1080/01621459.1967.10482935

Smith, A. J., Clutton, R. E., Lilley, E., Hansen, K. E. A., & Brattelid, T. (2018). PREPARE: Guidelines for planning animal research and testing. *Laboratory Animals*, *52*(2), 135–141. https://doi.org/10.1177/0023677217724823

Sneddon, L. U., Halsey, L. G., & Bury, N. R. (2017). Considering aspects of the 3Rs principles within experimental animal biology. *Journal of Experimental Biology*, *220*(17), 3007–3016. https://doi.org/10.1242/jeb.147058

Spence, R., Gerlach, G., Lawrence, C., & Smith, C. (2008). The behaviour and ecology of the zebrafish, Danio rerio. *Biological Reviews of the Cambridge Philosophical Society*, *83*(1), 13–34. https://doi.org/10.1111/j.1469-185X.2007.00030.x

Wellcome. (2021). Use of animals in research policy. https://wellcome.org/grant-funding/guidance/use-animals-research-policy Retrieved October 4, 2021.

Whitlock, M. C., & Schluter, D. (2020). *The analysis of Biological Data* (3rd ed). W. H. Freeman.

Zuur, A. F., Ieno, E. N., Walker, N. J., Saveliev, A. A., & Smith, G. M. (2009). *Mixed effects models and extensions in ecology with R*. Springer.

Chapter 10

Genotyping methods in zebrafish

Claire Allen

Abstract

Genome editing now enables research to create site-specific genetic modifications in a robust, widely adopted and dependable way. Zebrafish facilities, worldwide, have been inundated with all manner of newly created genetic modifications, often lacking visible phenotypes. Genotyping is the method that all zebrafish research facilities must employ to manage the genetics of their lines. DNA is extracted, preferably without sacrificing the fish, analysed and the results are relayed back to the identified fish that will be housed according to their genotype, and to the needs of research. Accuracy in the molecular biology assays undertaken to decipher genotypes is paramount for secure line management. The classic method used to obtain a satisfactory DNA sample is to fin-clip. Fin-clipping is swiftly losing its appeal for ethical and financial reasons, and in addition it fails to be sufficiently high throughput for certain endeavours. Most importantly it is widely accepted to cause adverse effects that can no longer be ignored. Research has a responsibility to mitigate these effects wherever possible and to refine the methods used. Alternatives to fin-clipping do exist, including swabbing adults and embryonic methods. Embryonic methods, however, remove the need to grow unwanted fish to adulthood and cause the least impact on fish welfare. This chapter reviews all the genotyping methods available for research and compares them.

Claire Allen and Jean-Philippe Mocho (eds) **Zebrafish**
DOI: 10.1079/9781800629431.0010, © CAB International, 2024

Keywords

zebrafish, genotyping, refinements, high-throughput, embryonic, fin-clipping

Introduction

Genotyping is the gateway tool for the identification and subsequent management of the laboratory genetically modified zebrafish. There has been an explosion in bioengineering with the development of the CRISPR-Cas system, which has revolutionized genome editing. The CRISPR-Cas system targets and creates site-specific genetic modifications with a Cas-9 endonuclease and guide RNAs (gRNAs) that combine to form a ribonucleoprotein (RNP) complex (reviewed by Allen et al., 2021a). This system is so robust that research potential is now almost limitless, allowing previously elusive modifications to be created and studied even within the injected zebrafish embryo (Hoshijima et al., 2019; Wu et al., 2018). For instance, our facility has seen a significant increase in the variety of lines that we house rather than an increase in the numbers of fish per line. This is alongside a clear shift from predominantly phenotypic genetically altered animals (GAA) to genotype identifiable GAA (Figure 10.1A). Multiple alleles within one line are increasingly popular, these require more careful management (Figure 10.1B), particularly when genotyping. With this recent capability in line creation, demand has also grown for automated genotyping assays to enable high-throughput research, such as drug screening and novel phenotypic analysis and response.

Genotyping is so crucial that in a facility primarily using fin-clipping to identify over 800 genetically modified zebrafish lines, more than 10,000 fin-clips are performed on adult (usually 4-month) fish each year. Adult zebrafish fin-clipping is a robust method of DNA extraction but there is important rationale for moving away from it, namely ethics, cost and the need for higher-throughput genotyping. Ethically we must consider the 3Rs – replacement, reduction and refinement (see Chapter 8 for more information). There is

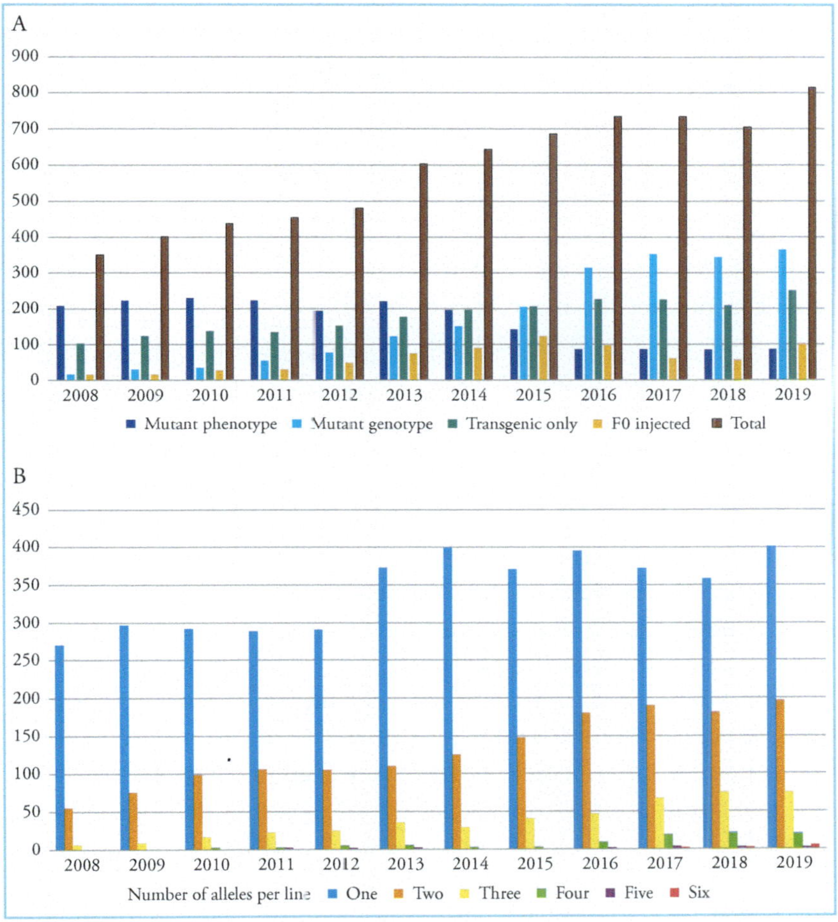

Figure 10.1 (A) Graph of line history; (B) graph of lines with numerous alleles. (A,B), *x* axis represents the years 2008–2019, *y* axis the number of lines. (B) The Biological Services Aquarium (BSA), The University of Sheffield (TUOS).

growing evidence that fin-clipping of adult zebrafish causes adverse effects (De Lomaert et al., 2017; Deakin et al., 2019a,b; Schroeder & Sneddon, 2017; Thomson et al., 2020; Tilley et al., 2020; White et al., 2017) and the compulsory isolation of each fish post-clip, while the DNA is analysed, is also considered a general welfare issue with specific husbandry demands. Armed with the knowledge that fin-clipping of adults has a negative impact on well-being, we have a responsibility to refine the method, reducing pain, suffering, distress and lasting harm. We also have an obligation to reduce the number of animals that we use, although zebrafish breeding schemes must incorporate out-crossing because inbreeding impacts the health of the line, reducing viability and fecundity (Allen et al., 2021b). Genetic heterogeneity should be maintained as a priority, but this means that excess fish are raised, many of which were never required. Mendelian rules of inheritance from any known parent cross can predict the number of fish with the correct genotype and those that are simply wild type. Because these breeding requirements bring with them significant cost implications for research, in order to maintain multiple complex lines cost effectively, different methods of genotyping, such as embryonic testing, should be considered. Finally, there is a drive for faster turnaround times for data collection and as we create more and more complex combinations of genetic modifications, so research needs to free itself of the genotyping bottleneck that fin-clipping poses. Solutions are emerging to resolve these issues, but each must be weighed up and considered for specific research goals (Figure 10.2A,B).

Zebrafish have the capacity to carry multiple genetic modifications which, in combination, maximize the research potential of a line (Figure 10.1B). Tracking and recording of a line's history and allelic accumulation within branches of a family can be managed through analysis of genetic material extracted from the individual fish in question. Genomic DNA is extracted from tissue samples, purified to varying extents and subsequently analysed by one or more biological assays, for example, polymerase chain reaction (PCR) amplification. Many of these methods can be repeated if the first analysis fails or fish become mixed although some methods are more invasive than others, needing particular life stages and specific equipment (Figure 10.2A). Genetic modifications can be identified through DNA sequencing, which is accurate but is costly and probably has to be done by a specialist service. Alternatively, knowledge of changes within the DNA code and how these relate to restriction enzyme digestion sites enables faster and often easier in-house analysis options. Advances in molecular biology continue to offer new and adaptable methods for genotype discovery, such as high-resolution

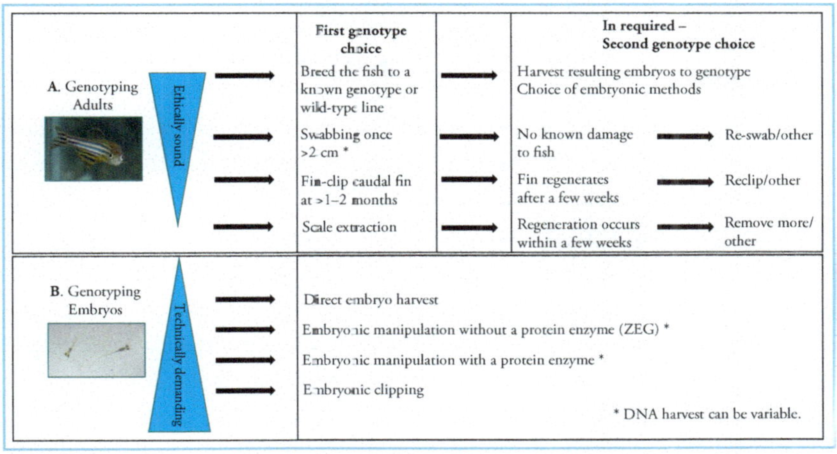

Figure 10.2 (A) Comparison of four ways of genotyping adult zebrafish; embryo analysis, swabbing, fin-clipping and scale removal, ordered by descending ethical suitability. (B) Comparison of four ways of genotyping embryonic zebrafish; direct harvesting, ZEG, protein enzyme and embryo-clipping, ordered by ascending technically demanding methods.

melting assay (HRMA) which, if available, is highly efficient and requires minimal effort (Wittwer, 2009). Whole-genome analysis is also possible in the identification of DNA markers, such as single-nucleotide polymorphisms (SNPs). SNPs create genetic variation that is increasingly important for breeding applications, and are commonly identified with a method called kompetitive allele-specific PCR (KASP–PCR; He et al., 2014).

Genotyping of DNA samples is not always required to decipher the genotype of a fish: it can be deduced by phenotype and/or known genetics within the parents. For example, in-crossed heterozygous (het) parents with recessive embryonic phenotypic traits can be identified by Mendelian inheritance in the appearance of phenotypes as the offspring grow and develop. Known homozygous (hom) fish can be out-crossed with a guarantee that all their offspring will be het. Reporter lines expressing visible indicators of their presence within a line, such as fluorescent proteins, help identify the carriers and specific parentage, be they het or hom, which helps to build the picture of the genotypes within the cohort. In fact, the first collection of genetic alterations in zebrafish (Haffter et al., 1996) created for cohesive research were all selected from phenotypically significant characteristics. The genetic knowledge of each allele was initially unknown, or at best anticipated. A combination of characterization and developmental analysis initiated the zebrafish as a model organism and, as molecular biology techniques evolved and developed, so did the genetic knowledge of each line. The opposite is now true of the phenotype–genotype relationship for research, and many modern genetically engineered lines have targeted gene alterations with no phenotype. Consequently we are increasingly dependent on genotyping in order to manage the line at each generation (Allen et al., 2021b).

Genetic material can be obtained through a number of methods (Table 10.1). By far the most popular and robust method has always been fin-clipping, whereby a portion of fin is cut. The clip is usually taken from the end of the caudal fin, due to its robust regenerative properties. The sample is then placed in an Eppendorf or multi-well place, which is fully labelled to match back to the individually housed fish. The fin can then be processed for DNA extraction, amplification and analysis of the desired genetic modification/s. The alternative methods to fin-clipping currently available include scale extraction (Kumar et al., 2007; Wasko et al., 2003; Yue & Orban, 2001), embryo sampling (Meeker et al., 2007; Parant et al., 2009), embryo clipping (Kosuta et al., 2018; Wilkinson et al., 2013; Xing et al., 2014), adult swabbing (Breacker et al., 2017; Tilley et al., 2020), embryo shaking (Lambert et al., 2018; Zhang et al., 2020) and embryonic fin scratching (Venditti et al., 2022). Some of these non-fin-clip methods work better than others, offering choice and refinement in regard to the adverse effects of fin-clipping (Figure 10.2A). For instance, if fin-clipping is too invasive for the survival of the fish due to their phenotype, or the cost implications are too high to grow fish to clippable sizes, embryonic methods are preferable (Figure 10.2B). If genotype is required from the point of protection to raise genetically distinct shoals, then embryonic methods are essential.

Successful genome editing and genotyping, although crucial in line creation and management, are not a research result and

Table 10.1 Techniques available to genotype zebrafish.

Technique Adult Embryo	Publications	Age range	Technical level	Chorion removal required	Proteinase	Anaesthesia	Equipment level	Viability	Genotyping success
Fin clipping >120 hpf	Akimento *et al.*, 2003 Poss *et al.*, 2003 Gemberling *et al.*, 2013 Cardiera *et al.*, 2016	Any	Medium	N/A	No	Yes	Low	100%	100%
Scales	Yue & Orban, 2001 Wasko *et al.*, 2003 Kumar *et al.*, 2007	Adult	Medium	N/A	No	Yes	Medium	100%	100%
Swabbing	Breacker *et al.*, 2017 Tilley *et al.*, 2020	>2 cm	Low	N/A	No	Optional	Low	100%	100% as per publication. *problems in practice.
Whole embryo	Meeker *et al.*, 2007 Parant *et al.*, 2009	>24 hpf	Low	No	No	No	Low	0%	100%
Embryo clipping	Wilkinson *et al.*, 2013 Kosuta *et al.*, 2018	72 hpf	High	Yes if <72 hpf	No	Yes	High	100%	100%
Embryo digestion with viability	Zhang *et al.*, 2020	>24 hpf <96 hpf	High	No	Yes	No	Low	>90%	90%
Chorionic fluid chip (no maternal DNA)	Samuel *et al.*, 2015	1–2 dpf	Very high	No	Yes	No	High	100%	78%
Fin clip chip	Samuel *et al.*, 2015	1–2 dpf	Very high	Yes	No	No	High	100%	100%
Embryo multi-plate shaking	Zhang *et al.*, 2020	>24 hpf <96 hpf	High	Yes	Yes	No	Average	>90%	>95%
Zebrafish embryo genotyper (ZEG)	Lambert *et al.*, 2018	<120 hpf	Medium	Yes	No	Yes	High	>90%	>90% for 1 allele. *problems in practice.

take up significant amount of bench time. The essential molecular biology techniques require a competent level of expertise, particularly to perform the subsequent genetic analysis. These facts can cause problems both for the implementation of the work and management of the line through the generations. For the time-pressured researcher, outsourcing might be preferable from a logistical viewpoint. It represents an investment in accuracy, reliability and reproducibility, and several international companies offer fast, efficient and accurate analysis from your tissue samples. The benefits to this approach include: elimination (or at least significant reduction) of human error; acceleration of research by bypassing this laborious task, which needs to be trained to competency; reduction in solitary tank time; and reduction in tank costs. One concern, however, is that sometimes samples are declined by these companies because they cannot process tiny specimens, such as those generated by swabbing or embryo shaking. Nonetheless in many cases, with adequate sample size, outsourcing genotyping can be cost effective for research and, perhaps more importantly, it gives peace of mind. It also ensures that the correct genotypes are within the tank, which is pivotal to all subsequent research endeavours that utilize the shoal and future generations bred from it.

Methods used for adults: fin-clipping

Originally used to distinguish farmed fish from their wild counterparts, fin-clipping is one of the oldest and most basic methods performed on fish. The fin is cut with a sharp blade and part, or all, of the chosen fin is removed. For zebrafish, the method was first suggested in the appropriately named *Zebrafish Book* (Westerfield et al., 1994), which is now in its 5th edition (Westerfield et al., 2007).

The regenerative properties of the zebrafish (Gemberling et al., 2013), particularly its fins (Akimenko et al., 2003), make surgical fin amputation/fin-clipping/finectomy a robust and plentiful source of genomic DNA. In the past it has often been referred to by many in the zebrafish research community as equivalent to experiencing a haircut. This lack of ethical consideration for fish has been historically commonplace in aquaculture and fisheries, but in a research setting we need to be more exacting with our research model (Sneddon, 2017). The evidence is growing that fin-clipping is an adverse event and far from a painless trim (Sneddon, 2020). The caudal fin is the first fin that the zebrafish develops (Siomava et al., 2018) and is the target fin of most fin-clippers. Anatomically this fin is known to comprise segmented bony rays, with inter-ray mesenchymal tissue between, is covered in epidermis and is found to be both vascularized and innervated. Finectomy (surgical fin amputation) initiates a regenerative process with three distinct stages: wound healing, blastema formation and regenerative outgrowth (Akimenko et al., 2003; Gemberling et al., 2013; Poss et al., 2003). A differentiation gradient is established along the proximal–distal axis, which signals a variety of cells including osteoblasts (Cardeira et al., 2016). The structural evidence thus indicates that this technique is invasive and is, therefore, one that should be refined. Further evidence of adversity comes from research into behavioural changes on fin-clipped zebrafish (Deakin et al., 2019a,b; De Lomaert et al., 2017; Schroeder & Sneddon, 2017; Tilley et al., 2020; White et al., 2017). Findings show that the freshly fin-clipped zebrafish does not simply respond with a withdrawal reflex and then return to normal behaviour, but displays distinct and longer-lasting effects for around 5–6 hours post-clip (Deakin et al., 2019a). These may not be seen in the first few moments after the clip, since the anaesthetic from the clip has not worn off and its analgesic properties are still working. Immediate apparent adverse effects to look for include bleeding from the gills, regurgitation of stomach contents, failure to recover from the anaesthesia and reduced buoyancy. Regurgitation can be prevented by withholding food from the fish for a number of hours before they are scheduled to be clipped. Visual indicators of the adverse effects to look for once the anaesthetic has worn off include seeing the fish spending increased time at the bottom and corner of their tank, erratic slower swimming and wafting their remaining tail (Schroeder & Sneddon, 2017). The final visible indicator of longer-lasting adverse effects, following visible recovery, is a lack of appetite. The more subtle, but nevertheless significant, adverse effects are only observable with the help of fairly specialist behavioural analysis

(Deakin et al., 2019b). These analyses are, unfortunately, quite difficult to reproduce without the correct training, equipment and facilities. Additional evidence of adversity arises from molecular studies carried out to compare skin swabbing and fin-clipping, which have shown that cortisol, the stress indicator, is increased post-clip (Tilley et al., 2020). Interestingly, Tilley et al. (2020) also proposed that sexual maturation could be inhibited by fin-clipping, something not generally observed or reported as a problem in post-clipped fish.

The application of an analgesia post-clip is a good solution to minimizing adverse effects and responses, and is beginning to be adopted by the community (see Chapter 13 for more information). Indeed, analgesia is considered ethical and necessary for fin-clipped adults (Deakin et al., 2019a) and is now embedded in UK project applications.

Additional considerations need to be made to reduce the impact of handling the fish. Netting is the stress inducer that cannot be prevented, and thus fish will show signs of stress as a result of being moved. This cannot be fully mitigated, but all methods need to prevent repeated netting of the fish, which leads to stress following gasping events (Tilley et al., 2020). Analgesia administration should not include additional netting events and, with a specially designed genotyping rack, this should not be a problem. The design is such that isolated fish never leave their tank or exit the water. If a genotyping rack is not available, protocols with no additional netting events are available (Schroeder et al., 2021; see Chapter 13 for more information). It should be noted, however, that we simply do not know the extent to which fish are impacted by being netted out shortly with gasping, or the level of discomfort post-clip. The question remains: are analgesia protocols where fish are netted out, which creates stress, still beneficial? This requires further investigation.

Research models of disease, such as the zebrafish, emulate the characteristics of the illness, and deleterious phenotypes are emerging that do not tolerate anaesthesia and invasive techniques. Two examples of this have emerged from the Noel group (E. Noel, Sheffield, 2019, personal communication), who are working on heart morphogenesis during zebrafish development and congenital heart disease. The Noel group has developed two new lines, which are currently unpublished and are thus named genes A and B here. Both gene A and gene B were raised to adulthood in a shoal of genetically mixed fish (including het, hom and wild-type siblings). The aim was to identify and segregate the hom fish by genotyping with fin-clipping. Gene A fish were raised to adulthood and looked normal, although the occasional fish would temporarily stop moving and drift. Some fish remained undersized at 1 month and were split out into a second tank to prevent them being outcompeted for food by larger fish. Fin-clips occurred on these fish one at a time and the fish were observed to recover. After one hour a number of fish had started to spin or stop moving and had to be euthanized. Post-mortem genotyping revealed that only hom GAA fish were affected. The working theory with this line is that the adverse effects were due to anaesthesia-induced arrhythmia, which is known to occur in susceptible humans with these genotypes (Figure 10.3; E. Pollitt and E. Noel, Sheffield, 2019, personal communication). Gene B is known to affect heart development in zebrafish, and adult hom fish were indistinguishable from het and wild-type siblings. Following fin-clipping, some fish were observed to have bleeding from their gills during anaesthesia recovery (this is a known but rare adverse event). These fish also took much longer than normal to recover from anaesthesia, requiring minutes rather than seconds until they resumed movement. Similarly, only hom fish were affected with an undefined cause, possibly low blood pressure or poor gill integrity (Figure 10.3; E. Pollitt and E. Noel, Sheffield, 2019, personal

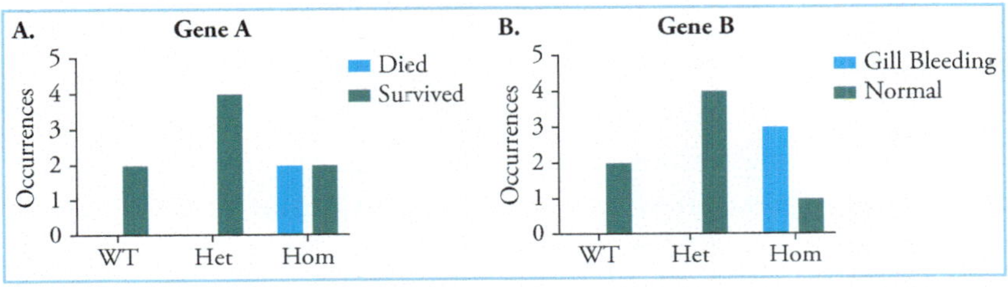

Figure 10.3 Adverse effects during and after fin-clipping. (A) Gene A; anaesthesia is thought to induce arrhythmia, which occurs in susceptible humans. (B) Gene B; undefined cause, possibly low blood pressure/poor gill integrity. Credit: Pollitt and Noel, personal communication, 2019.

communication). For both genes, embryonic genotyping resolved these issues and enabled the raising of hom tanks in isolation.

Fin amputation, according to the *Zebrafish Book*, briefly involves anaesthesia to restrain the fish, resting the fish on a small spoon and clipping around 50% of the caudal fin with scissors (Westerfield et al., 2007). It reports that further fin removal can jeopardize regeneration success. It also reports that regeneration is not guaranteed if the anal and dorsal fins are amputated and the pectoral and ventral fins may regenerate well but are logistically harder to clip. No alternative to fin-clipping is mentioned in the *Zebrafish Book* (Westerfield et al., 2007), which perhaps hinders global adoption of refinements that are now well known within the community. If we look at the *Journal of Visualised Experiment*, their fin-clip is done with a sterile razor blade on absorbent tissue and approximately 50% of the caudal fin is removed (https://www.jove.com/v/20184/dna-extraction-from-tail-clip-a-method-used-in-zebrafish-genotyping; retrieved 5 April 2022). Both of these protocols suggest that 50% of the tail is a popular proportion to take (Figure 10.4); this is perhaps too much, however, since the genetic material extracted from a 5–10% clip is known to give sufficient DNA for genotyping of multiple alleles. Indeed even embryonic clipping is known to produce more than adequate DNA for multiple allelic analyses (Wilkinson et al., 2013). Clipping of <10% of the whole fin may only be a partial amputation but it reduces regeneration time and does not cut through as many anatomical structures in the tail as does a 50% clip (Figure 10.4). If we make a simple visual comparison of the behaviour of fish with large clips (>50%) with those having had around 1/10 to 1/15 taken, we see more favourable behaviour and minimal adverse effects with the latter (personal observation). However, only detailed behavioural analysis would confirm this observation. Unfortunately, armed with the knowledge that regeneration occurs posterior to the point where the scales end, researchers have been known to carry out total fin amputations causing visible bleeding from the wound. This is extremely likely to cause pain, increases the chances of infection and reduces the ability of the fish to regenerate the fin and ultimately recover. An anti-vivisection social media post reported 50 zebrafish (in a

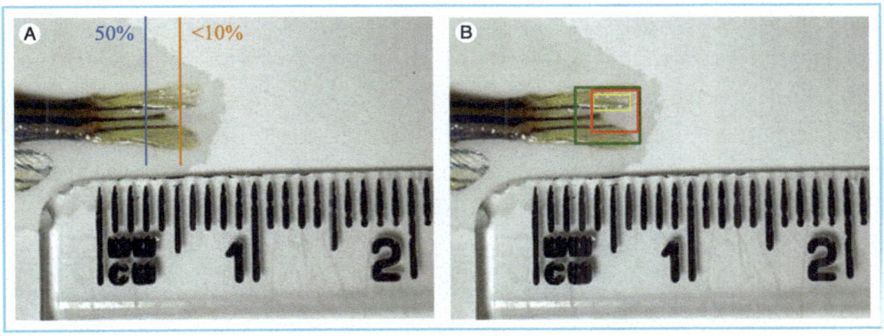

Figure 10.4 (A) Caudal fin-clip options: blue line, 50% cut line; orange line, <10% cut line. (B) Red box, 3 × 3 mm^2 or 40% of caudal fin; green box, estimated 50% area; yellow box, approximately 10% at the tip of the fin.

research setting) that had to be culled due to excessive caudal fin removal; unsurprisingly this was not seen as satisfactory treatment of zebrafish in the hands of researchers. More considerate, or at least knowledgeable and fully trained researchers, with thousands of fish that need clipping could choose to create a library in longfin (TL) wild-type strains. This strain has additional caudal fin tissue and so clipping can be reduced to <5%, which potentially reduces the impact on their ability to swim post-clip. We can only hypothesize that this reduces adverse effects.

Transection of the caudal fin requires a fresh sterile razor blade that is both clean and sharp. Scissors have a more shearing type of action that creates adverse tissue damage. The cutting implement needs to be cleaned with water to remove debris and then disinfected in alcohol between each clip (taking care that all the ethanol has evaporated before use on the fin). Alternatively, a fresh razor blade could be used each time but that is rather wasteful. The anaesthetic agent must be optimized to prevent deep and irreversible anaesthesia, and fish should be observed going to sleep on an individual basis. Once the operator feels confident, small groups of fish can be anaesthetized and recovered in one set. Pretreatment with analgesics is also becoming a popular step in the protocols of several facilities. Considerations to bear in mind with anaesthetic agents are their health and safety implications for the operator and reproducibility and adverse effects for the fish, which can last for days post-anaesthesia (see Chapter 13 for more information). Once anaesthetized, the operator must work fast on the fish. To prevent cross-contamination the fish should be positioned and/or placed on a sterile surface that will be replaced, or cleaned between each fish. Parafilm, or a clean tissue, are both ideal and easily obtainable but neither minimizes the amount of time spent by the fish out of water. To reduce the time the fish is exposed to air and gasping we designed a specific 3D-printed mould that allows water to cover the gills while the tail remains beyond the end of the mould (Figure 10.5; instructions on how to 3D print a 'fish cradle' are available on request in standard format (stl)). This helps with the controlled clipping of tiny quantities of caudal fin. Sterile fine tweezers can help to transfer the fin from the cutting implement into the sterile test tube (usually an Eppendorf or multi-well plate). Pigmented fish allow visualization of the clip on a surface or instrument, reducing loss of samples. This can also be

Figure 10.5 A 3D printed mould 'fish cradle' used to enable fin-clipping <10% of the caudal fin while keeping the gills submerged.

confirmed visually in the test tube holding the clips. Extra care must be taken when fin-clipping pigmentless zebrafish, since the fin will be less visible to the naked eye.

The use of multi-well plates, such as a 96-well plate (any higher and the fin will not fit and/or is difficult to insert), and scaling up the liquid-handling options means that higher-throughput fin-clipping can be achieved. Time must be spent understanding the 96-well plate coordinates and the locations of the fish in their recovery tanks so that mixing of samples is prevented. Isolation of fish for recovery and subsequent individual identification, while genotyping occurs, is essential and is discussed later in more detail. Labelling both the location of each fin and the original fish in a systematic way is paramount. To complement the 96-well format, manufacturers have designed both fish grids and genotyping tanks. These grids have small compartments each with a space of around 200 ml for the fish (Figure 10.6), and the genotyping tanks are designed to fit onto specialist racks (not shown). These all have coordinates that readily relate back to multi-well formats and house the fish within transparent sides. The genotyping racks are designed for large-scale fin-clipping. One design has trays of 24 fish with readily removable tanks, dividers and lids. These are all coded with a unique letter or number to allow rapid processing of

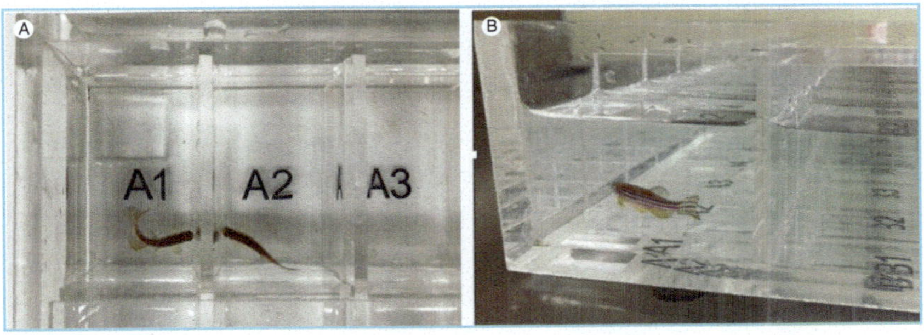

Figure 10.6 Fish in a fish grid with 48 compartments labelled to coordinate with a 96-well plate. Fish have at least 200 ml of water, which should be refreshed every 24 hours if fish are fed, or every 48 hours if withholding of food is permitted. (A) Two fish interacting across a transparent barrier. (B) A fish viewed from the side of the grid.

fish and to facilitate addition of analgesia to the water for the first few hours post-clip. The temptation to put pairs of fish together should be avoided (White et al., 2017). It is easy to believe that space can be saved and that planning to divide the sexes can be achieved once you have identified their genotypes. Unfortunately, pairs quickly form aggressive behaviours as a consequence of dominance hierarchy and extreme violence can lead to deterioration in health, breeding potential and, at the extreme end of the spectrum, death.

Although fin-clipping does not generally reduce survival or growth rate (Eriksen et al., 2011), variability in techniques, operators, environmental conditions and the health status of a facility could all contribute towards adverse effects for the individual fish. Studies show variability, presumably as a consequence of these factors, but it seems increasingly clear that we cannot ignore the adverse effects resulting from this method (Sneddon, 2021).

Methods used for adults: extraction of DNA from scales

The sampling of scales for several fish species, whereby detached scales are collected for DNA extraction, is described as a non-invasive, gentle method (Kumar et al., 2007; Wasko et al., 2003; Yue & Orban, 2001). This is highly debatable, and we hypothesized that there are adverse effects associated with scale removal even though it does not affect their survival or growth (Sire et al., 2000). Consider this: the teleost scale (elasmoid scale) is a highly regenerative, membranous bone (dermal bone) embedded in the skin. The scales are attached to the body and their removal leaves a scale pocket in the dermal cells with removal of the outermost layer of the epidermis (Iwasaki et al., 2018). Thus it is highly likely that scale removal will be adverse and, as such, is already a regulated procedure in many countries.

The zebrafish has a distinct set of several hundred scales distributed over the body, which develop and regenerate in controlled orientations (Iwasaki et al., 2018). Adult laboratory zebrafish are found to have approximately 7.5% regenerated scales (when housed on a sand substrate), which indicates that some natural loss does occur, presumably through their environment, netting and aggression (Sire et al., 2000). Regeneration occurs in approximately 2 weeks (Mariotti et al., 2015) and regenerating scales are morphologically distinct from the originals (Sire et al., 2000). Although removal is often reported to be carried out with forceps, the use of a glass slide to obtain loose scales by gently stroking the body surface is generally considered less traumatic. Scales are translucent and 1–2 mm in diameter, so removal is technically demanding and must be done under a microscope. Four zebrafish scales are found to yield sufficient DNA for analysis (Yue & Orban, 2001).

Scale extraction has not been widely adopted by the zebrafish community, presumably due to the extraction procedure, which is more onerous than fin-clipping. Veterinary opinion suggests that scale removal leaves the fish prone to infection, with more clinical significance than a fin-clip, because the scale is far closer to internal vital organs than the caudal fin (J. P. Mocho, London, 2021, personal communication).

Methods used for adults: swabbing

The method of swabbing zebrafish to harvest DNA from mucous membrane was first published by Breacker et al. (2017). This protocol has great potential as an alternative and refinement to fin-clipping, particularly because it can be done without anaesthesia and is considered non-invasive. The method requires that adult zebrafish (approximately or greater than 2 cm in body length) be netted and restrained in order to rub a sterile rayon-tipped swab over the scales from 'head to toe' (operculum to caudal fin), ensuring no touching of the eyes. This mucus-harvesting swab is repeated five times along the same direction to prevent disruption to the scales. The published protocol does not indicate whether both sides of the body can be swabbed but the authors suggest that, if consistency with one side is upheld for all swabs, the un-swabbed side is left as a repeat swab option (C. Tilley, Leicester, 2021, personal communication). The swab reportedly becomes covered in sufficient DNA from surface mucus and sloughed epithelial cells for extraction with commonly and widely used molecular biology methods. Breacker et al. (2017) did not anaesthetize the swabbed fish, which was deemed to benefit them. Interestingly, when some facilities have tried swabbing with anaesthesia this seems to interfere with the protocol, rendering it ineffective (C. Tilley, Leicester, 2021 personal communication). The reasons for this problem are unknown and probably circumstantial rather than directly related. Tilley et al. (2020) confirmed the potential of this method as a refinement through a comprehensive analysis of operculum beat measurements, cortisol measurements, mRNA profiles and behavioural tests. Needless to say, swabbing does cause some long-term behavioural changes (analysed at 1 month) but these are less averse when compared with fin-clipping. Indeed the need to net the fish and expose them to air during such procedures triggers stress responses (Tilley et al., 2020). A decision must be made regarding the impact of anaesthesia, which helps to restrain the fish, versus the conscious experience of swabbing with restraint and potential gasping. At present the general uptake of swabbing across the zebrafish research community is below expectations, probably due to technical issues not in the swabbing method itself, but rather in the biological assays and DNA analysis that follow. Of note, Tilley et al. (2020) did not extract DNA from the swabs as part of their work to confirm this method as a refinement, which is unfortunate since some facilities, including my own (L. Van Hateren, Sheffield, 2023, personal communication), report low DNA yields.

When we trialled this technique in house we proposed swabbing with the head of the fish submerged in system water to prevent suffocation in air (Figure 10.7A,B). This requires the use of anaesthesia to prevent failures in manual handling of the fish and induction of stress responses. In actual fact, once the fish is restrained in the wet net with its eyes covered, it is very easy to obtain a swab. Yield is a problem, and it would also be interesting to see whether swabbing from both sides of the fish improves DNA yield without impacting health. We also propose that the addition of a manual scraping step, along the mucus-loaded swab into a collection tube

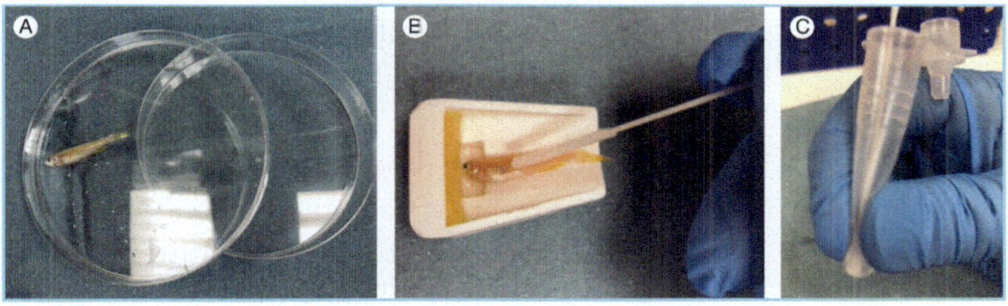

Figure 10.7 (A) Petri dish on a slant to hold fish with gills submerged. (B) 3D-printed fish cradle to hold fish with gills submerged while swabbing is in progress. (C) Scraping the swab into a soft-walled test tube.

(Figure 10.7C), will improve extraction from the surface of the swab. Several publications have now addressed the issue of DNA yield. Venta et al. (2020) successfully used swabbing and extracted DNA with a 'hot-shot' procedure (Meeker et al., 2007) for DNA profiling, and Jang et al. (2022) performed DNA extraction immediately following a caudal fin-swabbing event (the trunk of the fish was avoided to prevent damage to internal organs). This second method incorporated a DNA collection buffer containing proteinase K, thereby improving DNA yield. Additional steps to increase DNA yield have been devised whereby the loaded swabs are put directly on ice in a sodium hydroxide solution until DNA extraction (N. Shriever, K. Camargo Sosa and R. Kelsh, Bath, 2023, personal communication).

It seems that the initial issues and inconsistency reported by the zebrafish community establishing this as a mainstay genotype method may possibly have been resolved. It remains to be seen whether every allele and, just as importantly, multiple alleles, can be genotyped using this refined method.

Methods used for adults: gamete extraction for specific genetic analysis

Certain research scenarios may require the maternal or paternal genotype, in which case the parent fish can be squeezed for their gametes

(Hagedorn & Carter, 2011; Wasden et al., 2017). Samples can be processed for DNA analysis, and information on the genotype of the parentage will be revealed. Meeker et al. (2007) demonstrated this method with sperm.

Methods used for adults: conclusion

There is a significant caveat with adult genotyping, namely the cost and unnecessarily prolonged life of GAA in research facilities. We can refine adult methods and carefully manage the adverse effects as we discover more about the experiences of zebrafish pre- and post-procedure. By continuing to use a method for adults, however, we fail to address the reduction criteria of the 3Rs. We predict a future where adult fin-clipping has become a less popular technique due to advances in other methods, which will also work out cheaper and more suitable for high-throughput genotyping. This will require a concerted effort to change general perceptions and processes. There is great potential in genotyping embryos which, of course, are not protected, since they are not deemed to experience pain, suffering, distress or lasting harm. These methods will now be discussed.

Methods used for embryos: reporter lines

As mentioned above, the simplest genotyping method of embryos is by sight, where a GAA phenotype or reporter is visible microscopically (see Chapter 11 for more information). Subtleties in individual lines may take time to identify, depending on the detection level required and training in scope use. Because fluorescent reporter lines are often localized and transient, although the researcher will have expectations in regard to the structures and stages with expression, unusual patterns are not uncommon. It is, therefore, advised that a thorough developmental assessment is conducted, which may require dechorination, pigment removal with phenylthiourea (PTU) and anaesthesia to aid visualization of all structures (again see Chapter 11 for more information). Mendelian inheritance can assist with the expected number of GAA individuals to be identified. This can become complicated though; for example, crossing three different het GAA (one double het parent (two alleles) and one single het parent (one allele) yields 8 genetic variations of offspring). Therefore, if the numbers do not tally, parents require either retesting with other potential carriers or direct genotyping themselves. These methods are primarily embryonic and no DNA analysis is required, simply a microscope. However, the parents do need to be housed separately until the embryos' inherited genetic modifications are determined.

Methods used for embryos: embryo sampling

Embryo DNA extraction, whereby you are able to raise the fish with the correct genotype without causing them damage, incurring complex care procedures or demanding high research costs, is perhaps the perfect genotyping method. Utilization of embryos with self-sufficiency in their yolk stores and their microscopic size are two valuable attributes in embryonic genotyping. Embryonic methods do require liquid handling and microscope work to extract samples, and some of these requirements slow the process down and are not totally optimal for high-throughput demands.

Genetic analysis on the embryonic zebrafish is limited only by the quantity and quality of the DNA extraction. Using the whole embryo was first suggested by Kawakami and Hopkins (1996), who were aiming to identify a transgene in a first generation of fish. Embryos can be readily harvested and processed for DNA extraction and analysis (Meeker et al., 2007; Parant et al., 2009). Breeding by pair-mating of two parent fish with unknown genetic modifications, followed by sampling a proportion of their progeny, can identify known aspects of the genetic modification possessed by the parents. This can be a complicated process if the parents have different or multiple genetic components. Breeding by pair-mating of two parent fish where one has a known genotype simplifies the identification of the desirable genotype in the target fish but requires more pairings. This method may be useful if the parents have already been genotyped by another method and regulations prevent additional invasive procedures. For instance, when contamination of a particular genotyped line occurs, caused by mistakes in the mixing of fish, you can deduce whether particular alleles are present or absent. This is an option that should be considered in certain research scenarios where more invasive methods are detrimental to the fish (Figure 10.3).

Whole-embryonic genotyping is also utilized in clutch verification of newly injected embryos, whereby a proportion of the fish are sacrificed (e.g. 10%) for genomic evaluation. This confirms whether integration of the genetic modification into the genome has been successful. If the clutch is identified as having the correct genome edit/s, they are then grown up (G0/F0 generation) and the next generation should be analysed for germline verification. Because each fish is a potential founder (F1 generation), it is out-crossed to known genotypes and the offspring can be assessed for integration into the germline and subsequent family breeding. Molecular biologists are starting to streamline this process because of the demand for high-throughput, null-mutation analysis whereby multiple potential genetic modifications are designed, injected, assessed and categorized for further research endeavours. Advances in CRISPR-Cas9 in combination with duplex guide RNA (dgRNA) are now demonstrating the potential to generate G0/F0 embryos that are no longer mosaic and inconsistent in their integration but where all cells are shown to carry the desirable mutations (Hoshijima et al., 2019). Rapid screening of novel and true protein-null mutations in G0/F0 eliminates the need to raise lines lacking a phenotype that can be studied. This development is good news in regard to compliance with the 3Rs, reducing the number of fish bred and refining their genotyping experience. Genotyping bottlenecks are removed and screening of embryos pinpoints significant avenues of interest for further research. Subsequent generation genotyping, however, will revert back to the usual methods and considerations.

Methods used for embryos: embryo genotyping

Early attempts of embryo sampling without sacrificing the embryo utilized the protein-digesting enzyme proteinase K (Kawakami & Hopkins, 1996). Unfortunately, in their pursuit of obtaining sufficient usable genomic DNA, survival suffered whereby at 60 dpf only 59–87% of the fish remained. This level of survival is unacceptable, for both ethical and efficiency reasons. Historically the use of protein enzymes has been a risky business, marred by failure and concern over the welfare of enzyme recipients. One recent publication has carefully optimized the use of protein enzymes (Zhang et al., 2020) with >90% survival. Their comparison of pronase and proteinase K showed the increased viability of the latter, and consequently Zhang et al. (2020) optimized the protocol for proteinase K. Sensitivity was further developed by optimizing the buffer, shaking frequency and timings for extraction to release sufficient DNA from the skin of the embryo without long-term damage to the fish. This well-considered study went on to demonstrate that there were no significant adverse effects associated with the method. It instils confidence in the use of proteinase K as a robust method for high-throughput embryo genotyping. In-house optimization of this method is still advised, and careful analysis of fish well-being following treatment is always recommended.

Methods used for embryos: embryo clipping

Embryonic clipping methods are described as high-throughput methods of genotyping zebrafish (Kosuta et al., 2018; Wilkinson et al., 2013; Xing et al., 2014), but whether they are or not is totally dependent on your opinion of what high-throughput should be. Microscopic fin-clips of the developmental caudal fin are taken from embryos between 72 and 120 hpf, before fish are considered to be able to feel pain. It is deemed possible to clip from 48 hpf, but these embryos are smaller and therefore harder to sample and more inaccessible because they are still living within their chorion. From 72 hpf the fish has dechorionated itself naturally and so manual removal of the chorion, which adds technical steps to the protocol, is not required. Digestion of the chorion by pronase might be considered as obviating the need for dissection with tweezers, but pronase also digests embryos if left to work for too long. Pronase is known to wound weakened embryos and left-over traces of pronase may impair healing, which could lead to death or poor development. In such cases careful washing and monitoring is essential (F. Van Eeden, Sheffield, 2021, personal communication).

Embryonic fin-clipping must be performed with anaesthesia and under a dissecting microscope. DNA extraction and genotype identification are performed while the embryos recover in a multi-well plate. Each well must hold at least 200 microlitres of medium for each embryo, plus an air gap for oxygen exchange. Embryos can be kept for up to 2 days in these vessels, after which media changes are recommended if the genotyping processes are not completed (Kosuka et al., 2018). This method takes advantage of the yolk sac used by each embryo to supply nourishment, which also means that the water quality parameters within individual containers can be easily maintained. The addition of food would have a negative impact upon health and well-being because it pollutes with water. Care needs to be taken when handling the embryos, and technical dexterity minimizes loss. The anaesthetic is essential because it helps to immobilize the fish for dissection. Once the embryo is aligned and clearly in focus under the light microscope, the fin can be gently cut within the pigment gap (Wilkinson et al., 2013). Cutting in this specific site does not damage the notochord and is distal to any locally developing blood vessels. Kosuka et al. (2018) report that, when genotyping pigmented embryos, the clips are visible to the naked eye as small black dots, which helps sample processing. The use of TWEEN®20 on the tweezers, used to pick up the cut fin, prevents it from sticking and helps with transfer into the sample tube. Care must be taken, therefore, with embryos lacking pigment because the pigment gap is not present and the clips will not be visible once removed. In this situation a very conservative clip must be taken by a researcher trained to competency on pigmented embryo clipping. Regeneration of the fin begins immediately after with the formation of a blastema, and there is no evidence that this alters their development into adults. The amount of DNA harvested with this method is calculated to be sufficient to genotype as many at 1,800 alleles, which is adequate for most applications (Wilkinson et al., 2013). This method is initially laborious but, with training and practice, speeds of 96 clips in 2 hours can be achieved.

Methods used for embryos: microfluidic techniques to assist embryo genotyping

Microfluidic devices have been widely developed to assist numerous fields of zebrafish research. Two such devices have been developed to aid the genotyping of zebrafish embryos: the chorionic fluid chip and the fin-clip chip. The former involves controlled digestion of the chorion of 1–2 dpf embryos with pronase followed by pulsing pressure in a specifically designed disposable tube that culminates in the release of both an unharmed embryo and chorionic fluid. Interestingly, chorionic fluid only contains DNA from the embryo rather than maternal DNA, which could be important in certain research endeavours. Viability was 100% with this method but genotyping success was only 78% with HRMA, which indicates that normal PCR product analysis by gel electrophoresis would have lower success rates. The second device, the fin-clip chip, was designed to utilize the reflex motor response to optimally position the embryo for clipping. This means that anaesthesia is not used, which could be an advantage in certain research endeavours (Figure 10.3). Both processes are somewhat complicated but were developed as potentially harmless methods for high-throughput genotyping of embryos. Samuel et al. (2015) reported that, although genetic material was obtainable, it was variable, inconsistent and required pushing the PCR to limits that introduced false positives. It does not appear that either method has been adopted or adapted further, although Lambert et al. (2018) investigated the potential of chorionic fluid when developing a new embryo-shaking technique.

Methods used for embryos: embryo shaking

Two recently published methods involve shaking embryos to release DNA. Both have been mentioned previously, one using proteinase K (Zhang et al., 2020) and the other utilizing a rough-surfaced slide (Lambert et al., 2018). The second is now commercially available and is called the zebrafish embryo genotyper (ZEG). The ZEG has been developed to utilize microfluidic harmonic oscillations in combination with individual compartmentalized roughened wells on a glass slide. Each slide is covered in hydrophobic tape with 24 cut-out circles to separate samples. The hydrophobic tape creates an open-top droplet system for easy loading of 3 dpf dechorionated embryos. Again there is an advantage in waiting for the fish to naturally dechorionate (>72 hpf), because it requires less work and is more high throughput. Each embryo is bathed in around 10 µl of E3 medium, mounted on the ZEG device and vibrated at 1.4 V for 10 min. The glass slides are single use, to prevent DNA contamination and, when tested, no evidence was found of cross-contamination in the design of the wells. Care needs to be taken when the slides are out of date because the hydrophobic tape detaches and allows cross-contamination between samples.

The ZEG offers great potential for higher-throughput embryonic DNA extraction, although it does have cost implications: investment is required in the equipment and ongoing supply of the unique slides. Our rough calculations estimate that the ZEG is 64% cheaper over 3 months for 24 fish compared with fin-clipping as adults. Using the ZEG for genotyping is more laborious than indicated by the initial publications, with each allele requiring careful testing for efficacy. The DNA extracted currently enables the genotyping of only one allele. This limitation is frustrating, particularly since so many lines now have multiple genetic modifications (Figure 10.1B). This is a new technology, and teething problems are being addressed by the manufacturers so that larger volumes of DNA-rich extract are obtainable.

Concerns over the disruption to structural integrity and stress responses through vigorous shaking of embryos on the ZEG have begun to be discussed within the zebrafish community. The worry is that, because genetic modifications cause deleterious effects, shaking of these embryos might impact and change both cellular and molecular processes. With the availability of the ZEG, we and others (Douek et al., 2021) envisaged researchers 'Zegging' fish with phenotypes, carrying on with their research and having skewed results because of the impact of the ZEG. Our concerns focused on the impact this has on the developing heart and we then investigated the Southpaw (Spaw) line (C. Derrick and E. Noel, Sheffield, 2019, personal communication). Spaw is crucial for both early and late aspects of cardiac left–right asymmetry (Long et al., 2003). Genotyping Spaw embryos with the ZEG made no discernible difference to hom embryos (Figure 10.8) and the fish were genotyped without issue. Fish were subsequently raised and appeared as normal adults. At the same time another group, working on muscle tissue, began to investigate the structures that they felt could be affected by the ZEG process, namely skeletal muscle, spinal cord and motor neurons (Douek et al., 2021). They found that there

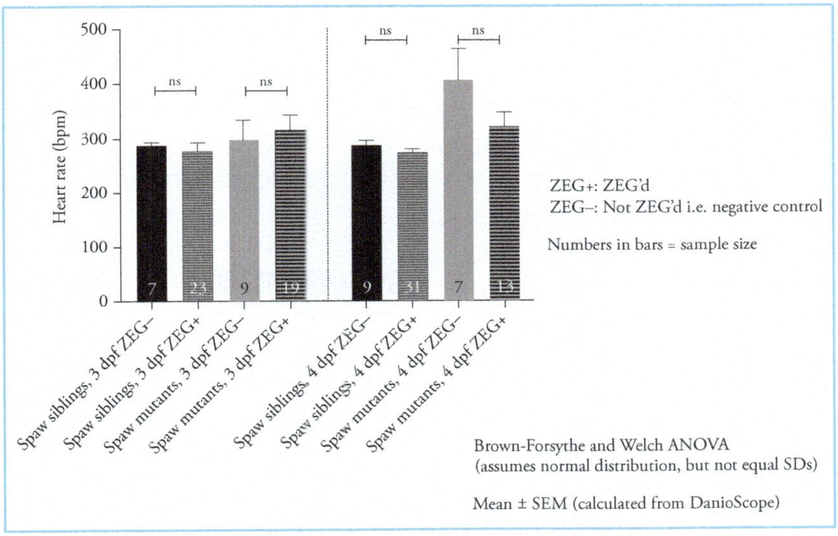

Figure 10.8 Impact of 'Zegging' on the Spaw embryo heartbeat, as monitored by DanioScope. *x* axis represents the genotype and age of embryos post ZEG; *y* axis represents heart rate (bpm). Ns, not significant. Credit: C. Derrick, E. Noel, Sheffield, 2019.

was a low-level, acute stress response but no other long-term effects were observed, concluding that post-ZEG investigations should wait >24 hours after the procedure to allow the embryos time to recover.

Methods used for embryos: conclusion

The potential to genotype embryos is as old as the zebrafish model, and the ability to genotype them has evolved alongside our knowledge of genomics. Embryos are not deemed to feel pain under the point of protection (5 dpf at 28.0 ± 0.5 °C) because mechanisms of response are not developed. There is, however, a very fine time line in regard to their ability to perceive adverse events. Remember that this species develops rapidly, an advantage exploited for the benefit of research. Mechanisms of pain response are evident in larvae around the point of protection, because they respond to a noxious challenge in a similar way to adult zebrafish and which can be alleviated by analgesia (Lopez-Luna et al., 2017). Wherever possible it is advisable to work well below this point when genotyping embryos, partly to avoid adverse effects but also because the fish are ready to be fed from 5 dpf (at 28.0 ± 0.5 °C). If there are delays in procedures, the fish will have to be fed. Feeding embryos in multi-well plates is not only extremely technically challenging but also detrimental to the fish, due to subsequent water quality deterioration.

The zebrafish embryo contains sufficient DNA for analysis and, if harvested, can reveal the genotype. The first challenge in embryo genotyping is being able to extract sufficient DNA, without detrimental effects, while allowing the fish to be raised. Whether sampling, clipping, shaking or treating the embryo with enzyme, the technique you choose requires optimization within your facility for your specific lines. The second challenge lies in the molecular biology assays that must be employed when the DNA sample is low both qualitatively and quantitatively.

Solitary confinement

Once genetic samples have been harvested from the fish and, in the absence of any other means of identification, each fish must be kept in isolation (usually a recovery tank for adults or multi-well plate for embryos). The time spent in solitary confinement should be minimized and is dependent on the results of DNA analysis. We know that prolonged isolation has a negative effect on fish (Anneser et al., 2020; Collymore et al., 2015; Fulcher et al., 2017; White et al., 2017). We also know that these detrimental effects can be alleviated quickly if the fish are returned to their original and familiar shoal. Currently, the best scenario for the isolated fish is to be able to see shoal mates through tank walls or dividers (Figure 10.7B).

Genotyped fish with matching, or at least known, genetic modifications can then be housed separately from non-carrier siblings as a genetically distinct population, ideally with mixed sexes. If the genotyped fish is one of a kind, and this is often the founder fish, then it must be outcrossed to secure the line before a decision is made about its housing situation. Unique individuals should, wherever possible, be given a companion shoal of visually distinct fish, preferable of a similar age, to create a small group with equal sex ratios. For instance, if the individual is a striped strain, house this fish with similarly sized males and females that lack pigment or have leopard spots. This allows visual identification of the founder at a later stage. The only caveat is that preferences exist for similarly pigmented conspecifics (Engeszer et al., 2007), so a watchful eye is required to detect hierarchical relationships and aggressive behaviours developing between fish. In certain circumstances visible implant elastomer (VIE) tagging could be utilized if a distinct genotype is required to live in a visibly similar shoal (Rácz et al., 2021).

Because optimal husbandry parameters are an ethical expectation for isolated fish, daily well-being checks and water quality management are required, regardless of age. The larger the volume of water, the more time is available before it is polluted: that is, oxygen levels are too low, pH varies by too much or ammonia or nitrite is present at too high a level (see Chapter 7 for more information). Usually, five adult fish per litre can remain out of recirculation for at least

48 hours before variation in water parameters becomes significant. Water is therefore changed every 48 hours for fish isolated in breeding tanks housing fish in standing water. One advantage of the genotyping rack is that water is shared, filtered and recirculated. This provides extra interaction between fish and allows easier care for isolated fish, which can then be fed daily with little risk. It is, however, important to remember that fish need sufficient space to move around while isolated. This space is often restricted within a genotyping rack but not when fish are singly housed in breeding tanks (off the system). Fish in breeding tanks can also be fed at a reduced ratio to avoid water pollution, but this can be technically challenging. All of these considerations are harder to assess and manage with embryos in small multi-well plates and so quicker turnarounds, preferably within a couple of days, are recommended.

Signs of deterioration relating to being in solitary confinement must be monitored, so regular checks of isolated fish are paramount. Long-term health is impacted more visibly for females than males, because females become egg-bound (see Chapter 4 for more information). Egg-binding in females arises when a female becomes unable to reabsorb her eggs due to the prolonged absence of breeding males (Nasiadka & Clark, 2012). A maximum of 1 month of solitude is recommended for adult female zebrafish (Niimi & LaHam, 1974), which gives ample time to genotype the fish by any method. Intervention with egg-bound females is possible and the fish may be gently 'squeezed', following a mock breeding to stimulate her to release her eggs. Anaesthesia is advised to aid gentle and controlled squeezing in the early stages of a female being egg-bound. There are no guarantees of success, however, and anaesthesia may be risky for severely egg-bound females. Squeezing without anaesthesia is possible if the female is extremely egg-bound and it must be done swiftly and gently, by restraint in a net, but is deemed to be stressful. Excess eggs may be degraded and, once removed, the body cavity becomes slackened. Following a period of rest, young, healthy, robust females should begin to breed again successfully. If an egg-bound female is left unmanaged this can become an irreversible state, rendering the fish unusable for ongoing research.

Conclusion

Genotyping your GAA zebrafish is a requirement for line management. The genotyping choices that you make in your facility will vary according to expertise and resources (Figure 10.9). All procedures, including genotyping methods, are performed more efficiently if the operator is trained to competency by proficient trainers. A directly observed practical skills (DOPS) document now exists for fin-clipping of zebrafish and is available online (https://etplas.eu/wp-content/uploads/2022/01/ETPLAS-DOPS-Model-1-Zebrafish-fin-clipping.pdf). All training should include ethical considerations and a clear understanding that working with animals in research is a privilege and not a right. If we reflect upon all the methods that are available, there are clear choices to both reduce numbers of fish raised and refine historical protocols. Change in management may be the biggest challenge in the adoption of alternative methods, but change we shall. There is no reasonable excuse for saying that, because a zebrafish is a lower neurophysiological model than other vertebrates used in research, it does not feel. Research has proven its worth time and again, so we cannot conveniently ignore the evidence that fin-clipping and solitary confinement, for instance, cause adverse effects. Even if changing from the classic and enduring fin-clipping method is met with resistance, consideration of the amount of fin taken (recommended <10%) and application of analgesia are good initial refinements. Legal frameworks should reflect these ethical considerations, by not being specieist against the needs of fish and giving tailored guidance on their preferences. Needless to say, science is categorically better if the health and welfare of the fish is upheld as a priority (see Chapter 7 for more information), and the successful uptake of the ARRIVE guidelines by >600 journals is evidence of the undeniable importance of this ideology (see Chapter 8 for more information).

We conclude that the future of genotyping zebrafish should focus on embryonic methods wherever possible. This aim completely removes the requirement for refinements to minimize pain, suffering, distress and lasting harm. It also reduces the number of zebrafish with undesirable genotypes required to be raised and maintained.

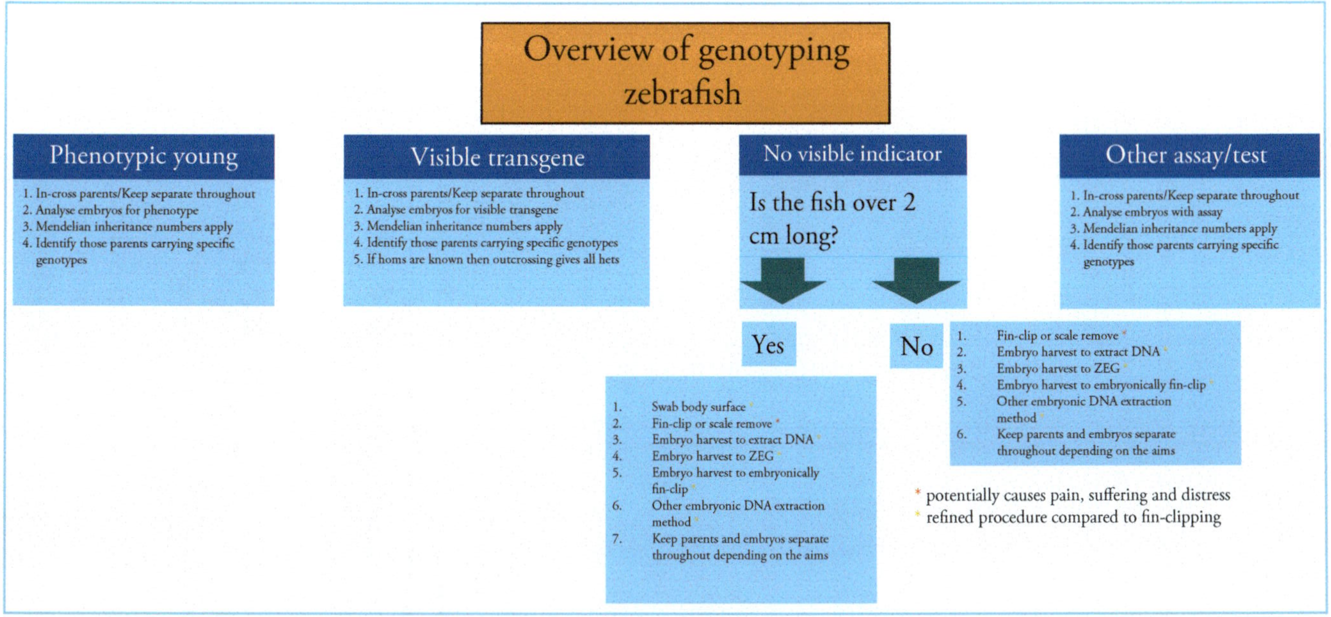

Figure 10.9 Overview of genotyping zebrafish.

Acknowledgements

I would like to thank the Biological Services Aquarium (BSA) team for all their hard work in managing the zebrafish facility at The University of Sheffield (TUOS), UK, particularly throughout the COVID pandemic of 2020–2021; with particular mention of Vickie Thomas-McArthur for her work on producing Figure 10.1, Yulia Nikolaeva on producing Figures 10.4–10.6 and Lisa Van Hateren for her dedicated work on refining the genotyping service at TUOS.

References

Akimenko, M. A., Marí-Beffa, M., Becerra, J., & Géraudie, J. (2003). Old questions, new tools, and some answers to the mystery of fin regeneration. *Developmental Dynamics, 226*(2), 190–201. https://doi.org/10.1002/dvdy.10248

Allen, C., Argenton, F., Van Eeden, F. J., & Tiso, N. (2021b). Transgenesis, mutagenesis, knockdown, and genetic colony management. In L. D'Angelo & P. de Girolamo (Eds.), *Laboratory fish in biomedical research* (pp. 139–152). Elsevier.

Allen, D., Rosenberg, M., & Hendel, A. (2021a). Using synthetically engineered guide RNAs to enhance CRISPR genome editing systems in mammalian cells. *Frontiers in Genome Editing, 2*, 35.

Anneser, L., Alcantara, I. C., Gemmer, A., Mirkes, K., Ryu, S., & Schuman, E. M. (2020). The neuropeptide Pth2 dynamically senses others via mechanosensation. *Nature, 588*(7839), 653–657. https://doi.org/10.1038/s41586-020-2988-z

Blaser, R., & Gerlai, R. (2006). Behavioral phenotyping in zebrafish: Comparison of three behavioral quantification methods. *Behavior Research Methods, 38*(3), 456–469. https://doi.org/10.3758/bf03192800

Breacker, C., Barber, I., Norton, W. H., McDearmid, J. R., & Tilley, C. A. (2017). A low-cost method of skin swabbing for the collection of DNA samples from small laboratory fish. *Zebrafish, 14*(1), 35–41. https://doi.org/10.1089/zeb.2016.1348

Cardeira, J., Gavaia, P. J., Fernández, I., Cengiz, I. F., Moreira-Silva, J., Oliveira, J. M., Reis, R. L., Cancela, M. L., & Laizé, V. (2016). Quantitative assessment of the regenerative and mineralogenic performances of the zebrafish caudal fin. *Scientific Reports, 6*, 39191. https://doi.org/10.1038/srep39191

Collymore, C., Tolwani, R. J., & Rasmussen, S. (2015). The behavioral effects of single housing and environmental enrichment on adult zebrafish (Danio rerio). *Journal of the American Association for Laboratory Animal Science, 54*(3), 280–285.

Deakin, A. G., Buckley, J., AlZu'bi, H. S., Cossins, A. R., Spencer, J. W., Al'Nuaimy, W., Young, I. S., Thomson, J. S., & Sneddon, L. U. (2019a). Automated monitoring of behaviour in zebrafish after invasive procedures. *Scientific Reports, 9*(1), 9042. https://doi.org/10.1038/s41598-019-45464-w

Deakin, A. G., Spencer, J. W., Cossins, A. R., Young, I. S., & Sneddon, L. U. (2019b). Welfare challenges influence the complexity of movement: Fractal analysis of behaviour in zebrafish. *Fishes, 4*(1), 8. https://doi.org/10.3390/fishes4010008

De Lombaert, M. C., Rick, E. L., Krugner-Higby, L. A., & Wolman, M. A. (2017). Behavioral characteristics of adult zebrafish (Danio rerio) after MS222 anesthesia for fin excision. *Journal of the American Association for Laboratory Animal Science, 56*(4), 377–381.

Douek, A. M., Klein, E. I., Kaslin, J., Currie, P. D., & Ruparelia, A. A. (2021). Cellular and molecular characterization of the effects of the zebrafish embryo Genotyper protocol. *Zebrafish, 18*(1), 92–95. https://doi.org/10.1089/zeb.2020.1958

Engeszer, R. E., Da Barbiano, L. A., Ryan, M. J., & Parichy, D. M. (2007). Timing and plasticity of shoaling behaviour in the zebrafish, Danio rerio. *Animal Behaviour, 74*(5), 1269–1275. https://doi.org/10.1016/j.anbehav.2007.01.032

Eriksen, T. B., Fraser, T., Gregersen, H., Kristiansen, M., Olufsen, M., Rajan, B., Ramirez, R., Ruangsri, J., Røn, J., & Sarmiento, M. G. (2011). *Should Fin Clipping Be Used as a Method for Identification of Fish?* http://www.norecopa.no Retrieved September 2021

Fulcher, N., Tran, S., Shams, S., Chatterjee, D., & Gerlai, R. (2017). Neurochemical and behavioral responses to unpredictable chronic mild stress following developmental isolation: The zebrafish as a model for major depression. *Zebrafish, 14*(1), 23–34. https://doi.org/10.1089/zeb.2016.1295

Gemberling, M., Bailey, T. J., Hyde, D. R., & Poss, K. D. (2013). The zebrafish as a model for complex tissue regeneration. *Trends in Genetics, 29*(11), 611–620. https://doi.org/10.1016/j.tig.2013.07.003

Haffter, P., Granato, M., Brand, M., Mullins, M. C., Hammerschmidt, M., Kane, D. A., Odenthal, J., van Eeden, F. J., Jiang, Y. J., Heisenberg, C. P., Kelsh, R. N., Furutani-Seiki, M., Vogelsang, E., Beuchle, D., Schach, U., Fabian, C., & Nüsslein-Volhard, C. (1996). The identification of genes with unique and essential functions in the development of the zebrafish, Danio rerio. *Development, 123*(1), 1–36. https://doi.org/10.1242/dev.123.1.1

Hagedorn, M., & Carter, V. L. (2011). Zebrafish reproduction: Revisiting in vitro fertilization to increase sperm cryopreservation success. *PloS One*. Public Library of Science One, *6*(6), e21059. https://doi.org/10.1371/journal.pone.0021059

He, C., Holme, J., & Anthony, J. (2014). SNP genotyping: The KASP assay. *Methods in Molecular Biology, 1145*, 75–86. https://doi.org/10.1007/978-1-4939-0446-4_7

Hoshijima, K., Jurynec, M. J., Klatt Shaw, D. K., Jacobi, A. M., Behlke, M. A., & Grunwald, D. J. (2019). Highly efficient CRISPR-Cas9-Based methods for generating deletion mutations and F0 embryos that lack gene function in zebrafish. *Developmental Cell, 51*(5), 645–657.e4. https://doi.org/10.1016/j.devcel.2019.10.004

Iwasaki, M., Kuroda, J., Kawakami, K., & Wada, H. (2018). Epidermal regulation of bone morphogenesis through the development and regeneration of osteoblasts in the zebrafish scale. *Developmental Biology, 437*(2), 105–119. https://doi.org/10.1016/j.ydbio.2018.03.005

Jang, J. Y., Liang, T., Kim, M. K., Kang, K. W., Lee, B., & Choi, S. Y. (2022). A rapid and noninvasive method that extracts polymerase chain reaction-ready genomic DNA from adult zebrafish. *Zebrafish, 19*(4), 160–164. https://doi.org/10.1089/zeb.2022.0015

Jaźwińska, A., & Sallin, P. (2016). Regeneration versus scarring in vertebrate appendages and heart. *Journal of Pathology, 238*(2), 233–246. https://doi.org/10.1002/path.4644

Kawakami, K., & Hopkins, N. (1996). Rapid identification of transgenic zebrafish. *Trends in Genetics, 12*(1), 9–10. https://doi.org/10.1016/s0168-9525(96)90083-9

Kosuta, C., Daniel, K., Johnstone, D. L., Mongeon, K., Ban, K., LeBlanc, S., MacLeod, S., Et-Tahiry, K., Ekker, M., MacKenzie, A., & Pena, I. (2018). High-throughput DNA Extraction and Genotyping of 3dpf Zebrafish Larvae by Fin Clipping. *Journal of Visualized Experiments*, 136.

Kumar, R., Singh, P. J., Nagpure, N. S., Kushwaha, B., Srivastava, S. K., & Lakra, W. S. (2007). A non-invasive technique for rapid extraction of

DNA from fish scales. *Indian Journal of Experimental Biology, 45*(11), 992–997.

Lambert, C. J., Freshner, B. C., Chung, A., Stevenson, T. J., Bowles, D. M., & Samuel, R. (2018). *Anautomated system for rapid cellular extraction from live zebrafish embryos and larvae development and application to genotyping* p. 13. Public Library of Science One.

Lidster, K., Readman, G. D., Prescott, M. J., & Owen, S. F. (2017). International survey on the use and welfare of zebrafish *Danio rerio* in research. *Journal of Fish Biology, 90*(5), 1891–1905. https://doi.org/10.1111/jfb.13278

Long, S., Ahmad, N., & Rebagliati, M. (2003). The zebrafish *nodal*-related gene *southpaw* is required for visceral and diencephalic left–right asymmetry. *Development, 130*(11), 2303–2316. https://doi.org/10.1242/dev.00436

Lopez-Luna, J., Al-Jubouri, Q., Al-Nuaimy, W., & Sneddon, L. U. (2017). Reduction in activity by noxious chemical stimulation is ameliorated by immersion in analgesic drugs in zebrafish. *Journal of Experimental Biology, 220*(8), 1451–1458. https://doi.org/10.1242/jeb.146969

Mariotti, M., Carnovali, M., & Banfi, G. (2015). Danio rerio: The Janus of the bone from embryo to scale. *Clinical Cases in Mineral and Bone Metabolism, 12*(2), 188–194. https://doi.org/10.11138/ccmbm/2015.12.2.188

Meeker, N. D., Hutchinson, S. A., Ho, L., & Trede, N. S. (2007). Method for isolation of PCR-ready genomic DNA from zebrafish tissues. *BioTechniques, 43*(5), 610–614. https://doi.org/10.2144/000112619

Nasiadka, A., & Clark, M. D. (2012). Zebrafish breeding in the laboratory Environment Institute. *ILAR Journal, 53*(2), 161–168. https://doi.org/10.1093/ilar.53.2.161

Niimi, A. J., & LaHam, Q. N. (1974). Influence of breeding time interval on egg number, mortality, and hatching of zebrafish Brachydanio rerio. *Canadian Journal of Zoology, 52*(4), 515–517. https://doi.org/10.1139/z74-063

Parant, J. M., George, S. A., Pryor, R., Wittwer, C. T., & Yost, H. J. (2009). A rapid and efficient method of genotyping zebrafish mutants. *Developmental Dynamics, 238*(12), 3168–3174. https://doi.org/10.1002/dvdy.22143

Poss, K. D., Keating, M. T., & Nechiporuk, A. (2003). Tales of regeneration in zebrafish. *Developmental Dynamics, 226*(2), 202–210. https://doi org/10.1002/dvdy.10220

Rácz, A., Allan, B., Dwyer, T., Thambithurai, D., Crespel, A., & Killen, S. S. (2021). Identification of individual zebrafish (*Danio rerio*): A refined protocol for VIE tagging whilst considering animal welfare and the principles of the 3 Rs. *Animals (Basel)*, *11*(3), 11(3):616. https://doi.org/10.3390/ani11030616

Samuel, R., Stephenson, R., Roy, P., Pryor, R., Zhou, L., Bonkowsky, J. L., & Gale, B. K. (2015). Microfluidic-aided genotyping of zebrafish in the first 48 h with 100% viability. *Biomedical Microdevices*, *17*(2), 43. https://doi.org/10.1007/s10544-015-9946-9

Schroeder, P. G., & Sneddon, L. U. (2017). Exploring the efficacy of immersion analgesics in zebrafish using an integrative approach. *Applied Animal Behaviour Science*, *187*, 93–102. https://doi.org/10.1016/j.applanim.2016.12.003

Schroeder, P., Lloyd, R., McKimm, R., Metselaar, M., Navarro, J., O'Farrell, M., Readman, G. D., Speilberg, L., & Mocho, J. P. (2021). Anaesthesia of laboratory, aquaculture and ornamental fish: Proceedings of the first LASA-FVS Symposium. *Laboratory Animals*, *55*(4), 317–328. https://doi.org/10.1177/0023677221998403

Siomava, N., Shkil, F., Voronezhskaya, E., & Diogo, R. (2018). Development of zebrafish paired and median fin musculature: Basis for comparative, developmental, and macroevolutionary studies. *Scientific Reports*, *8*(1), 14187. https://doi.org/10.1038/s41598-018-32567-z

Sire, J. Y., Girondot, M., & Babiar, O. (2000). Marking zebrafish, Danio rerio (Cyprinidae), using scale regeneration. *Journal of Experimental Zoology*, *286*(3), 297–304. https://doi.org/10.1002/(sici)1097-010x(20000215)286:3<297::aid-jez8>3.0.co;2-x

Sneddon, L. U. (2017). Pain in laboratory animals: A possible confounding factor? *Alternatives to Laboratory Animals*, *45*(3), 161–164. https://doi.org/10.1177/026119291704500309

Sneddon, L. U. (2020). Can fish experience pain? In T. Kristiansen, A. Fernö, M. Pavlidis & H. van de Vis (Eds.). *The Welfare of Fish*, *20*, 229–249. Springer. https://doi.org/10.1007/978-3-030-41675-1_10

Thomson, J. S., Deakin, A. G., Cossins, A. R., Spencer, J. W., Young, I. S., & Sneddon, L. U. (2020). Acute and chronic stress prevents responses to pain in zebrafish: Evidence for stress-induced analgesia. *Journal of Experimental Biology*, July 22, *223*(14). https://doi.org/10.1242/jeb.224527

Tilley, C. A., Carreño Gutierrez, H., Sebire, M., Obasaju, O., Reichmann, F., Katsiadaki, I., Barber, I., & Norton, W. H. J. (2020). Skin swabbing is a refined technique to collect DNA from model fish species. *Scientific Reports*, *10*(1), 18212. https://doi.org/10.1038/s41598-020-75304-1

Tunbak, H., Vazquez-Prada, M., Ryan, T. M., Kampff, A. R., & Dreosti, E. (2020). Whole-brain mapping of socially isolated zebrafish reveals that lonely fish are not loners. *eLife*, *9*. https://doi.org/10.7554/eLife.55863

Venditti, M., Pedalino, C., Rosello, M., Fasano, G., Serafini, M., Revenu, C., Del Bene, F., Tartaglia, M., & Lauri, A. (2022). A minimally invasive fin scratching protocol for fast genotyping and early selection of zebrafish embryos. *Scientific Reports*, *12*(1), 22597. https://doi.org/10.1038/s41598-022-26822-7

Venta, P. J., Nguyen, A. K., Senut, M. C., Poulos, W. G., Prukudom, S., & Cibelli, J. B. (2020). A 13-plex of tetra- and penta-STRs to identify zebrafish. *Scientific Reports*, *10*(1), 3851. https://doi.org/10.1038/s41598-020-60842-5

Wasden, M. B., Roberts, R. L., & DeLaurier, A. (2017). Optimizing sperm collection procedures in zebrafish. *Journal of the South Carolina Academy of Science*, *15*(2), 7.

Wasko, A. P., Martins, C., Oliveira, C., & Foresti, F. (2003). Non-destructive genetic sampling in fish. An improved method for DNA extraction from fish fins and scales. *Hereditas*, *138*(3), 161–165. https://doi.org/10.1034/j.1601-5223.2003.01503.x

Westerfield, M. (1994). *The zebrafish book. A guide for the laboratory use of zebrafish (Danio rerio)* (1st ed). University of Oregon Press.

Westerfield, M. (2007). *The zebrafish book. A guide for the laboratory use of zebrafish (Danio rerio)* (5th ed). University of Oregon Press.

White, L. J., Thomson, J. S., Pounder, K. C., Coleman, R. C., & Sneddon, L. U. (2017). The impact of social context on behaviour and the recovery from welfare challenges in zebrafish, *Danio rerio*. *Animal Behaviour*, *132*, 189–199. https://doi.org/10.1016/j.anbehav.2017.08.017

Wilkinson, R. N., Elworthy, S., Ingham, P. W., & van Eeden, F. J. (2013). A method for high-throughput PCR-based genotyping of larval zebrafish tail biopsies. *BioTechniques*, *55*(6), 314–316. https://doi.org/10.2144/000114116

Wittwer, C. T. (2009). High-resolution DNA melting analysis: Advancements and limitations. *Human Mutation*, *30*(6), 857–859. https://doi.org/10.1002/humu.20951

Wu, R. S., Lam, I. I., Clay, H., Duong, D. N., Deo, R. C., & Coughlin, S. R. (2018). A rapid method for directed gene knockout for screening in G0 zebrafish. *Developmental Cell, 46*(1), 112–125.e4. https://doi.org/10.1016/j.devcel.2018.06.003

Xing, L., Quist, T. S., Stevenson, T. J., Dahlem, T. J., & Bonkowsky, J. L. (2014). Rapid and efficient zebrafish genotyping using PCR with high-resolution melt analysis. *Journal of Visualized Experiments: JoVE, 84*(84), e51138. https://doi.org/10.3791/51138

Yue, G. H., & Orban, L. (2001). Rapid isolation of DNA from fresh and preserved fish scales for polymerase chain reaction. *Marine Biotechnology, 3*(3), 199–204. https://doi.org/10.1007/s10126-001-0010-9

Zhang, X., Zhang, Z., Zhao, Q., & Lou, X. (2020). Rapid and efficient live zebrafish embryo genotyping. *Zebrafish, 17*(1), 56–58. https://doi.org/10.1089/zeb.2019.1796

Chapter 11

Imaging of zebrafish

Karishma Chhabria

Abstract

The advance of zebrafish as a model organism with which to study a wide variety of biological processes stems in part from its amenability to a variety of imaging techniques. While visualization of gene expression or protein localization and dynamics in fixed tissues is widely used in a variety of organisms and tissues, zebrafish offer an unparalleled opportunity to observe, document and quantify these phenomena in a living vertebrate. This chapter will provide guidance for both live- and fixed-tissue imaging in zebrafish embryos and adults, covering aspects such as microscope considerations, sample preparation and maintenance, systems for high-throughput (HT) imaging as well as tips for optimization of visualization. While general imaging protocols and considerations are widely applicable to broad needs, specific protocols will vary depending on the specific requirement, and these protocols should be therefore used as guidelines and adapted to achieve the desired imaging outcome. In general, this chapter aims to make available zebrafish imaging protocols, techniques and guidance to researchers moving into mechanistic work using zebrafish and laboratories working with other model organisms who wish to start zebrafish experiments.

We start by giving a description of the various standard microscopy techniques with specifications of optics, followed by descriptions of methods and protocols used for different types of imaging

Claire Allen and Jean-Philippe Mocho (eds) **Zebrafish**
DOI: 10.1079/9781800629431.0011, © CAB International, 2024

including live imaging of embryos/adults and fixed-tissue imaging. Finally we give a brief excerpt of imaging applications focusing on state-of-art techniques.

Keywords

imaging, microscopy, high-throughput, optics, zebrafish

Microscopy considerations

With the advancement in optics providing us with high-numerical aperture (NA) lenses and lasers, there has been a revolution in imaging systems. There are various kinds of microscope that can be adapted to specific imaging needs. This section briefly discusses the basis of standard microscopes used for imaging in vivo, ex vivo and in vitro specimens, focusing on underlying optics. It is our hope to give the readers a basic idea of the standard microscopes available to make the right selection for specific imaging requirements.

Wide-field microscopy

Wide-field (WF) imaging, also called brightfield imaging, is the oldest and most basic microscopy technique, and the small size of the zebrafish allows for imaging of the entire specimen when exposed to a light source. A standard WF microscope consists of a light/fluorescent lamp, an objective lens and an eyepiece (Figure 11.1A). Image capture is facilitated by conventional camera-based photodiodes, those most used including charge-coupled devices (CCDs) and the complementary metal oxide semiconductors (CMOSs). Depending on the specifications required for the specimen of interest (such as signal-to-noise, frame rate and capture time), either CMOS or CCD can be used. There are two kinds of WF microscope predominantly used in zebrafish research worldwide: upright

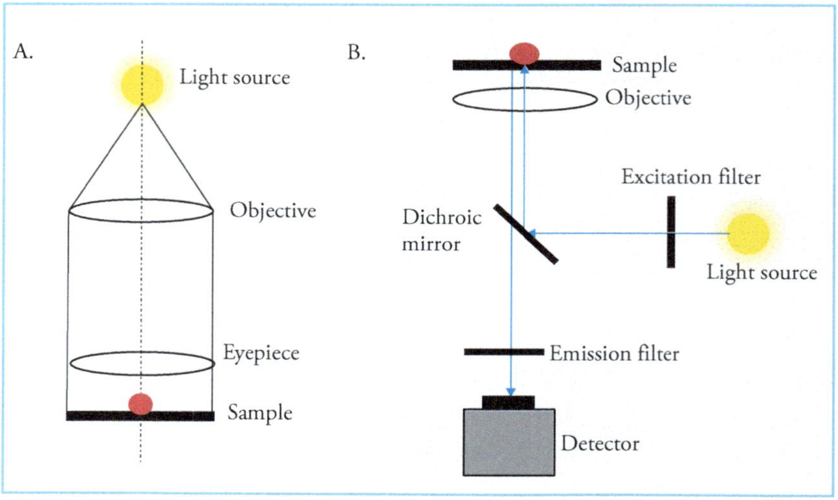

Figure 11.1 Schematics of the optics. A standard WF (A) and an epifluorescence microscope (B).

and inverted WF microscopes. Upright microscopes are useful to image fixed samples mounted and coverslipped on glass slides; inverted WF microscopes are used to image living cells commonly grown in liquid media.

Using a filter set with the same objective lens to both illuminate and collect the light emitted by the specimen, a WF microscope can be used as an epifluorescence microscope (Figure 11.1B). In a typical epifluorescence microscope the light travels from the laser source, passes through the filter set and a dichromatic mirror that acts as a beam splitter and excites the specimen using the shorter-wavelength excitation light. The reflected light from the specimen is collected by the objective and passes through the dichromatic mirror, eventually to be collected at the eyepiece forming an image by the camera.

A variation of WF microscopy deploying contrast methods such as phase contrast is called differential interference contrast (DIC) microscopy and is often used to image the morphology of structures in a specimen. A typical DIC microscope comprises a light source similar to a WF microscope and two birefringent prisms that act as beam splitters (or beam combiners). One of the beam splitters is positioned near the condenser lens (eyepiece) aperture to align and shear incident polarized wavefronts into two orthogonal components. The other beam splitter is located on the rear front of the objective lens and recombines the sheared and deformed wavefronts into linear and elliptically polarized light, which then passes through an analyser (polarizer) located in an intermediate tube above the objective. The polarized light components emerging from the analyser recombine through constructive and destructive interference at the image plane into the eyepiece/camera to form the contrasted image. There are various configurations of beam splitters and analysers used, depending on the manufacturer.

Current DIC microscopes position the beam splitters and analyser in strategic locations with respect to the field lens, condenser, objectives and observation tubes. The beam splitters are mounted in a revolving turret condenser that can accommodate three or more beam splitters and other auxiliary components such as phase contrast annular rings or dark field light stops, depending on the manufacturer. Furthermore, each beam splitter prism should be matched to the range of objective numerical apertures such that a typical prism, which works for only one or two objectives (10× and 20×). In general, each of the prisms can be customized (designed by the manufacturer) to image across the spectrum of dry- and oil-immersion objectives with varying numerical aperture.

Confocal microscopy

The confocal laser scanning microscope is one of the oldest fluorescent high-resolution microscopes used for acquiring serial optical sections. It provides high optical resolution and contrast due to the spatial pinhole that blocks out the off-focus light reflected from the specimen. The pinhole size is adaptable as a function of the NA of the objective lens and wavelength of the laser used for scanning (Figure 11.2). Commercially produced confocal microscopes (e.g. Leica) currently deploy a bi- or multi-laminar iris design pinhole

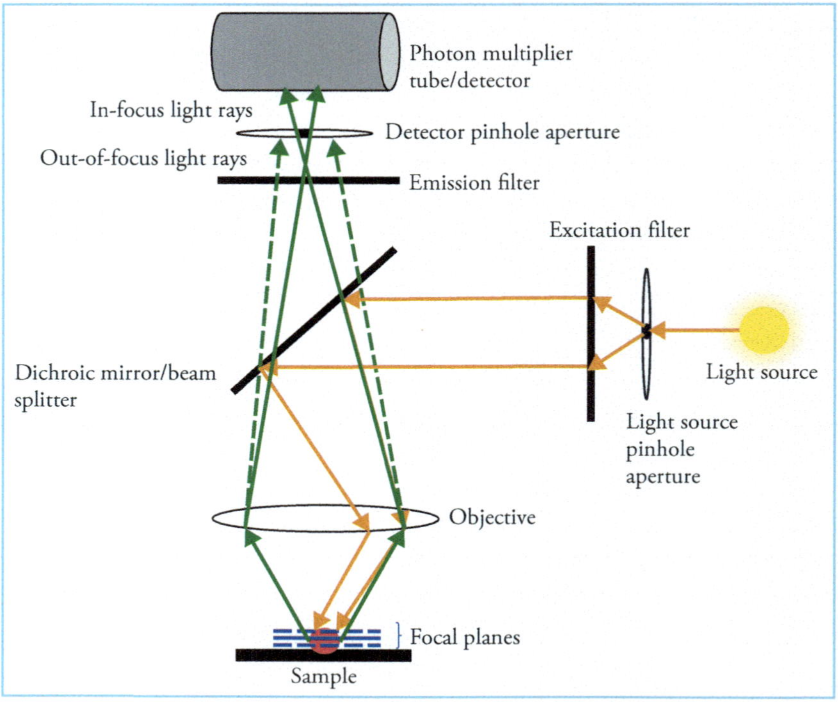

Figure 11.2 Schematic of optics used in a standard confocal microscope. The light source used here is derived from different lasers, depending on the excitation wavelengths needed for a particular fluorophore.

that automatically adapts the pinhole diameter depending on the NA of the objective, wavelength of the laser and internal magnification of the optical elements in the microscope. A conventional confocal microscope deploys parallel scanning patterns across scanning motors. The parallel scanning method allows for tuneable scanning speed. In general, lower scanning speeds are preferred because they provide high signal to ratio and hence higher contrast. This causes a 'time bottleneck' for scanning of a full frame. However, recent advances allow for fast scanning using resonant mode scanners giving 8–10 times the speed of conventional scanners, enabling acquisition of serial two-dimensional (2D) images at varying depth that are then amenable to three-dimensional (3D) rendering and reconstruction.

Lightsheet microscopy

Selective plane illumination microscopy (SPIM), also called light-sheet fluorescent microscopy, has become a popular method for imaging diverse specimens, ranging from single cells to organoids and multicellular organisms, in developmental biology. Unlike other methods of microscopy, SPIM illuminates and detects the specimen along two orthogonal horizontal directions. The emitted light is then collected with a standard microscope objective and projected onto an imaging sensor such as a CCD, CMOS camera or an electron/photon-multiplying CCD (Figure 11.3). This allows for

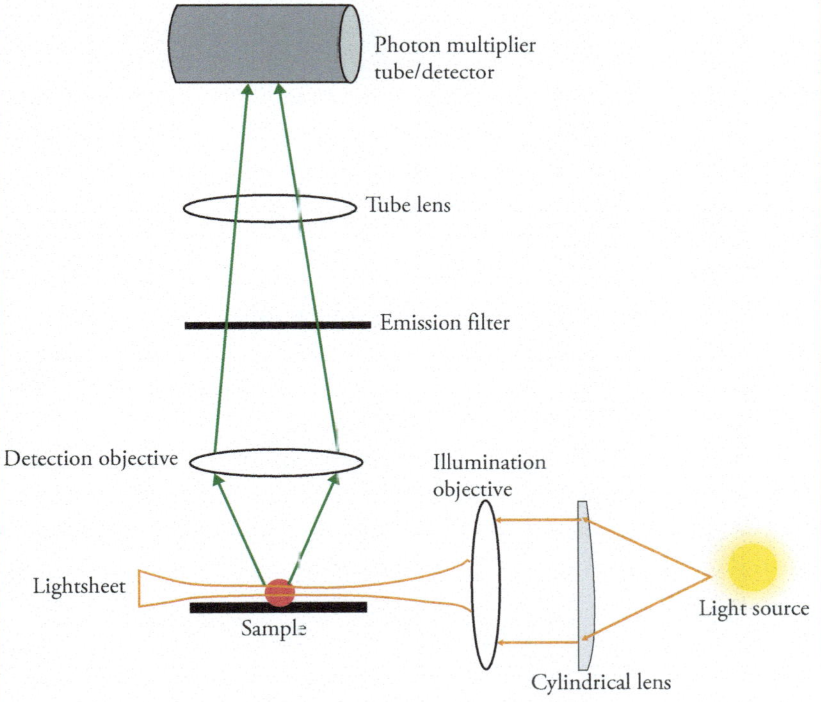

Figure 11.3 Schematic of optics used in a standard lightsheet microscope. The light source used here is derived from different lasers depending on the excitation wavelengths needed for a particular fluorophore. A cylindrical lens is needed to generate the lightsheet as opposed to the pinhole aperture in a confocal microscope.

only part of the specimen to be exposed to the laser being imaged, which gives SPIM its two major advantages: non-invasive and longitudinal imaging. Due to planar illumination, an entire section can be recorded in one shot with a high-speed digital camera, without the need to scan through every pixel as in a confocal microscope.

Multi-photon imaging

Multi-photon imaging, including commonly used two-photon imaging, has been gaining traction due to its superior ability for 3D optical sectioning without absorption at greater depths compared with other microscopy techniques such as confocal microscopy. Two-photon excitation arises from simultaneous absorption of two photons in a single quantized event. According to the Planck–Einstein equation, the energy of a photon is inversely proportional to its wavelength and hence the two absorbed photons must have a wavelength roughly twice that required for one-photon excitation as in a confocal microscope. The penetration at greater depths is achieved due to the use of higher-wavelength excitation laser beams that undergo less scattering than shorter-wavelength blue light. This reduced scattering thereby contributes to fewer detrimental effects of photobleaching and phototoxicity as observed in confocal microscopy. A further advantage of reduced scattering is that there is increased excitation light reaching the focal plane deep within the intact tissue.

The optics of a two-photon microscope are very similar to those of a confocal microscope, except for the lasers and the fact that no pinhole aperture is needed (Figure 11.4). Mode-locked pulsed lasers are used for generating high photon densities, which are required for a significant number of two-photon absorption events. In general, photon density should be a million times more than that needed for one-photon absorption events. Furthermore, the absence of a pinhole allows greater flexibility in detection geometry.

Although two-photon microscopy is the most prevalent multiphoton imaging technique, another nonlinear optical process, three-photon excitation, has been shown to have potential biological applications. Three-photon excitation is very similar to two-photon

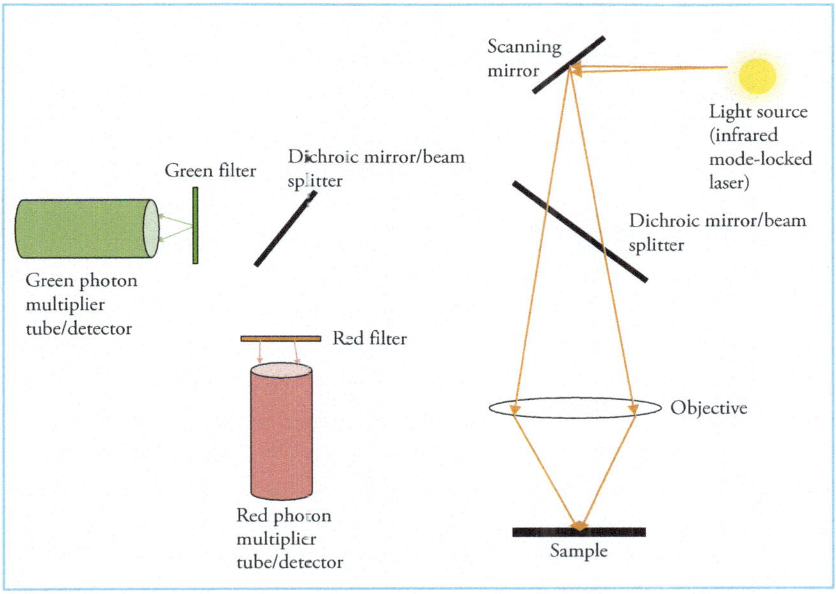

Figure 11.4 Schematic of optics in a standard two-photon microscope. Note that a scanning mirror is used before the objective dichroic to modulate the scanning speed of the tissue. As opposed to other microscopy techniques, multiple fluorophores can be scanned in parallel at the same scanning speed.

excitation, the photon density required being only tenfold more than for two-photon due to the non-linear quantum mechanical properties of fluorescence absorption. An infrared laser (1,050 nm) can produce three-photon excitation of an ultraviolet-absorbing fluorophore (~350 nm) while causing two-photon excitation of a green-absorbing fluorophore (~500–530 nm), providing greater scope in multi-modal imaging. Other higher-order optical processes, such as four-photon absorption, have been shown to have adverse heating effects and hence are unlikely to have potential future biological applications.

Methodologies

Different imaging needs are catered for by different microscopies and sample preparation methods. In this section we describe the specifics of live- and fixed-tissue imaging while also focusing on sample preparations for different microscopy needs. Table 11.1 highlights the key differences between live and fixed imaging, illustrating example applications wherever possible.

Live-tissue imaging

The ability to image zebrafish live, either as adults or embryos, allows us to investigate a wide variety of biological processes and phenomena, with relatively simple experimental set-ups. Because embryogenesis occurs relatively rapidly when compared with higher vertebrates, coupled with the advance of transgenesis in allowing visualization of specific organs or proteins using fluorescent proteins, we can observe in real time, and at high resolution, multiple processes of interest. Broadly speaking, live-imaging experiments fall into two categories – high-resolution imaging of cellular processes requiring sample immobilization, and lower-resolution organism-wide morphology; behavioural analyses will outline specific examples of each in this section.

Live imaging of embryos

High-resolution/quality imaging of small numbers of samples, requiring precise manual orientation and mounting, can be used to image organ- or cell-specific processes at tissue- or cell-level resolution. Conversely HT experiments, where large numbers of usually whole embryos are imaged, is better suited for screening well-defined and clear read-outs, such as levels of gene expression/reporter transgene, without the need for high spatial resolution or high-quality images. Both experiments require slightly different experimental set-ups and considerations, which we will outline below:

For many applications involving low throughput and hence high-resolution imaging, embryos need to be immobilized. In such applications, embryos can be mounted in low-percentage, low-melting point (LMP) agarose. The percentage of agarose used

Table 11.1 Summary of key aspects of live- and fixed-tissue imaging with corresponding examples.

	Live-embryo imaging	Fixed-tissue imaging
Mounting considerations	1. Low-melting-point-agarose-based methods are deployed to immobilize larvae. The percentage of agarose varies among imaging paradigms and ranges between 0.3 and 4% (w/v). 2. For longer dynamic imaging of vascular, neuronal and muscle development, a layered agarose method can be used. This method helps with compensation for distortions due to larval growth (Upadhyay et al., 2020). 3. Methylcellulose (2–3% w/v) is an alternative to low-melting-point-agarose, often used to provide sufficient rigidity to image and manipulate live embryos. 4. Most live imaging requires anaesthetics inclused in the embedding media to aid in immobilization of larvae, especially for time-lapse imaging. Tricaine is commonly used at 4.2 mg/ml for most imaging paradigms, except for certain task-based neuro-imaging paradigms. 5. For high-throughput imaging paradigms, both an agarose-based microfluidic device and a more accurate 3D-printed agarose cast (Kleinhans & Lecaudey, 2019) silicone cast-based method can be used (Masselink et al., 2014).	6. Fixed tissue mounting is traditionally performed using 100% glycerol on a glass slide preparation post fixation, as used for whole-mount in situ hybridization. 7. For fixed-section imaging, agarose mounting/embedding is the method of choice. Various kinds of casting mould have been suggested, including triangular, radial and shoe moulds depending on the desired position of the fish for sectioning (Copper et al., 2018). 8. A custom glycol methacrylate plastic-based method has been shown to be optimal to preserve fluorescent signal (inclusive of photo-activatable fluorophores) without the need for antibodies (Oralova et al., 2019).
Microscope considerations	1. Confocal microscopy is by far one of the most widely used microscopy tools due to the detailed structural information it provides. It has been used mostly to image 2D + t stacks, but also for 3D imaging and rendering. However, it has the disadvantages of photo-bleaching and photo-damage and hence is not ideal for longitudinal imaging.	6. Conventionally, a compound or stereomicroscope suffices for imaging sections with anti-digoxygenin staining or whole-in situ hybridization. The main disadvantage of whole-in situ hybridization is the lack of 3D information. However, imaging software is available to help with the 3D reconstruction of anti-digoxygenin-stained sections, while 3D rendering has its own limitations.

Table 11.1 (continued)

	Live-embryo imaging	Fixed-tissue imaging
	2. Multi-photon microscopy is a choice when considering imaging of dynamic signals in live animals. With the advent of genetically encoded calcium indicators, two-photon microscopy has been employed to acquire whole-brain maps at single-neuron resolution. Lightsheet fluorescent microscopy is a preferred technique owing to its requirement for low excitation intensities and hence minimal photo-toxicity. More advanced versions, such as thin-sheet laser-imaging microscopy, offer a wider field of view while compensating for the penetration artefacts of conventional lightsheet microscopy (Santi, 2011). 3. Other microscopes for consideration, based on low bleaching while offering the ability to image for a longer duration and commonly used for macromolecular structural imaging include the second harmonic generation microscopy (Sun et al., 2004). 4. Recent advancements offer the ability to visualize the dynamics of biomolecular processes in a label-free system using confocal Raman spectroscopic imaging (Høgset et al., 2020). 5. Other microscopy methods include micro-computerized axial tomography, magnetic resonance imaging and high-frequency ultrasound.	7. Confocal microscopy is the method of choice for 3D imaging requirements. 8. To avoid the need for 3D rendering, fluorescent in situ hybridization is the procedure of choice because fluorescent microscopy techniques such as multi-photon and light sheet fluorescent microscopy can be used for volume imaging.
Imaging and analysis considerations	**Imaris, Meshlab, Solidworks, Fiji, MATLAB, IGorplot**	**Fiji, MATLAB, µManager, Virtual brain explorer (ViBE-Z)**

should be sufficiently high to ensure that the sample remains immobilized during the experiment; however, too high an agarose percentage will result in inability of the embryo to continue to grow and undergo morphogenesis. Therefore, careful consideration must be taken to ensure that the optimum concentration is used, and typically this will range between 0.25 and 1%. LMP agarose should be used because this allows sufficient time for embryo orientation while the agarose is still liquid, but at a temperature that is not detrimental to the health of the embryo. Anaesthetic (e.g. tricaine) must be added to the liquid agarose to ensure that the embryo does not move during the imaging process. Wherever possible, embryos mounted in agarose should be kept covered with E3 medium under anaesthesia and, if imaged in a dish, should be kept covered. This will prevent minor changes in temperature of the sample associated with excessive evaporation, and is particularly important when imaging temperature-sensitive processes such as comparative analysis of heart rate.

Changes in temperature and evaporation during imaging can also cause the mounting medium to expand or contract. For long-term imaging experiments this can cause the sample to shift (particularly in the *z*-plane) during imaging and may result in the sample moving out of the imaging field. It is therefore vital to equilibrate the sample to the temperature of the imaging chamber prior to starting the imaging experiment, to reduce as much as possible the possibility of the sample moving out of the field of view during the experiment.

Protocol 1: using agarose to mount embryos

1. Heat 0.25–1% LMP agarose in E3 medium until the agarose is fully dissolved.
2. Aliquot the liquid agarose into 1 ml aliquots and allow to cool to approximately 40–50 °C (can be placed in a heat block at 42 °C at this point).
3. Once the agarose has cooled sufficiently, add 40 µl of 4g/l tricaine stock solution (at a final concentration of 100 mg/ml) to the agarose.

4. Maintain the agarose in a heat block at 42 °C while mounting the embryos.
5. For mounting, remove an aliquot of agarose from the heat block.
6. Drop the embryo(s) into the agarose, transferring as little medium as possible.
7. Wait for the embryo to sink to the bottom of the agarose and then transfer the embryo to the mounting chamber in a sufficient amount of agarose.
8. For mounting on an inverted microscope, transfer the embryos to a coverslip/coverslip-bottomed imaging dish in a drop of agarose.
9. Orient the embryo in the desired position and wait for the agarose to solidify.
10. Once the agarose has solidified, cover the embryo with a layer of E3 medium (with tricaine at 4%) to prevent the agarose from drying out.
11. For mounting embryos in light-sheet imaging capillaries, transfer them into a nine-well dish.
12. Using the capillary-plunger provided with the lightsheet microscope, suck up the agarose containing the embryo into the capillary.
13. Rotate the capillary to orient the embryo into the desired position and lay the capillary flat to allow solidification. This can then be transferred directly to the mounting chamber of the lightsheet.

Documenting the gross morphology of live zebrafish embryos is incredibly straightforward. While it is possible to image embryos/larvae in a Petri dish in E3/system water, it can be difficult to orient embryos precisely and reproducibly. The significant amount of yolk present in younger stages can prevent the embryo from lying flat for lateral views and prevents orientation for dorsal imaging. Similarly, once the swim bladder has developed, aligning larvae laterally is problematic. Methylcellulose is an excellent medium (at 2–3%) for accurate orientation of embryos/larvae – its viscosity allows specific alignment and orientation of the embryo, and the inclusion of tricaine in the medium will allow the embryo to remain immobile during imaging. Methylcellulose solutions may be used for short-term live-imaging experiments as well as single snapshots.

Protocol 2: using methylcellulose to mount live embryos.

1. Standard protocols use a 3% methylcellulose solution in E3 medium.
2. Allow the solution to mix gently overnight to ensure that it is fully dissolved, and to remove bubbles from the medium.
3. To dissolve quickly, freeze E3 with the powder mixing regularly as it cools.
4. When handling methylcellulose solutions take care not to introduce bubbles.
5. Methylcellulose (3%) can be aliquoted and stored at –20 °C, and working solutions at 4 °C.
6. Place anaesthetized embryos in a drop of 3% methylcellulose on a depression slide and use a pipette tip or embryo poker to orient them.

HT screening

Zebrafish, owing to their various advantages, are popular models for being amenable to HT analysis. Commonly employed methodologies include phenotype-based visualization used as a mode for preclinical drug screening. Plate reader methodology is the oldest implemented HT screening method for behavioural assays. A commonly used toolbox is DanioScope, which is used for larva tracking and performing non-invasive measurements. Its automatic larval recognition software can recognize embryos within the chorion, and measure their activity such as burst movement/duration/frequency and inactivity. Among the most frequently used applications are blood and gut flow measurement, larval morphology and cardiology measurement. This system can be used in combination with the Zeiss-based stereo discovery platform and is best suited for acquisition speeds >60 frames/second for a range of different video platforms. Functional assays assessing fluorescence represent a relatively recent approach to phenotypic analysis. Automated reporter quantification in vivo (ARQiv) using fluorescent microplate readers can scan volumes >50,000 units/day (Walker et al., 2012). ARQiv is an economical off-the-shelf instrumentation method that can be used for drug discovery due to its ability to quantify changes in cell number, cell signalling and metabolism. A simple yet potent screening-only method, called complex object parametric analyser and sorter (COPAS), can be used for a sample size of 10–1,500 um (Liu et al., 2018). A more sophisticated method developed by Pardo-Martin et al. (2010), vertebrate automated screening technology (VAST), is a popular method because it can automatically load, position and rotate zebrafish larvae. It has the capacity to acquire high-profile imaging of each zebrafish embryo within 9.6 seconds, thereby ramping up screening accuracy and imaging capability. Further advancement in VAST has recently increased its utility in 3D reconstruction. Using the voxel residual volume (VRV) maximization problem, Guo et al. (2017) have shown accurate volume and surface area calculations of acquired samples. A real-time HT screening method proposed by Liu et al. (20178) extends the potential of HT by imaging embryos flowing through a capillary tube. Using a cylindrical lens to generate a sheet of light for imaging alongside a linear charge-coupled device (Lc-CCD), it can scan at a frame rate of up to 66 kHz. Another significant advantage of this method includes sorting of dead and live embryos based on grey-level distribution.

Live imaging: adults

While there is a repertoire of methods available for larval HT screening, there have also been substantial advances in imaging of multiple adult zebrafish. Bang et al. (2002) have shown an imaging rate of 108 adult zebrafish per hour while quantifying responsiveness to auditory stimulation; this can be used to quantify various components of the auditory system, including swim bladder, Weberian ossicles, perilymphatic spaces and sensory maculae (Bang et al., 2002). This can be coupled with various image-processing techniques for the detection of startle response to visual, auditory and proprioceptive stimulation. Light-emitting diode (LED)-based fluorescent microscopy is a low-cost method for screening transgenic animals and a method recommended for simultaneous imaging of up to 30 adults (Blackburn et al., 2011). The salient optical feature of this method includes the availability of between four and six different LED lights and a selection of up to three optic filters. Overall, this method has a high detection rate at an economical cost and with an easy build.

Behavioural imaging is another dominant sphere in the utilization of adult zebrafish models to investigate swim patterns, through the use of video tracking of individual location and posture. This versatile and common method involves placing the fish in shallow water and filming at around 1,000 Hz from above/below, allowing tracking of location, orientation, tail curvature, horizontal eye position and pectoral fin motion (Easter & Nicola, 1997). Several image-processing algorithms and open-source software packages have been used for automatic kinematic variables from video recordings. Various software packages now also provide ways to measure complete trajectories of fish and have a diverse set of applications, including the study of anxiety, sleep, exploration or socialization. Most of the tracking systems developed for individual tracking have now developed algorithms to extract fingerprints for multiple individuals, aiding in the tracking of individuals in groups. With these high-dimensional data, machine learning algorithms can be applied to identify specific behaviours associated with specific individuals in the group (Kabra et al., 2013).

Fixed-tissue imaging

Analysis of gene expression through mRNA in situ hybridization (ISH), or protein localization by immunohistochemistry, can be carried out on either entire embryos (whole mount) or those that have been sectioned (generally using a cryostat).

ISH: whole mount

Imaging of whole-mount embryos usually requires clearing of embryonic tissue and yolk as much as possible to enhance the signal within the specimen. Clearing of embryos is typically done through equilibration into either a 50% glycerol solution (directly made from phosphate buffered saline with Triton X100 (PBST)) or a 100% mixture of benzyl alcohol and benzyl benzoate (BABB) solution (following dehydration into methanol (MeOH)). Both systems have their advantages and disadvantages. BABB is an excellent clearing medium, giving unparalleled signal resolution over background and is particularly effective at clearing background staining in the yolk. Although embryos imaged in BABB may be stored long term in this solution, they may be subject to loss of signal, particularly in weakly stained samples. Alternatively, embryos may be stored long term in MeOH before and after imaging in BABB, resulting in better preservation of signal. It should be noted that BABB can melt some plastics including PET, so care should be taken when storing embryos and working with BABB. Embryos in BABB exposed to light/heat are also prone to the rapid development of background in the embryonic tissue and pink staining in the yolk, and thus should be protected from strong light as much as possible during imaging. To this end, embryos stored in MeOH/BABB must therefore be stored in the dark at 4 C. Finally, certain substrates used for the chromogenic reaction in the ISH protocol, for example INT/BCIP, dissolve in BABB resulting in a loss of staining and therefore BABB is not universally suitable for imaging whole-mount ISH.

Glycerol is more commonly used as a medium for both storage and imaging in whole-mount ISH, typically at a concentration of around 50%. Glycerol is an adequate clearing reagent although can be less successful at removing background staining in the yolk (treating embryos with ethanol prior to imaging in glycerol can help to reduce background staining in the yolk for difficult probes). Glycerol provides the added advantage that embryos can be stored long term in glycerol at room temperature, negating the need for specialist storage, and is compatible with a wide variety of chromogenic substrates.

Thisse and Thisse (2008) described another standard protocol for ISH. Commonly used labels such as digoxigenin and fluorescein are detected using alkaline phosphatase-conjugated antibodies that catalyse a chromogenic reaction resulting in blue or red fluorescence. This allows for simultaneous multiple tagging of two different genes. Protocols for synthesis of digoxigenin- and fluorescein-labelled probes from linearized plasmid or PCR products are available, but beyond the scope of this chapter (Moens, 2008).

Protocol 3: whole-mount ISH

1. Dechorionated embryos are fixed with 4% paraformaldehyde (PFA), either for 5 h at room temperature or overnight at 4 °C. Embryos are then equilibrated five times with PBST at room temperature as a preparatory step before permeabilization.
2. Proteinase K at room temperature is conventionally used as the permeabilization reagent. The duration of incubation varies depending on age, ranging from 1 to >10 minutes for stage 1–4 somite embryos to 3 days post fertilization (dpf). Care should be taken not to over-digest because embryos can disintegrate during subsequent steps.
3. Embryos are then re-fixed in 4% PFA for 1 h at room temperature, followed by rinsing five times with PBST as a preparatory step before hybridization.
4. Embryos are first pre-hybridized with the hybridization solution for at least 1 h at 65 °C. Digoxigenin-labelled RNA probe dilution must be derived empirically, starting from 1:200 in the hybridization solution.
5. Once the appropriate dilution is determined, the prehybridization buffer is replaced with a warmed hybridization buffer with the probe and incubated overnight at 65 °C.
6. Hybridization solutions inclusive of the probe can be removed after use and stored at –20 °C, then reused several times without loss of signal. Samples are subsequently washed in 2× saline sodium citrate (SSC) buffer (2:1), 2× SSC (1:2), 2× SSC, 0.2× SSC, 0.1× SSC, 0.2× SSC:PBST (2:1) and 0.2× SSC:PBST (1:2), each for 5 min at 65 °C.

7. Finally, embryos are washed with PBST for 5 m at room temperature followed by incubation in blocking solution for 1 h at room temperature.
8. Embryos are then incubated with AP-conjugated anti-digoxigenin antibody (1:5,000) for 2 h on a shaker at room temperature. Alternatively, they can be left overnight at 4 °C; however, this has the potential to lead to higher background noise.

For coloration, specific prefabricated coloration buffers are used. Embryos are first equilibrated in the coloration buffer four times for 5 min each, followed by the addition of NBT/BCIP solution to the coloration buffer.

The colour should become visible within 0.5–4 h, depending on the probe used.

The speed of colour development can be manipulated by changing the temperature.

Once the desired intensity of colour has developed, coloration is stopped by washing the samples with sterile water. For two-colour ISH, the coloration process is carried out sequentially.

Embryo orientation and imaging

Depending on the organ or tissue of interest, whole-mount ISH embryos can be mounted in a variety of ways to optimize imaging resolution.

To visualize anteroposterior axial structures – for example, the somites – younger embryos may benefit from being flat mounted. This involves removal of the yolk using forceps or needles then tape to generate a slightly raised border around a central cavity on a slide, placing the embryo flat within this cavity and putting a coverslip on top of the tape, sandwiching the embryo within the imaging window – ideally in a flattened orientation so that entire A–P axis is levelled to a single focal plane (Figure 11.5A–C). Flat mounting can also be useful for high-resolution imaging of lateral structures – for example, the eyes or ears in the early stages of development.

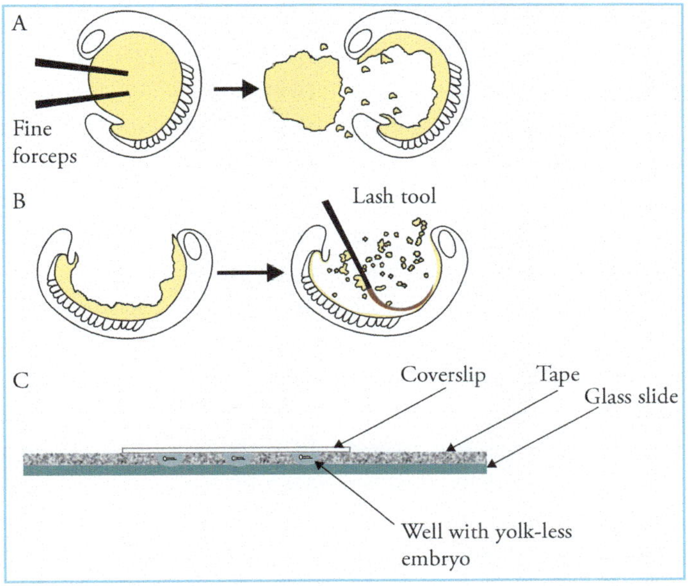

Figure 11.5 Flat mounting of embryo for fixed-tissue imaging. (A,B) The process of removing the yolk sac for flat-mount imaging (adapted from Cheng et al. (2014)). (C) Schematic of glass slide with well cast for flat-mount, un-yolked embryo.

Unfortunately, certain structures or tissues in the embryo may be disrupted by removal of the yolk – for example, the heart, intestines and developing fin buds. In these circumstances, flat mounting the embryos may be detrimental because the embryo must be imaged as intact as possible.

To image whole embryos, imaging slides can be prepared containing a significantly deeper 'imaging window' than that used in the flat-mounting protocol. Coverslips (22 × 22 mm^2) can be glued to either end of a standard slide – and the number of cover slips required can vary depending on the age of the embryos – typically anywhere between three and five coverslips per stack. The embryos to be imaged can then be placed within a droplet of mounting medium in the middle of the slide between the two stacks. A 22 × 50 mm^2 coverslip can then be placed on top of the stacks, over the droplet containing the embryos. Gentle pressure may be required to allow the droplet on the slide to form a liquid column with the coverslip. Once this liquid 'bubble' has been formed, the coverslip on top can be gently slid around on top of the coverslip stacks – this will gently rotate the embryo within its liquid bubble and allow the researcher to accurately orient it. If the tail of the embryo is not particularly of interest, it may be beneficial to remove the tail with a scalpel prior to mounting – this will provide more orientation options and is particularly important when imaging structures such as the heart, which may be obscured by the head in a simple dorsal view.

Depending on the structure being imaged, either DIC or bright-field optics may be suitable for imaging of whole-mount ISH. In circumstances where it is preferable to see cell outlines (for example, when visualizing a gene expressed in only part of a tissue), DIC optics may help to outline the organ of interest and orient expression within the tissue. However, in situations where the tissue of interest lies near the yolk, brightfield may be more suitable to prevent light refraction from the granular yolk obscuring or distorting the signal from the gene of interest. Furthermore, the use of DIC optics often requires a high intensity of light to be concentrated on the sample – if embryos are being imaged in BABB this will significantly increase the chance of background developing in the sample, and it may be preferable to use brightfield instead.

Fluorescent ISH

Fluorescent in situ hybridization (FISH) is a method of choice over ISH due to its amenability to high-resolution imaging such as confocal imaging and lightsheet microscopy. The main advantage of FISH over ISH is labelled RNA probe-based detection of genes. The first steps in the FISH protocol are identical to the standard Thisse protocol for ISH described above (Thisse & Thisse, 2008). The protocol diverges at the point of application of fluorescent substrate, and uses an amplification step to obtain high signal-to-noise ratio for visualization. Amplification is achieved by the tyramide signal amplification (TSA) technique (as described in Welten et al., 2006) that provides amplification of Cy3/Cy5 signal. Complementary to the NBT/BCIP detection substrates, horseradish peroxidase is used to catalyse deposition of a fluorophore-labelled tyramide amplification reagent at the site of probe binding. Even weak probes and small gene expression domains can be visualized using this approach for TSA detection. Another advantage of this technique is the ability to use nuclear counterstains such as SYTOX green or orange. Anti-fading agents such as Gelvatol-DABCO can be used before mounting. Care should be taken to ensure the use of fluorescent protection media for sections while mounting before imaging. A wide range of fluorescent mounting media are available, such as VECTASHIELD and Mowiol.

Immunohistochemistry

Immunohistochemistry has been gaining in popularity recently due to advancements in imaging technologies. Whole-mounted samples stained immunohistochemically can be imaged using lightsheet microscopy, and cross-sectional samples can be imaged using wide-field/confocal microscopy depending on the imaging resolution needed. A standard whole mount or section immunohistochemistry includes the steps described in Protocol 4 (below) except for mounting of those sections needed for section immunohistochemistry described alongside.

Protocol 4: whole-mount immunohistochemistry

1. Larvae are fixed with 4% paraformaldehyde overnight at 4°C, followed by 5-min washing with 1× PBS. For sections, larvae are equilibrated in a gradient concentration of sucrose (or commercially available anti-freeze) before the next steps.
2. Whole larvae can be resuspended in 100% methanol for long-term storage at –20°C.
3. Samples are rehydrated from methanol with three 10-min washes with PBST under gentle agitation. Samples are then suspended in 150 mM Tris-HCl (pH 9.0) for 5 min and subsequent heating at 70°C for 15 min. A general norm is to equilibrate the larvae for two rounds of (10 min) PBST and two rounds of (5 min) washes with distilled water.

4. The duration of the next step of permeabilization varies for larvae versus sections. For larvae, permeabilization with ice-cold acetone is done at –20°C for 20 min, followed by two rounds of washes with distilled water and equilibration with PBST. For cross-sections, 10 min of permeabilization should suffice.
5. Samples are initially blocked with a blocking buffer for 3 h at 4°C followed by incubation in the primary antibody buffer for 3 days at 4°C with gentle agitation.
6. Residual primary antibody is washed off with three hourly washes in PBST at room temperature, followed by three 10-min washes with PBST and equilibration in PBS-TS.
7. Lastly, samples are incubated in secondary antibodies at custom dilutions (commonly 1:200–1:500) for 2.5 days at 4°C in the dark, with gentle agitation. As a general practice, larvae can be stored at 4°C in the short term prior to imaging, with a brief washing with PBST.

Additional considerations

Temperature regulation

Temperature regulation is crucial both for in vivo and ex vivo imaging. Temperature can affect various intra-vital cellular processes including cell proliferation, cell morphology and migratory behaviours. For imaging procedures, temperature is regulated mainly at three levels (Johnson et al., 2001).

Room level

Most imaging regimens allow for imaging at room temperature; however, to image above or below room temperature, the building's heating ventilation and air conditioning (HVAC) systems can be used to modulate the temperature. Alternatively, microscopes are placed in a laboratory/department-specific hot or cold room. This is a method of choice over an HVAC control system because it minimizes temperature variations that could affect the sample and/or optomechanical stability, due to the possibility of air draughts produced by HVAC equipment.

Box level (local contained level)

A more common technique for more precise heat regulation is to enclose the microscope in a box whose temperature can be regulated independently of room temperature. This mimics the whole room control but is more consistent and quicker to control due to the smaller space involved. Furthermore, an enclosed system provides protection from electric draughts in a laboratory.

Sample level

The most common way to enable rapid equilibration and temperature control is via local and direct control of sample temperature regulation. Devices such as resistive heaters, Peltier coolers and water heater/cooler jackets are attached to either a sample holder or the objective. A major disadvantage of this system is the potential for poorly controlled temperature cycling, with consequent image instability and inconsistent sample temperature. As a general practice, a combination of regulation systems is preferred depending on the experimental set-up and economic factors.

Anaesthesia

There is no 'one size fits all' anaesthetic regimen for zebrafish imaging; rather, the choice of anaesthetic is scenario/experimental need dependent. Anaesthetics work through depression of the central nervous system, accompanied by differential levels of analgesia (absence of pain) and muscle relaxation. It is important to gauge various parameters before selecting a particular anaesthetic, such as duration of procedure, age of zebrafish and functional assay of interest. Tricaine is one of the most widely used anaesthetics due to its minimal safety concerns and a standard working dose. A 0.4% w/v concentration is the preferred dose used for various experiments involving larval zebrafish. However, in cases where higher exposure times and dose are required, various growth-adverse effects such as aversion, epidermal and corneal lesions, hypoxaemia, decreased heart rate and death should be considered. Anaesthesia up to a duration of 50 min can be induced using benzocaine with no substantial effects on physiology (Olt et al., 2016). Anaesthesia becomes a critical factor in adult zebrafish (see Chapter 13 for more information).

Inhibition of pigment formation

The relative transparency of zebrafish compared with other models used in research makes them a popular choice. However, visualization can be hampered by melanization, the formation of a black pigment. Phenylthiourea (PTU) inhibits melanogenesis by binding to copper ions and interacting with side chains in the active site (Klabunde et al., 1998). A concentration of 0.003% (w/v) or 0.2 mM (commonly referred to as 1× PTU) is used as a standard for inhibition of melanization between 12 h post fertilization (hpf) to shortly before 24 hpf. Some studies have also shown that a lower concentration, of 0.075 mM or 0.375× PTU at about 25 hpf, can inhibit melanization (Karlsson et al., 2001). Due to the versatile nature of PTU treatment, it remains a primary method for inhibition of melanization. However, the use of PTU should be considered in respect to studies showing adverse molecular and physiological effects, such as activation of cytochrome P4501A1 (Wang et al., 2004), suppression of retinol-binding protein 4 (Tingaud-Sequeira et al., 2006) and suppression of thyroid hormone (T4) immunoreactivity in thyroid follicles (Elsalini et al., 2003). Alternatively, the use of genetic mutants such as *nacre* (Lister et al., 1999), *sandy* (Das et al., 2003) and *casper* (double mutant of nacre and *roy*, an iridophore mutant), is a popular method of choice for suppression of melanization. Recent advances in CRISPR-Cas9-based mutagenesis have provided novel methods to generate optically clear and pigment-free zebrafish embryos and larvae in any transgenic and/or mutant-carrying background. A recently published method shows significant improvement in the targeted cutting CRISPR-Cas9 procedure of mutagenesis, with minimal concerns about off-target effects (Davis et al., 2021).

Longevity of imaging (embryonic health maintenance during live imaging)

The transparent nature of zebrafish larvae allows for visualization of development in different organ systems. Because the timescale of organ development is in the order of several hours to days, the concept of imaging for long periods of time is desirable. Nevertheless, there are various factors to consider that influence the success and longevity of imaging; the key factor is the choice of microscope. Microscopes, such as the confocal, can be used for short-term fluorescence imaging because the power per unit area causes bleaching after a few hours of imaging and has potential side effects on the physiology of zebrafish due to heat produced at every focus. The effects are so detrimental to the fish that only one imaging session is tolerated and subsequent imaging attempts cause adverse effects, so longitudinal studies are not an option. A lightsheet fluorescence microscope, with its low power-per-unit area and lower bleaching, is ideal for several hours of imaging and longitudinal studies. Various studies imaging organogenesis in zebrafish embryos between 10 hpf and 3 dpf used 0.8–1% low-melting-point agarose as a mounting medium for lightsheet microscopy (Keller, 2013). Anaesthesia of 0.4% tricaine is commonly added to E3 medium in the chamber of lightsheet-based fluorescent microscopy (LSFM) for long-term imaging, to avoid both motion artefacts and stress that could affect the physiological development of zebrafish. Some studies use concentrations of >1% low-melting-point agarose combined with <0.4% tricaine to avoid the non-specific side effects of long-term use of anaesthetics, especially blockage of ion channels (Chhabria et al., 2020; Godinho, 2011). Higher concentrations (>1%) of agarose could have potential side effects such as suffocation due to low oxygen diffusion through the agarose. An ideal method would be to use a lower concentration of agarose that would not affect physiology substantially; however, an additional analytical method to correct for motion artefacts would be needed (Dunn et al., 2016).

However, this regimen of low agarose concentration is not ideal for larvae >3 dpf because the anaesthetic concentration might not be sufficient to yield complete muscle relaxation for a longer duration (>24 h) of imaging. Furthermore, higher concentrations of agarose are known to result in growth defects. There is a thus a balance to be struck between restraint and optimal physiological conditions. Some studies use a bungarotoxin to immobilize larvae with lower concentrations of agarose and mild <0.4% tricaine to image post 3 dpf. This is not a common method of choice because the larvae would be partially conscious throughout and paralysis without full anaesthesia and analgesia is not ethical (Swinburne et al., 2015).

Recent studies describe novel methods for long-term lapse imaging using adaptive optical gating combined with sophisticated computational methods to compensate for motion correction (Taylor et al., 2019). Recent advances in the development of different mounting techniques for imaging in LSFM provide novel ways to maintain correct orientation, allowing for freedom of movement for the outgrowth of the posterior body without affecting its normal development (Hirsinger & Steventon, 2017). Studies by Kaufmann et al. (2012) suggested fluorinated ethylene propylene (FEP) filled with low concentrations of agarose or methylcellulose to provide an optimal balance between sufficient confinement of the living embryo in a physiological environment over 3 days and optical clarity suitable for fluorescent imaging (Kaufmann et al., 2012).

Example experimental set-ups

Zebrafish are popularly used for imaging at various levels of organization due to their small size, transparency, availability of a repertoire of transgenics and HT properties. If money is no object and one facility could house all the microscopes on offer, the options for imaging would be extensive. The researcher could analyse whole systems such as the circulatory system and then focus on specific organs, structures and move onto cells and even protein-level imaging. This section primarily covers some important examples of zebrafish imaging model systems that have led to deeper understanding of biological phenomena. The examples below are constrained by larval zebrafish applications. While quantification and imaging at the organ level can be done at the level of the adult zebrafish, detailed imaging of sub-organ-level applications is a work in progress. In particular, state-of-art imaging technologies being deployed have been emphasized with a brief outline of the corresponding optics. This can be treated as a guide for the reader to understand the applications of the microscopy techniques described earlier in the chapter.

Structural behaviours

Biomolecular-level behaviours

Fluorescent microscopy is not adequate to deliver optimal aid in imaging of complex biomolecular interactions and dynamics. Recent developments in label-free imaging using Raman spectroscopy are becoming a popular method of choice to visualize and analyse local biomolecular heterogeneity in different live zebrafish embryo models. One of the first versions of Raman spectroscopy imaging (Müller et al., 2016) lacked the desired spatial resolution in live embryos but is a method of choice for 'label-free fixed-tissue imaging'. The Raman technique can be used in combination with lightsheet microscopy or confocal microscopy to provide spatially resolved hyperspectral, label-free imaging of zebrafish embryos (Høgset et al., 2020). Using an optically transparent mutant line lacking iridophore and melanophore pigments, Høgset et al. (2020) demonstrated imaging of biomolecules including lipids, proteins and nucleic acids with a 63× objective and 532 nm laser in confocal Raman spectroscopic imaging (Høgset et al., 2020).

Organelle-level behaviours

With the advent of novel, non-invasive optical technologies such as lightsheet microscopy, it is now possible to image spatially complex sub-assemblies in a living cell. We elaborate on this with one specific organelle, the mitochondrion, which is one of the most studied and extensively imaged organelles, in vivo, in zebrafish models. Due to their highly dynamic nature and importance in cellular energetics, various diseases are associated with mitochondrial dysfunction. Mitochondrial movement and transport can be imaged in vivo up to 48 hpf in the afferent axons of posterior lateral line nerves using simple episcope and kymograph analysis (Marra et al., 2015) (Figure 11.6). At the same time, Dalgin and Prince (2015) used a novel transgenic line with a *neurod* promoter driving expression of cytochrome C oxidase attached to a fluorescent reporter. A 63×, NA 1.2 objective with a 100 × 100 um^2 field of view was used on an LSM800 microscope. Various mitochondrial transport parameters, including distance, velocity and direction, from 3D + t data

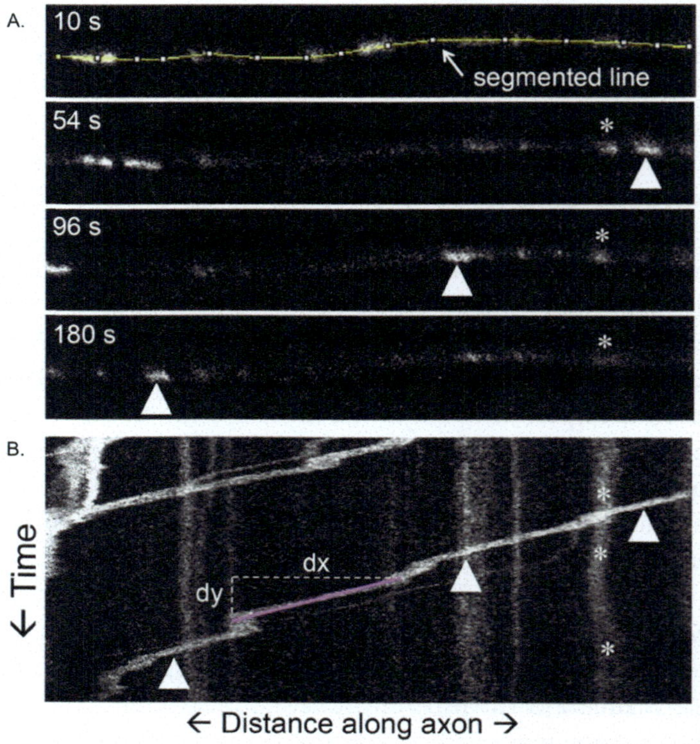

Figure 11.6 Mitochondrial movement analysis in the lateral line nerves of a 48 hpf embryo. Adapted from Marra et al. (2015).

obtained from fluorescent microscopy were extracted using kymograph analysis via either MATLAB- or Python-based codes, or with software such as MetaMorph (Biovision).

Single-cell mitochondrial dynamics can be investigated using a wide-field fluorescent microscope with 100× water-immersion dipping-cone objective and NA 1.0 coupled to a cooled CCD. The open-source microscopy software μManager is a tool of choice to analyse data (Stuurman et al., 2007) whereas, for bulk mitochondrial assays, wide-field fluorescent microscopy is often used in conjunction with a 40×/NA 0.8 water-dipping-cone objective. More recently, Liu et al. (2018) reported a further advancement in the method of imaging sub-cellular processes using lattice lightsheet microscopy with aberration-correcting adaptive optics in live zebrafish embryos. The various processes that can be imaged range from nanoscale diffusion of clathrin-coated pits, organelle remodelling during mitosis and growth cone dynamics during spinal cord development to axon-guided motility.

Organ- and cellular-level behaviours

A prime goal of physiology is to understand the basis of different bodily behaviours and understand how different organ systems interact to bring about homeostasis. Larval zebrafish are an attractive model for investigation of different organ systems and their behavioural repertoire. With the advent of various cellular reporters, in particular calcium reporters, there has been an upsurge in the study of cellular-level dynamics in zebrafish.

Organ-level behaviours

In the neuroscience domain, zebrafish models are an indispensable tool to understanding the mechanisms of various behaviours. Whole-brain functional imaging has been made possible with the use of volume imaging of calcium activations of all neurons using lightsheet microscopy, although in a restraint/head-fixed larval zebrafish (Ahrens et al., 2012). With recent advances in imaging of free-moving animals, it is now possible to conduct volumetric neuronal calcium imaging on freely behaving larval zebrafish (Cong et al., 2017).

The vertebrate retina forms another important organ for the study of neuronal circuits. The stereotyped laminar organization makes it an ideal model system to study neuronal circuit development. With the advancement in multi-photon imaging and readily available transgenics, there is an expanse of recent studies focusing on functional imaging of the developing zebrafish retina (Williams et al., 2013). For both confocal and multi-photon in vivo imaging of the retina, larvae need to be treated with PTU and can be immobilized in a simple, custom-designed imaging chamber (Figure 11.7). A repertoire of both commercial and non-commercial software packages is available to analyse 3D and/or 3D + t stacks such as ImageJ, Amira, Imaris and MetaMorph for processing and image visualization.

Imaging and accurate quantification of the heartbeat in zebrafish models is important for investigation of cardiovascular development and disorders. Heart rate can be quantified from high-frame-rate videos acquired, both at the single embryo level and in HT assays. Current methods of quantification require immobilization via mechanical constraint (agarose) and/or anaesthesia.

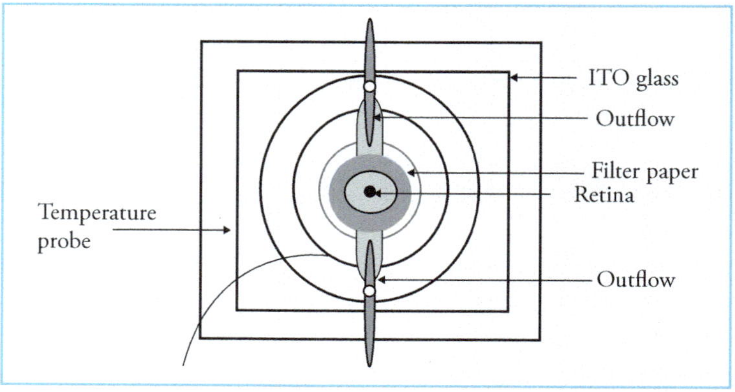

Figure 11.7 Schematic of custom-designed imaging chamber for live imaging of the retina in larval zebrafish.

Using two-photon microscopy, organogenesis during development has been studied extensively by visualization of polarized epithelial cells stained with Laurdan dye (Owen et al., 2010).

Cellular-level behaviours

The zebrafish is used extensively in cancer research owing to the availability of numerous genomic tools such as retroviral-based insertional mutagenesis, the LexPR system, Talens, CRISPr and the miniCoopR shuttle vector system. Studies have established various cancer pathways and mechanisms using several imaging modalities in zebrafish. Ridges et al. (2012) identified a selective inhibitor of lymphocyte proliferation using live imaging of zebrafish embryos with fluorescence microscopy. Combining novel mutant lines with confocal microscopy, the dynamics of tumour cells secreting human vascular endothelial growth factor have been investigated (Stoletov et al., 2007).

Kupffer's vesicle (KV) in zebrafish developmental stages is another important model system that extensively deploys imaging modalities to study the role of cilia in left–right patterning. DIC microscopy is a standard tool used to image KV in live embryos. To image fluid dynamics in KV, FluoSpheers® Fluorescent microspheres (Molecular Probes) are injected and imaged using DIC with a 60× water-immersion objective lens (Okabe et al., 2008). Recent advancements in DIC microscopy allow for a combination of the microscope with a high-speed camera (such as an Andor iXon camera), with speeds ranging between 100 and 200 frames per second for cilial motility and 50 frames per second for bead movements. The videos subsequently obtained can be analysed using custom Fiji plugin or MATLAB. Sub-organ/subcellular vacuole movements can be tracked using a spinning-disk confocal microscope with a 40× water-immersion objective (e.g. Perkin Elmer UltraVIEW Vox). Recent developments in stimulated emission depletion (STED) imaging using a 93× 1.3 NA motorized collar objective allow for four-dimensional time-lapse imaging with higher signal-to-noise ratio compared with confocal microscopy (Freifeld et al., 2017; Zhang et al., 2019).

Conclusion

In this chapter we have given an account of various imaging techniques that can be used for imaging live or fixed zebrafish specimens. The fundamentals of standard imaging microscopes are highlighted, taking the reader through their optics to aid researchers and clinicians in making an informed decision in the choice of imaging set-up. In general, the chapter provides multiple imaging options depending on whether the goal is to image dynamical biological phenomena or the structural specifics of whole tissue/macromolecules. Once selected, the microscopy specifics can be further customized using state-of-art developments in optics such as different excitation beams, objective lenses, filter sets and detector configurations. We further provide a collection of standard protocols for both mounting of live embryos and sample preparation of fixed tissue, briefly covering the specifics of chemical reagents needed. While there are fewer considerations needed for fixed-tissue imaging, there are critical parameters in live-embryo imaging that require thorough assessment. These considerations are necessary for maintaining the optimal physiological state such that dynamical measurements are unbiased. Finally we provide our readers with novel and state-of-art imaging techniques for imaging biological processes at the organismic (and sub-organismic) level. For example, we give an account of published literature describing imaging at various biological levels such as whole organism, organ, cellular and organelle structure, and function. This account also points to the rich diversity of functions that can be studied using the zebrafish as model. Furthermore, it should be noted that most of the techniques described in the chapter emphasize the non-invasive and hence more physiological nature of the imaging options and set-ups that have been developed for zebrafish. This is important because there are current issues with other mammalian animal models with respect to imaging that involve invasive manipulations for sample preparations, especially for live imaging. Our hope is to make this repertoire of rich techniques available to all researchers to enable them to quantify biological processes in the most optimal physiological state.

Acknowledgments

We would like to thank all authors of the book for their constant feedback on the chapter: specifically, Dr Claire E. Allen and Dr J.-P. Mocho for their most useful comments on the content of the chapter.

Further acknowledgements go to Dr Ryan MacDonald and Dr Emily Noel for their initial input into the chapter.

References

Ahrens, M. B., Li, J. M., Orger, M. B., Robson, D. N., Schier, A. F., Engert, F., & Portugues, R. (2012). Brain-wide neuronal dynamics during motor adaptation in zebrafish. *Nature*, *485*(7399), 471–477. https://doi.org/10.1038/nature11057

Bang, P. I., Yelick, P. C., Malicki, J. J., & Sewell, W. F. (2002). High-throughput behavioral screening method for detecting auditory response defects in zebrafish. *Journal of Neuroscience Methods*, *118*(2), 177–187. https://doi.org/10.1016/s0165-0270(02)00118-8

Blackburn, J. S., Liu, S., Raimondi, A. R., Ignatius, M. S., Salthouse, C. D., & Langenau, D. M. (2011). High-throughput imaging of adult fluorescent zebrafish with an LED fluorescence macroscope. *Nature Protocols*, *6*(2), 229–241. https://doi.org/10.1038/nprot.2010.170

Cheng, C. N., Li, Y., Marra, A. N., Verdun, V., & Wingert, R. A. (2014). Flat mount preparation for observation and analysis of zebrafish embryo specimens stained by whole mount in situ hybridization. *Journal of Visualized Experiments: JoVE*, (89), 51604. https://doi.org/10.3791/51604

Chhabria, K., Plant, K., Bandmann, O., Wilkinson, R. N., Martin, C., Kugler, E., Armitage, P. A., Santoscoy, P. L., Cunliffe, V. T., Huisken, J., McGown, A., Ramesh, T., Chico, T. J., & Howarth, C. (2020). The effect of hyperglycemia on neurovascular coupling and cerebrovascular patterning in zebrafish. *Journal of Cerebral Blood Flow and Metabolism*, *40*(2), 298–313. https://doi.org/10.1177/0271678X18810615

Cong, L., Wang, Z., Chai, Y., Hang, W., Shang, C., Yang, W., Bai, L., Du, J., Wang, K., & Wen, Q. (2017). Rapid whole brain imaging of neural activity in freely behaving larval zebrafish (Danio rerio). *eLife, 6,* e28158. https://doi.org/10.7554/eLife.28158

Copper, J. E., Budgeon, L. R., Foutz, C. A., van Rossum, D. B., Vanselow, D. J., Hubley, M. J., Clark, D. P., Mandrell, D. T., & Cheng, K. C. (2018). Comparative analysis of fixation and embedding techniques for optimized histological preparation of zebrafish. *Comparative Biochemistry and Physiology. Toxicology and Pharmacology, 208*, 38–46. https://doi.org/10.1016/j.cbpc.2017.11.003

Dalgin, G., & Prince, V. E. (2015). Differential levels of Neurod establish zebrafish endocrine pancreas cell fates. *Developmental Biology, 402*(1), 81–97. https://doi.org/10.1016/j.ydbio.2015.03.007

DaniScope. https://www.noldus.com/danioscope Retrieved November 7, 2023.

Das, T., Payer, B., Cayouette, M., & Harris, W. A. (2003). In vivo time-lapse imaging of cell divisions during neurogenesis in the developing zebrafish retina. *Neuron, 37*(4), 597–609. https://doi.org/10.1016/s0896-6273(03)00066-7

Davis, A. E., Castranova, D., & Weinstein, B. M. (2021). Rapid generation of pigment free, immobile zebrafish embryos and larvae in any genetic background using CRISPR-Cas9 dgRNPs. *Zebrafish, 18*(4), 235–242. https://doi.org/10.1089/zeb.2021.0011

Dunn, T. W., Mu, Y., Narayan, S., Randlett, O., Naumann, E. A., Yang, C. T., Schier, A. F., Freeman, J., Engert, F., & Ahrens, M. B. (2016). Brain-wide mapping of neural activity controlling zebrafish exploratory locomotion. *eLife, 5,* e12741. https://doi.org/10.7554/eLife.12741

Easter, Jr., S. S., & Nicola, G. N. (1997). The development of eye movements in the zebrafish (Danio rerio). *Developmental Psychobiology, 31*(4), 267–276. https://doi.org/10.1002/(sici)1098-2302(199712)31:4<267::aid-dev4>3.0.co;2-p

Elsalini, O. A., & Rohr, K. B. (2003). Phenylthiourea disrupts thyroid function in developing zebrafish. *Development Genes and Evolution, 212*(12), 593–598. https://doi.org/10.1007/s00427-002-0279-3

Freifeld, L., Odstrcil, I., Förster, D., Ramirez, A., Gagnon, J. A., Randlett, O., Costa, E. K., Asano, S., Celiker, O. T., Gao, R., Martin-Alarcon, D. A., Reginato, P., Dick, C., Chen, L., Schoppik, D., Engert, F., Baier, H., & Boyden, E. S. (2017). Expansion microscopy of zebrafish for neuroscience and developmental biology studies. Proceedings of the National Academy of Sciences of the United States of America, 114(50), E10799–E10808. https://doi.org/10.1073/pnas.1706281114

Godinho, L. (2011). Imaging zebrafish development. Cold Spring Harbor Protocols, 2011(7), 879–883. https://doi.org/10.1101/pdb.prot5647.

Guo, Y., Veneman, W. J., Spaink, H. P., & Verbeek, F. J. (2017). Three-dimensional reconstruction and measurements of zebrafish larvae from high-throughput axial-view in vivo imaging. *Biomedical Optics Express*, *8*(5), 2611–2634. https://doi.org/10.1364/BOE.8.002611

Hirsinger, E., & Steventon, B. (2017). A versatile mounting method for long term imaging of zebrafish development. *Journal of Visualized Experiments: JoVE*, (119), 55210. https://doi.org/10.3791/55210

Høgset, H., Horgan, C. C., Armstrong, J. P. K., Bergholt, M. S., Torraca, V., Chen, Q., Keane, T. J., Bugeon, L., Dallman, M. J., Mostowy, S., & Stevens, M. M. (2020). In vivo biomolecular imaging of zebrafish embryos using confocal Raman spectroscopy. *Nature Communications*, *11*(1), 6172. https://doi.org/10.1038/s41467-020-19827-1

Johnson, D. S., Deneau, T., Toledo-Crow, R., & Simon, S. M. (2001). Microscope enclosure for temperature regulation and light isolation. *Applied Sciences*, *11*(15), 6812. https://doi.org/10.3390/app11156812

Kabra, M., Robie, A. A., Rivera-Alba, M., Branson, S., & Branson, K. (2013). JAABA: Interactive machine learning for automatic annotation of animal behavior. *Nature Methods*, *10*(1), 64–67. https://doi.org/10.1038/nmeth.2281

Karlsson, J., von Hofsten, J., & Olsson, P. E. (2001). Generating transparent zebrafish: A refined method to improve detection of gene expression during embryonic development. *Marine Biotechnology*, *3*(6), 522–527. https://doi.org/10.1007/s1012601-0053-4

Kaufmann, A., Mickoleit, M., Weber, M., & Huisken, J. (2012). Multilayer mounting enables long-term imaging of zebrafish development in a light sheet microscope. *Development*, *139*(17), 3242–3247. https://doi.org/10.1242/dev.082586

Keller, P. J. (2013). In vivo imaging of zebrafish embryogenesis. *Methods*, *62*(3), 268−278. https://doi: 10.1016/j.ymeth.2013.03.015

Klabunde, T., Eicken, C., Sacchettini, J. C., & Krebs, B. (1998). Crystal structure of a plant catechol oxidase containing a dicopper center. *Nature Structural Biology*, *5*(12), 1084–1090. https://doi.org/10.1038/4193

Kleinhans, D. S., & Lecaudey, V. (2019). Standardized mounting method of (zebrafish) embryos using a 3D-printed stamp for high-content, semi-automated confocal imaging. *BMC Biotechnology*, *19*(1), 68. https://doi.org/10.1186/s12896-019-0558-y

Lister, J. A., Robertson, C. P., Lepage, T., Johnson, S. L., & Raible, D. W. (1999). Nacre encodes a zebrafish microphthalmia-related protein that regulates neural-crest-derived pigment cell fate. *Development, 126*(17), 3757–3767. https://doi.org/10.1242/dev.126.17.3757

Liu, T. L., Upadhyayula, S., Milkie, D. E., Singh, V., Wang, K., Swinburne, I. A., Mosaliganti, K. R., Collins, Z. M., Hiscock, T. W., Shea, J., Kohrman, A. Q., Medwig, T. N., Dambournet, D., Forster, R., Cunniff, B., Ruan, Y., Yashiro, H., Scholpp, S., Meyerowitz, E. M., . . . Betzig, E. (2018). Observing the cell in its native state: Imaging subcellular dynamics in multicellular organisms. *Science, 360*(6386). https://doi.org/10.1126/science.aaq1392

Marra, M. H., Tobias, Z. J., Cohen, H. R., Glover, G., & Weissman, T. A. (2015). In vivo time-lapse imaging in the zebrafish lateral line: A flexible, open-ended research project for an undergraduate neurobiology laboratory course. *Journal of Undergraduate Neuroscience Education, 13*(3), A215–A224.

Masselink, W., Wong, J. C., Liu, B., Fu, J., & Currie, P. D. (2014). Low-cost silicone imaging casts for zebrafish embryos and larvae. *Zebrafish, 11*(1), 26–31. https://doi.org/10.1089/zeb.2013.0897

Moens, C. (2008). Whole mount RNA in situ hybridization on zebrafish embryos: Mounting. CSH Protocols, 2008, pdb.prot5038. https://doi.org/10.1101/pdb.prot5038

Müller, W., Kielhorn, M., Schmitt, M., Popp, J., & Heintzmann, R. (2016). Light sheet Raman micro-spectroscopy. *Optica, 3*(4), 452–457. https://doi.org/10.1364/OPTICA.3.000452

Okabe, N., Xu, B., & Burdine, R. D. (2008). Fluid dynamics in zebrafish Kupffer's vesicle. *Developmental Dynamics: An Official Publication of the American Association of Anatomists, 237*(12), 3602–3612. https://doi.org/10.1002/dvdy.21730

Olt, J., Allen, C. E., & Marcotti, W. (2016). In vivo physiological recording from the lateral line of juvenile zebrafish. *Journal of Physiology, 594*(19), 5427–5438. https://doi.org/10.1113/JP271794

Oralová, V., Rosa, J. T., Soenens, M., Bek, J. W., Willaert, A., Witten, P. E., & Huysseune, A. (2019). Beyond the whole-mount phenotype: High-resolution imaging in fluorescence-based applications on zebrafish. *Biology Open, 8*(5), bio042374. https://doi.org/10.1242/bio.042374

Owen, D. M., Magenau, A., Majumdar, A., & Gaus, K. (2010). Imaging membrane lipid order in whole, living vertebrate organisms. *Biophysical Journal, 99*(1), L7–L9. https://doi.org/10.1016/j.bpj.2010.04.022

Pardo-Martin, C., Chang, T. Y., Koo, B. K., Gilleland, C. L., Wasserman, S. C., & Yanik, M. F. (2010). High-throughput in vivo vertebrate screening. *Nature Methods, 7*(8), 634–636. https://doi.org/10.1038/nmeth.1481

Ridges, S., Heaton, W. L., Joshi, D., Choi, H., Eiring, A., Batchelor, L., Choudhry, P., Manos, E. J., Sofla, H., Sanati, A., Welborn, S., Agarwal, A., Spangrude, G. J., Miles, R. R., Cox, J. E., Frazer, J. K., Deininger, M., Balan, K., . . . Trede, N. S. (2012). Zebrafish screen identifies novel compound with selective toxicity against leukemia. Blood, 119(24), 5621–5631. https://doi.org/10.1182/blood-2011-12-398818

Santi, P. A. (2011). Light sheet fluorescence microscopy: A review. *Journal of Histochemistry and Cytochemistry, 59*(2), 129–138. https://doi.org/10.1369/0022155410394857

Stoletov, K., Montel, V., Lester, R. D., Gonias, S. L., & Klemke, R. (2007). High-resolution imaging of the dynamic tumor cell–vascular interface in transparent zebrafish. *Proceedings of the National Academy of Sciences of the United States of America, 104*(44), 17406–17411. https://doi.org/10.1073/pnas.0703446104

Stuurman, N., Amdodaj, N., & Vale, R. (2007). µManager: Open source software for light microscope imaging. *Microscopy Today, 15*(3), 42–43. https://doi.org/10.1017/S1551929500055541

Sun, C. K., Chu, S. W., Chen, S. Y., Tsai, T. H., Liu, T. M., Lin, C. Y., & Tsai, H. J. (2004). Higher harmonic generation microscopy for developmental biology. *Journal of Structural Biology, 147*(1), 19–30. https://doi.org/10.1016/j.jsb.2003.10.017

Swinburne, I. A., Mosaliganti, K. R., Green, A. A., & Megason, S. G. (2015). Improved long-term imaging of embryos with genetically encoded α-bungarotoxin. *Public library of Science one. PloS One, 10*(8), e0134005. https://doi.org/10.1371/journal.pone.0134005

Taylor, J. M., Nelson, C. J., Bruton, F. A., Kaveh, A., Buckley, C., Tucker, C. S., Rossi, A. G., Mullins, J. J., & Denvir, M. A. (2019). Adaptive prospective optical gating enables day-long 3D time-lapse imaging of the beating embryonic zebrafish heart. *Nature Communications, 10*(1), 5173. https://doi.org/10.1038/s41467-019-13112-6

Thisse, C., & Thisse, B. (2008). High-resolution in situ hybridization to whole-mount zebrafish embryos. *Nature Protocols, 3*(1), 59–69. https://doi.org/10.1038/nprot.2007.514

Tingaud-Sequeira, A., Forgue, J., André, M., & Babin, P. J. (2006). Epidermal transient down regulation of retinol-binding protein 4 and mirror expression of apolipoprotein Eb and estrogen receptor 2a during zebrafish fin and scale development. *Developmental Dynamics*, *235*(11), 3071–3079. https://doi.org/10.1002/dvdy.20921

Upadhyay, S., Vergara, L., Shah, P., Gustafsson, J. Å., Kakadiaris, I., & Bondesson, M. (2020). A layered mounting method for extended time-lapse confocal microscopy of whole zebrafish embryos. *Journal of Visualized Experiments: JoVE*, (155), 60321. https://doi.org/10.3791/60321

Walker, S. L., Ariga, J., Mathias, J. R., Coothankandaswamy, V., Xie, X., Distel, M., Köster, R. W., Parsons, M. J., Bhalla, K. N., Saxena, M. T., & Mumm, J. S. (2012). Automated reporter quantification in vivo: High-throughput screening method for reporter-based assays in zebrafish. *PloS One*, *7*(1), e29916. https://doi.org/10.1371/journal.pone.0029916

Wang, W. D., Wang, Y., Wen, H. J., Buhler, D. R., & Hu, C. H. (2004). Phenylthiourea as a weak activator of aryl hydrocarbon receptor inhibiting 2, 3, 7, 8-tetrachlorodibenzo-p-dioxin-induced CYP1A1 transcription in zebrafish embryo. *Biochemical Pharmacology*, *68*(1), 63–71. https://doi.org/10.1016/j.bcp.2004.03.010

Welten, M. C., De Haan, S. B., van den Boogert, N., Noordermeer, J. N., Lamers, G. E., Spaink, H. P., Meijer, A. H., & Verbeek, F. J. (2006). ZebraFISH: Fluorescent in situ hybridization protocol and three-dimensional imaging of gene expression patterns. *Zebrafish*, *3*(4), 465–476. https://doi.org/10.1089/zeb.2006.3.465

Williams, P. R., Morgan, J. L., Kerschensteiner, D., & Wong, R. O. (2013). In vivo imaging of zebrafish retina. *Cold Spring Harbor Protocols*, *2013*(1). https://doi.org/10.1101/pdb.prot072652

Zhang, W., Noa, A., Nienhaus, K., Hilbert, L., & Nienhaus, G. U. (2019). Super-resolution imaging of densely packed DNA in nuclei of zebrafish embryos using stimulated emission double depletion microscopy. *Journal of Physics. Part D*, *52*(41), 414001. https://doi.org/10.1088/1361-6463/ab3200

Chapter 12

Behaviour in zebrafish

Raquel R. Martins and Paul G. Schroeder

Abstract

The zebrafish is a popular model for behavioural studies and has been widely used to study several complex brain disorders, including cognitive-, anxiety- and drug-related conditions. As a gregarious species these fish display a wide range of social behaviours, including the ability to recognize, interact and learn from conspecifics. Zebrafish can perform short- and long-term memory tasks and they are capable of classical and operant conditioning. Stress responses and behavioural flexibility may be associated with individual personality traits and affective states, and can affect the motivational state and learning abilities of this species. Importantly, zebrafish behaviour can be highly influenced by housing conditions, including the presence of conspecifics and environmental enrichment. This chapter describes some of the most common behavioural assays used in zebrafish, with a practical focus and emphasis on key aspects of zebrafish ecology, welfare and behaviour that may impact on experimental design.

Keywords

zebrafish, social behaviour, stress responses, anxiety-like behaviour, cognition

Claire Allen and Jean-Philippe Mocho (eds) **Zebrafish**
DOI: 10.1079/9781800629431.0012,

Introduction

George Streisinger, the first researcher to bring zebrafish into the laboratory, realized that, in addition to its favourable embryologic and genetic features, the zebrafish is a suitable model for behavioural studies, namely for visual behaviour screens. Streisinger and colleagues were pioneers in observing the developmental visual system in zebrafish and using behaviour assays to evaluate visual defects in larvae (Clark, 1981). Since then, zebrafish behaviour has been widely studied and many behavioural paradigms have been developed or adapted for this vertebrate model in order to test social behaviour, stress responses, cognition, sensorial abilities and others (Kalueff, 2017; Kalueff & Cachat, 2011; Kalueff et al., 2013). The zebrafish has proved an appropriate model to illustrate key behaviourally mediated neural circuits (Roberts et al., 2013) and for the study of a wide variety of complex brain disorders including those related to anxiety or fear (Piato et al., 2011; Stewart et al., 2010), cognition (Engeszer et al., 2004; Roberts et al., 2013), attention deficit hyperactivity (ADHD; Lange et al., 2018) and drug-related conditions (Cleal et al., 2021).

At the same time, it may be problematic to evaluate this species' behavioural baseline in an artificial environment such as the laboratory. To overcome this challenge, and to reliably recognize and measure certain behaviours, it is essential to develop a detailed understanding of the zebrafish. Thankfully today there are glossaries (Kalueff et al., 2013), including illustrated ones (Kalueff, 2017), describing the different behaviours of zebrafish and respective terminology. It is also crucial to identify possible confounding artefacts that may result from everyday events that may occur either before or during the assay. Slight visual, olfactory and/or auditory disturbances, changes in temperature, fish handling, acclimatization, time of day and general environmental conditions of captivity are examples of possible variables that can influence experimental outcomes (Ninkovic & Bally-Cuif, 2006). Knowledge of zebrafish ecology is a prerequisite to identifying these possible artefacts and their natural history is, therefore, briefly discussed in this chapter. Finally, the design of adequate controls – a critical requirement in every experiment – can be a significant challenge but will contribute to achieving more robust and reproducible behavioural assays.

This chapter will explore the most common behavioural assays used in zebrafish and emphasize key aspects of zebrafish ecology, welfare and behaviour that may impact on experimental design and interpretation of results.

Zebrafish ecology and its influence on zebrafish behaviour

Danio rerio (also known as *Brachydanio rerio*; Hamilton-Buchanan, 1822) is a freshwater teleost in the Cyprinidae family (Spence et al., 2008). Its natural range is limited to South Asia, and the species can be found in India (Arunachalam et al., 2013; McClure et al., 2006; Sterba, 1962), Bangladesh (Spence et al., 2006), Thailand (McClure et al., 2006), Pakistan, Myanmar and Sri Lanka (Karmakar et al., 2008).

In the wild, the zebrafish abundantly inhabits ditches, ponds and shallow lakes, often associated with rice cultivation (Arunachalam et al., 2013; Engeszer et al., 2007; Lawrence, 2007; McClure et al., 2006; Spence et al., 2006). These water bodies have been described as warm (24–31°C), shallow (1–100 cm depth), slow-moving waters, rich in vegetation and having a silty bottom with a gravel, sand or boulder substratum (Engeszer et al., 2007; McClure et al., 2006; Spence et al., 2006). Less abundantly, zebrafish can also be found in fast-flowing hill streams and upper reaches of rivers (Arunachalam et al., 2013; Lawrence, 2007; Spence et al., 2006). The zebrafish is described as a gregarious species, living in shoals of varying size (from <12 to >300 individuals), with both male and female in similar numbers (Suriyampola et al., 2016). Both the varied habitat selection and shoaling behaviour work as protection against predators. Of the latter, the most common are the Indian leaf fish (*Nandus nandus*), snakeheads (*Channa* spp.) and freshwater garfish (*Xenentodon cancila*). Putative avian predators, including the Indian pond heron (*Ardeola grayii*) and the kingfisher

(*Alcedo atthis*), have also been identified in areas with *D. rerio* habitats (Spence et al., 2006).

These actualities are likely to affect zebrafish behaviour, not only in the wild but also in the laboratory setting. For zebrafish in captivity an optimum temperature of 28.5°C is usually quoted (Lawrence, 2007) as this favours breeding performance and growth rate, although a wider optimum range of 24–30°C has been proposed (Matthews et al., 2002). In addition, laboratory experiments have shown that zebrafish prefer to be in a shoal rather than being kept in isolation (detailed later in this chapter; Ruhl & McRobert, 2005; Spence et al., 2008), and spend more time in an enriched environment (i.e. a tank with objects such as artificial plants and/or gravel) than in a barren one, when given a choice (Schroeder et al., 2014). Environmental enrichment has not only been reported to increase spawning (Wafer et al., 2016), but also to influence sex differentiation (or sex ratios) in domesticated laboratory zebrafish strains (Ribas et al., 2017) lacking the primary sex-determining gene (Wilson et al., 2014).

As with other species, when it comes to mating and spawning, females and males differ in their motivation and resulting behaviour; males defend a spawning area while females aggressively target other females in mid-water (Magurran & Maciás Garcia, 2005). Therefore, zebrafish have been identified as a useful organism for investigation of teleost courtship behaviour and its development. Courtship can be divided into three consecutive stages – initiatory, receptive/appetitive and spawning, with the successful outcome best facilitated by pairing males and females of similar or identical age categories (Darrow & Harris, 2004).

Activity levels in adult zebrafish of both sexes are heavily affected by the light/dark cycle, with the overwhelming majority of animals most active during the light phase of the cycle in a 12h/12h light/dark cycle (Hurd et al., 1998). When the alternating light/dark pattern is replaced by constant conditions, locomotor rhythmicity is preserved in most individuals, both in constant light and the dark. However, this is heavily dependent on temperature, with the highest proportion of fish retaining rhythmicity at 21°C (Hurd et al., 1998). This suggests that the circadian clock regulating zebrafish activity is modulated by temperature and, because rhythmicity is most robust at 21°C, reconsideration of optimal physiological temperature conditions for this species may be pertinent.

A varied, small-invertebrate-based diet in the wild indicates great capacity to adjust to different environments (Engeszer et al., 2007). Nonetheless, studies on captive intraspecific variation in feeding behaviour in zebrafish have shown that risk taking associated with foraging behaviour depends on nutrient demand. Not only is this attenuated by aversive stimuli but it can also be strain dependent, indicating that intraspecific variation in foraging behaviour has a genetic basis and supporting the more general idea that behavioural responses to manipulations may be strain specific (Oswald & Robison, 2008).

Together, the principal abiotic water quality parameters (temperature, pH and salinity), housing (single versus grouped housed and enriched versus barren environment), light/dark cycle and diet have all been reported to affect zebrafish behaviour and physiology (DePasquale et al., 2016; Marcon et al., 2018; McClure et al., 2006; von Krogh et al., 2010). Hence, these need to be considered when designing a behavioural assay.

Key zebrafish behaviours and behavioural paradigms

Shoaling and social behaviour

The formation of shoals increases the likelihood of mating/reproduction, promotes foraging and confers protection against predators ('safety in numbers'; Kalueff, 2017). As a gregarious species, the zebrafish shoals from very early stages, more specifically from the tenth day post fertilization (dpf), known as the flexion stage. This behaviour rapidly becomes fully established, because from the 16th dpf (post-flexion stage) zebrafish shoal preferences do not differ from adults (Engeszer et al., 2007). Supporting the idea that this is an innate behaviour, even when reared in isolation, fish form shoals when in the presence of conspecifics (Spence et al., 2008).

There is evidence that the social setting experienced in the first days post hatching defines any subsequent preference behaviour. Zebrafish prefer to be next to conspecifics that have the same pigment pattern as the fish they were raised with (Engeszer et al., 2007). For example, when wild-type (WT) are cross-reared with *nacre* fish (lacking melanophore stripes), they prefer to be next to the phenotype with which they were reared in preference to their own phenotype (Engeszer et al., 2004). Moreover, fish reared in isolation seem to be unable to discriminate between conspecifics with different pigment patterns (McCann & Matthews, 1974) and therefore do not have a preference for a specific phenotype. This may be because exposure to kin cues during the first days of life is crucial for imprinting. Olfactory cues from kin individuals at 6 dpf have been shown to be crucial for imprinting, whereas larvae exposed to these cues before or after this stage do not seem to develop imprinting (Gerlach et al., 2008).

These data suggest that, while the tendency to form a shoal is an innate behaviour, the preference for a certain strain is acquired at an early age. Furthermore they show that zebrafish can recognize conspecifics, a feature of social cognition. Several paradigms have been developed to assess social preference and social cognition for individual zebrafish based on this species' repertoire of shoaling behaviours.

Social preference is usually tested by a 'social preference test' (Pham et al., 2012). Briefly, this assay involves transferring a fish into an experimental three-compartment tank where it is exposed to an empty compartment versus a conspecific or a shoal (Figure 12.1). The time spent next to each of the compartments is observed, and this measure is used to determine whether there is any preference between the choices. As an alternative, animated fish images can be displayed to mimic a shoal, evoking similar group preference (Saverino & Gerlai, 2008).

Shoal interaction (i.e. affiliate behaviour) can be influenced by multiple stimuli, including visual cues (Larsch & Baier, 2018; Nunes et al., 2020). One example is exposure to stress, which has been reported to promote shoal cohesion (Chakravarty et al., 2013; Piato et al., 2011) – see below. This has been validated by the observation

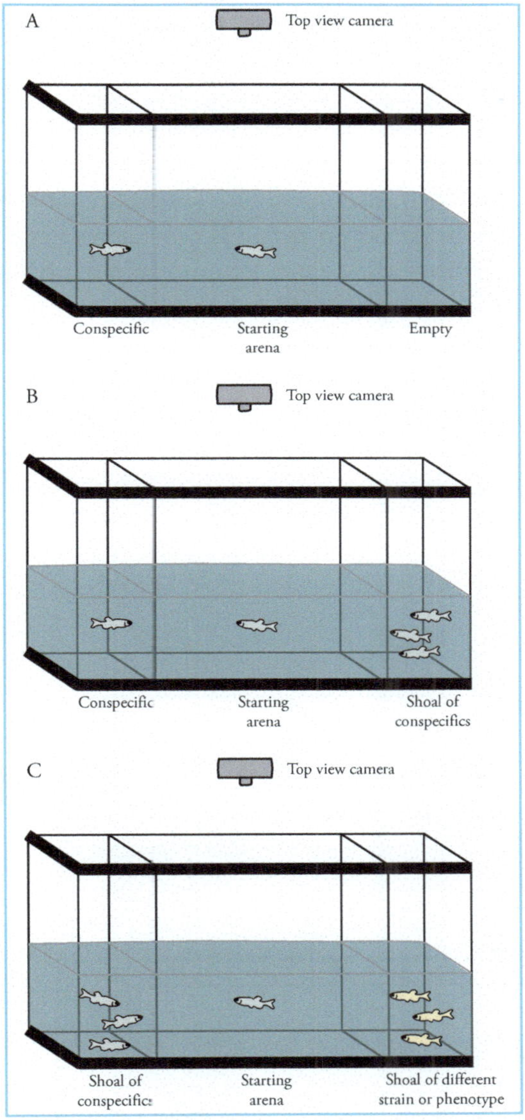

Figure 12.1 Schematic illustration of a social preference test. The tank is divided into three compartments, separated by transparent barriers. In the left- and right-hand compartments are presented the shoal stimuli: testing preference for (A) social isolation versus conspecific, (B) one conspecific versus a shoal of three conspecific individuals and (C) testing preference for shoaling with the same phenotype versus a different one. Adapted from Engeszer et al. (2007) and Ruhl and McRobert (2005).

that administration of anxiolytic drugs (e.g. scopolamine) leads to decreased shoal cohesion (Hamilton et al., 2017).

Affiliate behaviour can be assessed by a 'group behaviour test' (Pham et al., 2012; Piato et al., 2011), which consists of placing a small group of fish in a tank and observing the following: (1) vertical position in the tank (bottom dwelling suggesting anxiety), (2) locomotion (inhibition or excitement), (3) coloration (pale or chatoyant coloration suggesting fear or aggression/excitement) and (4) shoal cohesion (Piato et al., 2011). Shoal cohesion can be measured by the distance to the nearest neighbour, average of inter-individual distance and total area of the shoal (Figure 12.2; Miller and Gerlai, 2007, 2008).

Social hierarchies and the establishment of dominant–subordinate patterns are notable features of zebrafish behaviour (Larson et al., 2006; Spence & Smith, 2005). While some individuals tend to

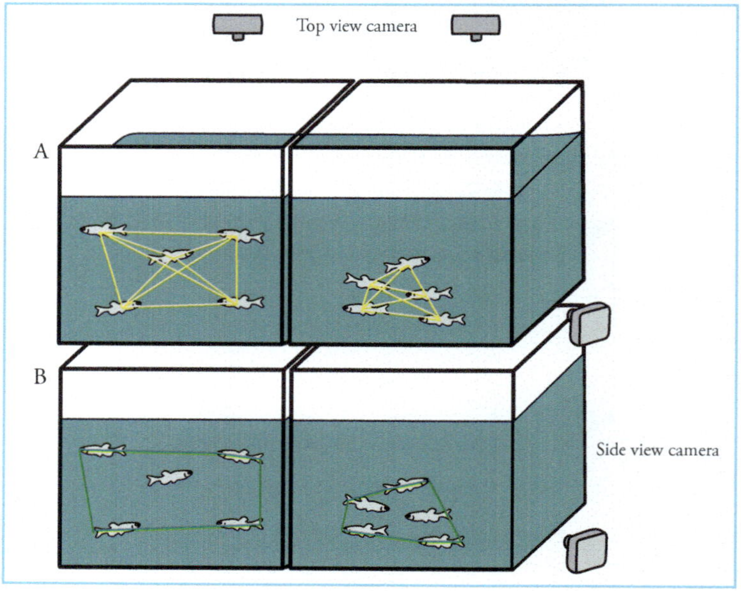

Figure 12.2 Assessment of affiliate behaviour in zebrafish. Schematic image representing a less cohesive (left) and a more cohesive (right) shoal. (A) Yellow lines represent inter-individual distance; (B) green lines represent the total area of the shoal. Adapted from Miller and Gerlai (2008).

exhibit aggressive behaviour (by biting, chasing and striking out at other conspecifics), others exhibit submissive behaviour, including freezing and fleeing responses (Teles et al., 2016). Based on this dichotomy, aggression has been studied in zebrafish in order to try to understand the neural mechanisms underlying this phenomenon (Carreño Gutiérrez et al., 2018; Jones & Norton, 2015).

Furthermore, more complex, shoal-related behaviours have been observed in zebrafish, including social recognition (i.e. ability to recognize conspecifics; Engeszer et al., 2007; Gerlach et al., 2008); social learning (i.e. ability to collect relevant information from other individuals (Abril-De-Abreu et al., 2015; Hall & Suboski, 1995; Madeira & Oliveira, 2017); and social buffering (i.e. faster recovery from a stressful event when in the presence of conspecifics (Faustino et al., 2017).

A wide array of zebrafish social behaviours have been described and studied, supplying the evidence base for this species' application as a relevant model for the study of conditions associated with abnormal social interactions, such as autism spectrum disorders (Gerlai & Gerlai, 2003), schizophrenia (Giacomotto et al., 2016), depression (do Nascimento et al., 2016) and anxiety (Sireeni et al., 2020).

Stress responses

Zebrafish display escape and/or avoidance behaviours from a very early age, with burst swimming (Budick & O'Malley, 2000; Buss & Drapeau, 2001) and startle responses (Colwill & Creton, 2011) identified from 2 dpf. Together with startle/escape responses evoked by strong acoustic or tactile stimuli (touch, sound, water flow or light; Colwill & Creton, 2011; Wolman & Granato, 2012), larvae also tend to swim towards the bright areas of the tank when threatened (Wolman & Granato, 2012). Moreover, physiological alterations are detected in larvae following acute stress exposure. For example, larvae exposed to either salinity (NaCl) or pH (HCl) alterations in their environment not only exhibit avoidance behaviour (vom Berg-Maurer et al., 2016) but also present increased cortisol levels in a dose-dependent way (vom Berg-Maurer et al., 2016;

Yeh et al., 2013). Taken together, this suggests that zebrafish present a strong stress response from an early age, involving behavioural alterations and activation of the hypothalamic–pituitary–inter-renal (HPI) axis.

Similar to larvae, stress exposure in adults also evokes behavioural alterations such as freezing, erratic movement (a stereotyped 'zig-zagging' movement) and bottom dwelling (Faustino et al., 2017; Fulcher et al., 2017; Marcon et al., 2018; Speedie & Gerlai, 2008). Moreover, these behavioural alterations can be accompanied by increased whole-body cortisol levels (Ghisleni et al., 2012; Ramsay et al., 2006). Restraint for 60 or 90 min (by keeping the fish inside a micro-centrifuge tube perforated to allow water flow) has been shown to induce a stress response with increased locomotion and higher numbers of turn angles, accompanied by increased whole-body cortisol levels (Ghisleni et al., 2012). Net chasing for 120 s (Giacomini et al., 2015), net handling twice with air exposure for 3 min with an interval of 3 min for rest (Ramsay et al., 2009) and overcrowding (40 fish/l; Ramsay et al., 2006) have all been reported to increase whole-body cortisol levels.

Predator exposure is another stimulus capable of triggering stress responses, causing an increase in freezing behaviour (Bass & Gerlai, 2008) and increased whole-body cortisol levels (Oliveira et al., 2017). Exposure to sympatric (i.e. same geographical origin) but not allopatric (i.e. foreign) predator cues, elicits these responses even in naïve captive zebrafish (Bass & Gerlai, 2008). In addition, animated images of the zebrafish sympatric predator Indian leaf fish (Gerlain et al., 2009), a bird silhouette (Luca & Gerlai, 2012a,b) or a large black dot moving on top of the tank (Luca & Gerlai, 2012b) evoke similar responses. Associated with predation avoidance, exposure to alarm substance – a chemical released from certain epidermal cells of fish when injured (Pfeiffer, 1977; Waldman, 1982) – also elicits stress behaviours, including erratic movement followed by freezing (Faustino et al., 2017; Speedie & Gerlai, 2008) and the formation of tighter shoals (Speedie & Gerlai, 2008).

Because the zebrafish is a gregarious species, the presence and behaviour of conspecifics can either exacerbate or attenuate stress responses. While the introduction of a stressed group of fish – but not of one individual fish – into a tank already occupied by a group of fish elicits stress responses in the resident shoal, leading to increased cortisol levels (Giacomini et al., 2015), the presence of non-stressed conspecifics decreases freezing responses in fish exposed to alarm substance (Faustino et al., 2017). This phenomenon seems to be mediated by both visual and olfactory cues, with the first showing a greater effect (Faustino et al., 2017). This suggests that zebrafish can modulate their stress behaviour in response to cues delivered by their conspecifics.

Several stressors have been combined for longer periods of time to induce chronic stress. The commonly called unpredictable chronic stress (UCS) protocol involves applying a stressor twice a day for several days, from one to five weeks (e.g. Fulcher et al., 2017; Song et al., 2018; Zimmermann et al., 2016). Table 12.1 summarizes the different stimuli that have been used to perform this protocol, with respective sources. The different stressors are usually presented in a randomized way and at different times of the day, to avoid predictability and subsequent habituation (Galhardo et al., 2011). The UCS protocol leads to both behavioural and physiological changes, resembling affective disorders in both rodents and humans (Bondi et al., 2008; Kalueff et al., 2014).

In juvenile zebrafish, exposure to a seven-day-long UCS protocol from 10 dpf led to increased locomotion in response to a dark–light transition, together with decreased exploration activity. However, it did not lead to the formation of tighter shoals (Golla et al., 2020). In adults, increased anxiety-related behaviours were identified after a period of 7 or 14 days of UCS administration, with higher levels of bottom dwelling in a novel tank test (see below for details; Fulcher et al., 2017; Marcon et al., 2016, 2018). A 5-week protocol elicited similar responses in a novel tank test, with stressed animals exhibiting decreased entries in the top part of the tank and increased time spent freezing (Song et al., 2018). Long-term stress has been reported to promote shoal cohesion and negatively affect cognition, leading to poorer results in an inhibitory avoidance task (Chakravarty et al., 2013; Manuel et al., 2014; Piato et al., 2011).

Besides behavioural alterations, the UCS protocol has been shown to effect physiological changes. In juvenile fish, chronic stress

Table 12.1 Stressors used in the UCS protocol in zebrafish.

Stressor	Protocol	References
Low water level	Lower the water level so that the dorsal fin of the fish is air-exposed:	
	– for 2 min	Piato et al., 2011; Chakravarty et al., 2013; Marcon et al., 2016, 2018; Zimmermann et al., 2016; Fulcher et al., 2017
	– for 15 min	Manuel et al., 2014
	– 20% of the regular amount of water	Song et al., 2018
Tank change	Transfer animals to a different tank:	
	– three consecutive times	Piato et al., 2011; Zimmermann et al., 2016
	– three consecutive times with 30 min intervals in between	Marcon et al., 2016, 2018
	– six consecutive times	Chakravarty et al., 2013; Fulcher et al., 2017
Net chasing	Chase the fish with a net:	
	– for 8 min	Piato et al., 2011; Chakravarty et al., 2013; Marcon et al., 2016, 2018; Zimmermann et al., 2016; Fulcher et al., 2017a
	– 2 × 3 min with 3 min interval in between	Ramsay et al., 2009
	– 2 × 5 min with 15 min interval in between	Manuel et al., 2014
	– for 20 min	Song et al., 2018
Air exposure	Elevate the net so that fish is exposed to the atmosphere air:	
	– for 2 min	Fulcher et al., 2017
	– 3 × 1 min with 5 min interval in between	Manuel et al., 2014
Water change	Replace the water of the tank for new water:	
	– for three consecutive times, with the animals inside	Piato et al., 2011; Zimmermann et al., 2016; Fulcher et al., 2017
	– once, without the fish inside	Manuel et al., 2014
Restraint stress	Keep the fish inside a 2 ml tube perforated in both extremities in order to allow water-flow:	

	– 60 min	Chakravarty et al., 2013; Fulcher et al., 2017
	– 90 min	Piato et al., 2011; Zimmermann et al., 2016
Social isolation	Maintain fish individually housed in a beaker:	
	– for 45 min	Piato et al., 2011; Marcon et al., 2016; Zimmermann et al., 2016
	– for 60 min	Chakravarty et al., 2013
	– for 8h	Song et al., 2018
Overcrowding	Maintain 67 fish/L for 50 – 60 min	Chakravarty et al., 2013
	Maintain 40 fish/L for 50 – 60 min	Piato et al., 2011; Manuel et al., 2014; Marcon et al., 2016, 2018; Zimmermann et al., 2016
	Maintain ten fish/L	Song et al., 2018
Heating water	Maintain fish in a tank with water at 33ºC for 30 min.	Piato et al., 2011; Chakravarty et al., 2013; Manuel et al., 2014; Marcon et al., 2016, 2018; Zimmermann et al., 2016
Cooling water	Maintain fish in a tank with water at 23ºC for 30 min.	Piato et al., 2011; Chakravarty et al., 2013; Manuel et al., 2014; Marcon et al., 2016, 2018; Zimmermann et al., 2016
Exposure to predator	Expose fish to a real predator, avoiding direct contact:	
	– for 50 min (e.g. *Archocentrus nigrofasciatus fish*)	Piato et al., 2011; Marcon et al., 2018; Zimmermann et al., 2016
	– for 24 h (*Trichogaster trychopterus* or gourami cichlids)	Song et al., 2018
	Expose fish an animated image of predator for 60 min (*Amatitlania nigrofasciata* and *Monocirrhus polyacanthus*)	Chakravarty et al., 2013
Exposure to a different strain	Expose fish to a visually different strain (pink glofish), for 24h	Song et al., 2018
Darkness	Expose fish to dark for 24 h	Song et al., 2018
Food deprivation	Starve fish for 24 h with exposure to alarm substance at the same time	Song et al., 2018
	Starve fish for 24 h	Manuel et al., 2014

impaired development growth, with the stressed fish displaying a smaller size than non-stressed fish (Golla et al., 2020). In adult fish, UCS was shown to influence the HPI axis by increasing the levels of corticotropin-releasing factor (CRF; Chakravarty et al., 2013) and whole-body cortisol levels (Marcon et al., 2018; Piato et al., 2011; Song et al., 2018).

Overall, these data suggest that the zebrafish is a robust model to study both stress responses and anxiety-like behaviour – which will be explored in the next section.

Validation of stress behaviour through the HPI axis

Whole-body cortisol is still the most popular method of sampling this hormone in both adults (Ramsay et al., 2009) and larvae (Yeh et al., 2013). Nonetheless, it is also possible to extract cortisol from blood collected from the adult tail vein (Pedroso et al., 2012). It is worth noting that both procedures are terminal. A non-invasive, non-lethal alternative is the extraction of cortisol from the water, based on the observation that fishes release free cortisol as steroid metabolite via their gills (Félix et al., 2013). Contrary to whole-body and blood collection from the tail vein, this method allows the measurement of cortisol in the same animal over time.

In terms of cortisol dynamics, the peak of cortisol levels in response to an acute stress (in an adult zebrafish) occurs 15 min after exposure to the stressor. After this period, levels start decreasing, returning to baseline approximately one hour after the stressor challenge (Ramsay et al., 2009).

Motivational states

As stated in the previous section, behavioural and physiological alterations are associated with affective states such as fear and anxiety. These are identifiable in both larvae and adult zebrafish. For this reason, several assays have been developed to assess anxiety-like behaviour in zebrafish, most of these based on the responses to novelty – which is thought to reflect the affective state of the animal as well as its behavioural adaptation to an unfamiliar environment

(Collier et al., 2017). There are two possible adaptive responses to novelty: exploration, which may lead to the discovery of food and/or mates, and anxiety (e.g. immobility), which may reduce predation risk (Pittman & Piato, 2017).

Methods used to assess anxiety-like behaviour

The open field test is a behavioural paradigm frequently performed in rodents (Katz et al., 1981; Prut & Belzung, 2003) that has been adapted for both larval (Ahmad & Richardson, 2013) and adult zebrafish (Stewart et al., 2012). This assay includes placing a fish into an empty, unfamiliar tank and observing its response to novelty (Figure 12.3A). This is a simple test used to assess not only locomotion (total distance travelled, speed, turning angles, number of movements, freezing and erratic movement) but also the degree of thigmotaxis (i.e. preference for proximity to the tank panels and avoidance of the more central open space; Champagne et al., 2010; Kalueff et al., 2013). Less anxious fish will explore the environment more and swim in the open central area whereas more anxious individuals will appear more passive and spend most of their time near the walls of the tank. There is increased thigmotactic behaviour after exposure to acute stress, such as change in illumination in larvae (Schnörr et al., 2012) or restraint stress in adults (Champagne et al., 2010). Contrarily, anxiolytic drugs such as diazepam (Schnörr et al., 2012) and scopolamine (Hamilton et al., 2017) reduce thigmotactic behaviour in both larvae and adults. In addition, fish have been found to adjust their locomotor activity to the size of the tank, swimming greater distances in a larger tank compared with a small one (e.g. a rectangular- versus a square-shaped tank). However, this does not appear to alter the spatiotemporal patterning of fish behaviour (Stewart et al., 2012), suggesting that the open field test is sufficiently robust and reproducible.

The novel object test is another novelty-based test used to assess anxiety-like behaviour (Stewart et al., 2012). The paradigm is similar to the open field assay, but also involves introducing an item to the middle of the tank – the novel object (Figure 12.3B). Besides

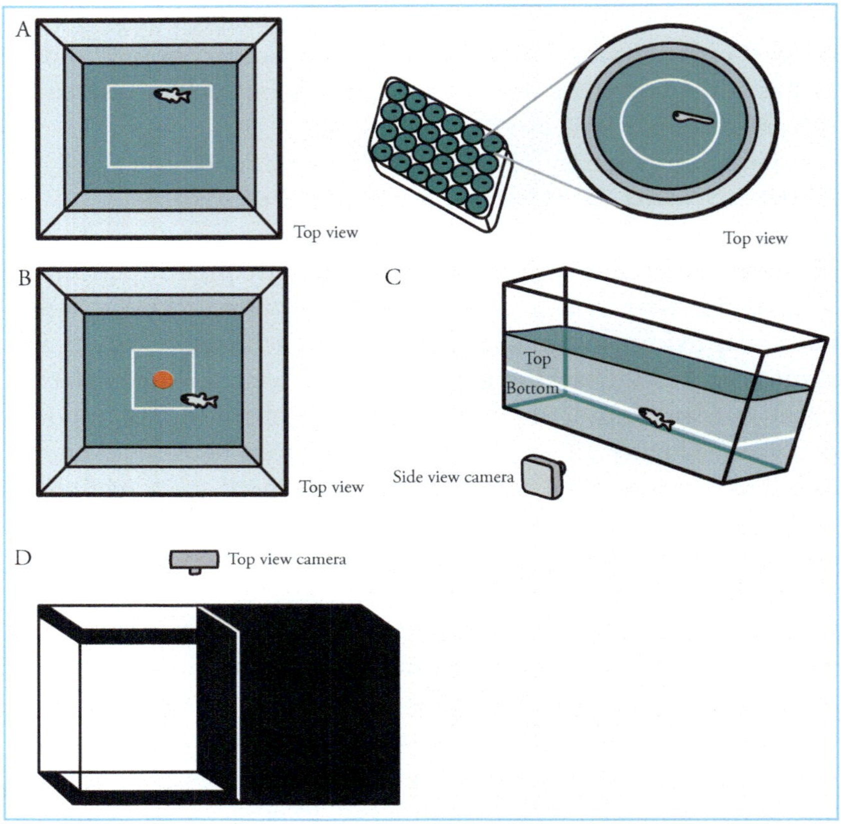

Figure 12.3 Measuring anxiety-like behaviour. (A) Example of apparatus used for the open field test in adults (left), adapted from Stewart et al. (2012); and in larvae (right), adapted from Schnörr et al. (2012). White lines represent the border between the inner and outer parts of the tank, in regard to measurement of thigmotaxic behaviour. (B) Novel object test with an object in the middle (red dot), adapted from Stewart et al. (2012). White line represents the area near the novel object used to measure exploration of that object. (C) Novel tank test, adapted from DePasquale and Leri (2018). White line represents the border between top and bottom parts of the tank, in regard to measurement of bottom dwelling. (D) Apparatus used in the light/dark test, where one half of the tank is dark and the other bright, adapted from Maximino et al. (2010a,b).

locomotion, this paradigm allows assessment of exploration by quantifying the latency to approach, and time spent near, the object. Just as in the open field test, in the novel object test the less anxious fish are expected to explore the object more while more anxious fish will avoid it (e.g. Kalueff et al., 2013).

The novel tank test is also employed as a general measure of anxiety, to assess the emotional state of the fish or study the effect of key anxiolytic drugs (DePasquale & Leri, 2018; Wong et al., 2010). In essence, this assay consists of placing a fish inside a novel/unfamiliar tank and, by using a lateral camera (Figure 12.3C), assessing the vertical position of the fish (Cachat et al., 2010). This is based on a behaviour known as geotaxis (i.e. the tendency to seek the bottom part of the tank, especially when threatened and/or anxious; Kalueff et al., 2013). Geotaxis is likely to be due to the fact that the upper water portion is more exposed to avian natural predators (Spence et al., 2008) and that greater proximity to the bottom provides better camouflage (Spence et al., 2006, 2008). Bottom dwelling is markedly increased following chronic stress experience (Chakravarty et al., 2013; Fulcher et al., 2017; Marcon et al., 2018; Song et al., 2018) and decreased in response to anxiolytics such as scopolamine (a muscarinic acetylcholine receptor antagonist; Hamilton et al., 2017), diazepam or fluoxetine (Gaikwad et al., 2011). The novel tank test is a robust test that assesses locomotion, exploration, freezing and erratic movement (Cachat et al., 2010). It has been reported that zebrafish can develop habituation to the novel tank test during the trial by increasing the number of visits to the top of the tank and decreasing freezing behaviour, especially after the first 10 min. These behavioural changes are also found in repeated exposures to unfamiliar tanks (6 min daily) over the course of 7 days, especially after the 4th day (Wong et al., 2010).

The light/dark preference test can also be used to assess anxiety-like behaviour. Zebrafish larvae prefer bright environments over darker ones, a behaviour for which the term 'scotophobia' has been proposed (Kalueff et al., 2013; Steenbergen et al., 2011). By contrast, adult zebrafish prefer dark environments, avoiding bright ones (Serra et al., 1999), a behaviour known as scototaxis (Maximino et al., 2010a,b). Thus, the light/dark preference test aims to assess the time spent in the dark versus the bright part of the tank, by placing the fish into a tank half white/half black (Figure 12.3D) and quantifying the time spent in each area, the number of entries in each zone and the total distance travelled in each zone (DePasquale & Leri, 2018; Maximino et al., 2010a,b, 2011). In the adult it is expected that more anxious fish will spend more time in the darker part of the tank. Indeed, following UCS exposure, fish presented more scototactic behaviour compared with non-stressed controls (Chakravarty et al., 2013). Accordingly, anxiolytic and antidepressant drugs such as chlordiazepoxide (Stewart et al., 2011) and fluoxetine (Collier et al., 2017) promote more entries and time spent in the white compartment. Moreover, contrary to the novel tank test, the outcome of the dark–light test is unaffected by habituation over time – no differences in time spent in the dark area of the tank were found in 15-min trials, even when repeated several times on the same day or over consecutive days (Maximino et al., 2010a,b).

Locomotion can be an ambiguous measure in the context of these paradigms, because it is an indicator of both physical condition and excitement. For instance, while acutely stressed fish may exhibit higher levels of locomotion (Ghisleni et al., 2012), animals subjected to chronic stress will display decreased locomotion, resembling a depression-like behaviour (Chakravarty et al., 2013; Piato et al., 2011). Moreover, in the absence of any stressor stimulus, fish living in an enriched environment may also display decreased locomotion compared with those living in barren tanks (von Krogh et al., 2010; see Chapter 2 for more information). Intermediate doses of ethanol increase locomotion while high doses markedly decrease it, leading to depression-like behaviour (Gerlai et al., 2000). With ageing, zebrafish generally display progressively lower activity levels (Martins, 2020; Yu et al., 2006). More specifically, old fish present a poorer swimming performance in terms of endurance (i.e. they reach fatigue earlier than young fish when exposed to a gradual increase in water velocity; Gilbert et al., 2014), and in distance swum (Martins, 2020; Yu et al., 2006; see Chapter 6 for more information).

Behavioural flexibility

The behaviour of individual fish is influenced by its personality traits and affective state. In this context, a bold–shy dichotomy has been proposed for zebrafish and other teleost species, where 'boldness' and 'shyness' are categorized in fish according to a range and strength of displayed behaviours. Bold fish generally are less averse to risk taking and may spend more of their time in the open (or in the upper sections of a test tank), perform a relatively higher frequency of swimming and show a quicker return to baseline behaviours following a stress response. Shy fish spend most of their time under cover (or at the bottom of a test tank) and display less exploratory behaviour (Thörnqvist et al., 2019). This binary paradigm has been associated with the HPI axis, manifested in increased expression of glucocorticoid receptors and genes regulating the synthesis of stress hormones in shy individuals (Oswald et al., 2012), as well as with the higher expression of delta opioid and dopamine D2 receptors in the brain of bold individuals, believed to be important for active stress coping (Thörnqvist et al., 2019).

The evaluation of an ambiguous stimulus is influenced not only by personality traits but also by the animal's affective state, which in turn is influenced by the animal's previous experiences (Mendl et al., 2009). This is known as cognitive or judgement bias. In practical terms, this means that an individual exposed to a set of aversive experiences, including algesiogenic (pain-producing) events, may show reduced success at associative learning when reward stimuli are offered only intermittently or alternated with negative stimuli (such as chasing or noise), thus rendering these cues ambiguous. Cognitive bias is commonly tested by training the animals to associate a positive location with a reward and a negative location with either an absence of reward or an aversive stimulus. After learning, the animals are presented with an ambiguous location (in the middle between the positive and negative locations). Animals that rapidly respond to the ambiguous location are considered optimistic, while animals that take a long time approaching this new location are considered to be in a more pessimistic affective state (Brydges et al., 2011; Harding et al., 2004; Mendl et al., 2010). An animal with an optimistic affective state tends to interpret an ambiguous stimulus as positive, and one with a pessimistic affective state as negative (Harding et al., 2004; Mendl et al., 2009). Although the cognitive bias paradigm has been long studied in mammals, it was only recently validated for zebrafish, using a half-arm radial maze (Espigares et al., 2021, 2022).

Post-nociceptive behaviour

Post-nociceptive behaviours are measured responses following the detection of potentially tissue-damaging stimuli. These responses include protective ('nocifensive') and escape behaviours and are critical for survival (Tracey, 2017).

Behavioural parameters in the context of nociception may be difficult to measure and interpret because different animals may present different outputs. They must be validated to determine whether they also exist under stressful but not painful conditions so that they can be robustly attributed to painful events. This can be achieved by using sham controls through which the behavioural effect of potentially stressful (e.g. netting, air emersion, chasing), but non-algesiogenic components, of a pain-inducing protocol can be accounted for (e.g. Schroeder & Sneddon, 2017) and meaningful baselines and consideration thresholds can be established. Investigations of post-nociceptive behaviours in zebrafish alone (Maximino, 2011) and zebrafish together with other fish species (Reilly et al., 2008) established that the type and magnitude of behavioural patterns following painful events can be species specific. These studies also showed that an increased ventilation rate was indicative of nociceptive stimuli in zebrafish and trout but not in carp. Zebrafish displayed a reduced swim rate, measured as direct movement longer than one body length, whereas rainbow trout showed an increase and no change was observed in common carp (Reilly et al., 2008). Injection of acetic acid into the lip elicited 'lip rubbing' in rainbow trout, an atypical behaviour thought to be analogous to humans rubbing painful areas, although this causality has not been accepted by everyone (Rose, 2003) and was not seen in control fish. This particular behavioural response has also been recorded in goldfish experiencing a painful event (Newby et al., 2009). A potentially analogous response has been observed in zebrafish in which, following acid injections near the tailfin, animals reacted with rapid fin movements described as 'tailfin fanning' (Maximino, 2011), with similar responses evoked following tailfin clipping of zebrafish (Schroeder & Sneddon, 2017). The latter also showed an array of other post-clipping behavioural changes, including reduced swimming behaviour – also observed for goldfish and Nile tilapia (Roques et al., 2010).

Post-nociceptive behaviour of zebrafish is also increasingly being analysed through fractal analysis (detailed later in this chapter), where the complexity and variability of three-dimensional (3D) zebrafish movement patterns are converted into a single value through video-processing software, showing that zebrafish display less complex and varied behaviour after being subjected to the algesiogenic stimuli of 'fin-clipping' (Deakin et al., 2019a) or acetic acid injection (Deakin et al., 2019a).

Cognition – memory and learning

Learning is a key aspect of animal behaviour, a prerequisite not only for social interactions but also navigation and crucial for foraging and predator avoidance (Engeszer et al., 2004). Zebrafish have been shown to have short- and long-term memory capacity, as well as complex learning abilities (Cheng et al., 2011).

Non-associative learning is a basic form of learning observed in both larvae (Best et al., 2008) and adult zebrafish through habituation (Wong et al., 2010). Habituation is the reduction of responses elicited by a certain stimulus following repeated exposures (Thompson & Spencer, 1966). This is exemplified in 7-dpf larvae displaying a reduced startle response to repeated acoustic stimuli (Best et al., 2008). Notably, habituation can be reversed, and this can be assessed by introducing a second stimulus known to induce the same response as the first (Thompson & Spencer, 1966). This was observed in larvae which, after habituation to acoustic stimuli, presented startle response again following a different stimulus (light pulse). Following this second stimulus, larvae reverted to displaying a startle response reacting to the acoustic signal (Best et al., 2008).

The zebrafish is also capable of more complex learning tasks. In addition to being able to discriminate between conspecifics (individuals of the same species) and heterospecifics (individuals of different species), *D. rerio* can recognize two different conspecifics through both olfactory and visual cues (Engeszer et al., 2007). Zebrafish are able to retain this information for 24 h (long-term memory; Madeira & Oliveira, 2017). Furthermore, adult zebrafish can recognize objects. Two independent studies showed that zebrafish discriminated between a familiar and a novel object after a one-trial test (Lucon-Xiccato & Dadda, 2014; May et al., 2016). Together, this suggests that zebrafish are capable of social and object recognition.

It is also clear that zebrafish are capable of classical conditioning and operant conditioning from an early age (Roberts et al., 2013). Classical conditioning is demonstrated through associative tasks – for instance, when learning to associate an innocuous stimulus, such as a visual cue (conditioned stimulus, CS), with a rewarding or aversive reinforcing stimulus (unconditioned stimulus, US). Following repeated CS–US pairing, the animal will start to respond to the CS as it would to the US. This is known as a conditioned response (Pavlov, 1927). Hinz et al. (2013) reported that a group of 6–8-day-old larvae (conspecifics) learned to associate a visual stimulus with a reward (Hinz et al., 2013). Following training the larvae spent more time next to the visual stimulus where the shoal used to be present, even in the absence of reward reinforcers. In addition, Valente et al. (2012) showed that 3–6-week-old larvae were able to associate a visual pattern in the bottom of the tank with an electrical shock (aversive stimulus), following training, by turning away from that part of the tank containing the CS (Valente et al., 2012). Such conditioning was also demonstrated for adult male zebrafish where animals learned the association of light stimuli with alarm substance but not with visual alarm cues, suggesting that chemical cues constitute a much more robust alarm signal (Pinho et al., 2020). Operant conditioning requires the performance of a voluntary action in order to receive a reward or avoid a punishment (Roberts et al., 2013). Zebrafish larvae are capable of operant behaviour, again using an electrical shock as aversive stimulus (Valente et al., 2012). Besides this, little is known about classical and operant conditioning in larvae and more studies are needed to address this question (Pritchett & Brennan, 2020).

Several studies have successfully implemented association learning in adult zebrafish. For example, adult zebrafish can learn to associate a reward (e.g. food or a shoal of conspecifics) with a visual cue (e.g. red card) by decreasing the latency to approach to the red card overtime (Al-Imari & Gerlai, 2008; Sison & Gerlai, 2010; Spence et al., 2011). In addition, zebrafish were reported to be capable of olfactory conditioning by associating unnatural odorant (phenylethyl alcohol) with food (Braubach et al., 2009). Spatial learning, which implies the establishment of a 'spatial map' of the environment (Karnik & Gerlai, 2012), has also been reported in adult zebrafish. For example, following training in a four-armed plus-maze, fish learned to associate a certain part of the tank (fixed location) with food, not by internal cues in the tank but by external cues around the maze (items in the room surrounding the tank; Gaikwad et al., 2011; Sison & Gerlai, 2010). Spatial learning has also been tested by using an aversive reinforcer. For example, in a tank divided into three chambers with two interconnecting slide doors, fish learned how to escape from being chased by a net by swimming towards a 'safe' compartment (Arthur & Levin, 2001).

Adult zebrafish have been reported to accomplish even more complex learning tasks, including ones featuring alternation and numerical elements. Fish were placed in a tank where an image of a shoal was presented on one of the sides but alternating from one side to the other. Fish were able to learn the sequence, spending more time near the side that was expected to present the image (Pather & Gerlai, 2009). In another study, food was presented on alternating sides of the tank containing a barrier in the middle to force the fish to swim underneath it to cross to the other side. Following training the fish learned the sequence, increasing the percentage of correct choices when deciding which side of the tank to swim to (Williams et al., 2002). In addition, it has been suggested that zebrafish can distinguish between two different geometrical figures with different numbers of objects on them (e.g. five versus ten or eight versus 12), although not all the fish were able to achieve this (Agrillo et al., 2012). Furthermore, a multiple-choice serial reaction-time task was also optimized for the first time to assess impulse control in zebrafish. In this paradigm the fish is rewarded with food after detecting a briefly presented light in one of the five distinct spatial locations of the tank, swimming towards that area. This may be a suitable assay to explore disorders related to impulsiveness, such as attention deficit hyperactivity disorder (ADHD) or Rett's syndrome (Parker et al., 2013).

Factors impacting cognitive behaviour

Cognitive abilities can be influenced by different variables. More complex environments appear to facilitate associative learning, at least in wild-caught zebrafish (Spence et al., 2011), while exposure to acute stress (alarm substance or predator) appears to impair associative and spatial learning (Gaikwad et al., 2011). Another variable likely to influence cognitive skills is age. Older zebrafish were shown to be slower in establishing an anticipatory response prior to food administration when in the presence of a CS. However, when a negative reinforcement cue (mild electric shock) was tested no differences between young and old were observed (Yu et al., 2006). This suggests that older fish can learn quickly but require stronger motivation to acquire a conditioned response. Moreover, in a different task, age turned out to be an important factor in developing generalization because the middle-aged and old cohorts did not develop generalization following a conditioned place preference (CPP) test while old fish exhibited faster extinction (Yu et al., 2006; see Chapter 6 for more information).

It is still not established reliably whether the presence of a shoal facilitates learning. Gleason et al. (1977) showed that learning in groups of five (but not in pairs) enhances the learning process, suggesting that the presence of a shoal may facilitate learning. A more recent study showed that zebrafish first trained in a shoal but subsequently tested as individuals did not show the same level of consistency in their learning success (Mcaroe et al., 2017). However, more studies are needed to further elucidate this subject matter.

Possessing vision with peak absorbance in ultraviolet (362 nm), blue (415 nm), green (480 nm) and red (570 nm; Spence & Smith, 2008), zebrafish have shown the ability to distinguish colours, namely red, blue, green and yellow (Avdesh et al., 2010), as have several other teleost species. While 5-dpf larvae show a preference for blue to the detriment of the other three colours (Park et al., 2016), the same colour has been shown to be aversive for adults. Adult zebrafish displayed equal preference for red, green and yellow (Avdesh et al., 2010). This should be considered when designing experiments and experimental set-ups.

Planning behavioural assays

Confounders associated with zebrafish husbandry

Housing conditions

It is now well established that environmental enrichment and social setting (grouped versus isolation) affect behaviour, physiology and cognition. Compared with fish living in a barren tank, those living in an enriched environment display decreased locomotion (von Krogh et al., 2010) and less anxiety when submitted to a novel tank test (DePasquale et al., 2016). Notably, environmental enrichment alone was enough to decrease cortisol levels in response to a UCS protocol (Marcon et al., 2018). In addition, fish from enriched environments present higher levels of proliferation in the telencephalon (von Krogh et al., 2010) and larger brains (DePasquale et al., 2016), together with enhanced cognitive skills in a spatial learning test (DePasquale et al., 2016; see Chapter 2 for more information).

Furthermore, stocking density can have a substantial impact on fish welfare. On the one hand, overcrowding has been reported to lead to a four-fold increase in whole-body cortisol levels (Ramsay et al., 2006), to decrease growth rate and negatively impact the immune system (Lawrence, 2007), compromising fish well-being. On the other hand, the zebrafish is known to exhibit antagonistic behaviour and form dominance hierarchies (Larson et al., 2006; Pavlidis et al., 2011), and aggression is thought to be inversely correlated with stock density (Schroeder et al., 2014; Spence & Smith,

2005). Taken together, this suggests that intermediate densities are the most favourable.

Impact of nutrition

There is growing evidence to suggest that nutrition and feeding patterns influence zebrafish behaviour. For example, food deprivation in larvae can alter risk-taking behaviour in foraging decisions, leading to poor predator vigilance and increased willingness to approach unknown objects. Feeding state is, therefore, likely to impact on an individual's interpretation and acceptance of ambiguous stimuli present in the environment (Filosa et al., 2016). Moreover, food composition itself is likely to affect fish behaviour. Likewise, probiotic-rich diets have been shown to alter shoaling and exploratory behaviour in adult fish (*Lactobacillus rhamnosus*; Borrelli et al., 2016), as well as locomotion and anxiety-like behaviour in larvae (*Lactobacillus plantarum*; Davis et al., 2016). Apart from food composition, feeding predictability (i.e. when food is supplied with the same frequency every day) has also been reported to increase exploratory behaviour in adult fish (Holley et al., 2014; see Chapter 3 for more information)).

Impact of pathogens

Several studies report that zebrafish infected with microsporidiosis (causative agent *Pseudoloma neurophilia*) show altered startle responses with increased reaction to fearful stimuli (Spagnoli et al., 2015), as well as increased shoal cohesion (manifested through reduced inter-fish distances; Spagnoli et al., 2017). These modified responses were interpreted as a parasite-induced increase in stress, fear and anxiety. In a more recent publication, sociability and aggression were reported to be unaltered yet hosts exhibited reduced activity, increased displays of freezing behaviour and decreased exploration following infection with *P. neurophilia*. More generally there is strong evidence that zebrafish behaviour may be impacted by a range of pathogens, which is particularly problematic since in most cases infection status is unknown. This emphasizes the importance of comprehensive monitoring and surveillance in the detection of subclinical infections in laboratory animals (Midttun et al., 2020).

A similar cautionary approach can be gleaned from a recent review of viral infections in zebrafish, which presents a range of naturally occurring viral pathogens with unknown subclinical or clinical manifestation (Crim, 2020), many of which go largely undetected as only zebrafish picornavirus is typically screened for (see Chapters 5 and 6 for more information).

Differences between strains

The Zebrafish International Resource Center (ZIRC) have several recognized breeding lines, including AB, Tübingen (TÜ), Tübingen Long Fin (TL), Wild Indian Karyotype (WIK) and others (Spence et al., 2008; www.zfin.org). There is growing evidence that there is variability in behavioural responses among different strains and populations of fish.

For example, a study comparing six different strains (AB, ABstrg, casper, EK, TÜ and WIK) showed inter-strain differences in locomotion throughout development, with casper displaying the highest variability compared with the other strains (Lange et al., 2013). A more recent and detailed study also explored the differences between AB and four other zebrafish strains (TL, WIK, pet store-purchased and golden zebrafish) in a wide variety of behaviours, including locomotor activity, exploratory behaviour and social interactions (Audira et al., 2020). In this study Audira et al. (2020) observed that, within the strains analysed, TL fish were the most similar to AB fish because no differences were detected between strains in terms of locomotor activity, exploratory behaviour or social interaction. However, when compared with AB, commercial (pet store-purchased) fish, widely used in behaviour assays, displayed increased locomotion activity and lower exploratory and social interaction behaviour (Audira et al., 2020). Finally, inter-strain differences have also been reported to account for variations in cognitive abilities, e.g. with slower learning in an associative task in the WIK strain when compared with wild-caught fish (Spence et al., 2011).

Apart from inter-strain differences, it has been suggested that some strains display greater variability in terms of behaviour than others. Audira et al. (2020) reported that AB and TL fish present a lower inter-individual variability than other strains, suggesting that the behaviours tested in these two strains are strongly robust and reproducible.

Differences between sexes

While there are generally relatively few differences in behavioural phenotypes between the two sexes of this species, male wild-type zebrafish are usually bolder in social encounters than females (Moretz et al., 2007) and display higher aggression levels when exposed to chronic stress (Genario et al., 2020); they also explore more in novel environments (Reolon et al., 2018). It is well established that the zebrafish displays sexual dimorphism in both physiology and anatomy (Conradsen & McGuigan, 2015). Some behavioural differences between sexes have been attributed to morphological inequities, for example in swimming performance where males were shown to be faster than females despite being significantly smaller. In this case, the more rotund body shape of the female was seen to be detrimental to swimming velocity (Conradsen & McGuigan, 2015).

Practical guidance

General rules

There are important considerations to make before starting a behavioural assay. If the fish are obtained from a commercial supplier, they require acclimatization to their new holding system for at least 15 days before the experiment. Note that Wong et al. (2010) reported a habituation to novelty over time, over the course of 7 days, when subjecting the fish to a novel tank test for 6 min every day. In case the fish are not maintained in the same room where the behavioural assays will be performed, one should allow them to acclimatize to the behavioural room for at least one hour before the experiment.

When preparing the set-up, bear in mind the scototactic behaviour in zebrafish (Serra et al., 1999) and avoid casting shadows in the experimental tank – because this can alter fish behaviour and lead to confounding results. Also, if covering the walls of the tank to avoid external cues, consider using an opaque paper/partition in order to avoid 'shoal effect'. In addition, during the behavioural assays it is essential to avoid any loud noise and/or vibrations since these can lead to startle responses in the fish and other behavioural alterations.

Furthermore, adequate controls should be carefully planned for each experiment. Fish subjected to a specific stressor challenge should always be compared with fish where that challenge has been omitted but that are otherwise handled in the same way and maintained in the same conditions during the experiment (commonly known as 'sham controls'). Finally, when performing an experiment, anything that might influence zebrafish behaviour should be controlled and recorded. Thus, handling stress should be minimized and kept at constant levels across all treatment conditions, while visual, tactile or olfactory cues during assays should be avoided in order to prevent alterations in behaviour.

Video analysis

Most behaviours can be quantified manually. For instance, by dividing the tank into multiple squares it is possible to achieve an approximate quantification of distance travelled by counting the number of squares crossed throughout the assay (Champagne et al., 2010). Also, by defining lines isolating key parts of the tank (e.g. bottom versus top part in a novel tank test, or inner versus outer part in an open field test), it is possible to count the number of entries and the time spent in each area. *The Observer XT* (Noldus, Netherlands; https://www.noldus.com/animal-behavior-research/products/the-observer-xt) is a software that facilitates this analysis by allowing the quantification of observational event-behaviours. With this tool, regions of interest in the videos can be drawn and special keys on the keyboard to register specific behaviours can be defined, while each behaviour can be captured and quantified. The software then allows the export of the data into a spreadsheet format for further analysis.

Other IT-based tools have been developed allowing reliable high-throughput analysis of zebrafish behaviour. Video-tracking software such as *EthoVision* (Noldus, Netherlands; https://www.noldus.com/animal-behavior-research/products/ethovision-xt) and *ViewPoint* (ViewPoint Behaviour Technology, France; http://www.viewpoint.fr/en/home) are examples of this. By keeping track of the coordinates of the fish throughout the video, these software applications automatically record a variety of measurements including speed, distance travelled, time spent in each area of the tank, number of entries in each part of the tank and the angles of any turns. Video-tracking therefore enables fast, complete and reliable analysis of fish behaviour.

Other than these tools, customized software have been developed to access more specific behaviour such as shoaling (Miller & Gerlai, 2007; Pérez-Escudero et al., 2014). One example is the idTracker, a multi-tracking algorithm developed by Gonzalo De Polavieja and colleagues that allows the discrimination of up to ten different fish based on their silhouette (Pérez-Escudero et al., 2014; Romero-Ferrero et al., 2019). It is important to note, though, that the idTracker analysis requires high-quality videos, which rely on sufficient camera resolution (1,080 × 1,080 pixels) as well as illumination of the set-up. The experimental set-up should therefore be optimized and tested for this type of analysis.

More recently the concept of fractal dimension (FD) was introduced to assess post-nociceptive zebrafish behaviour, calculating 3D trajectories to determine the effect of a range of algesiogenic stimuli on movement patterns (Deakin et al., 2019b). This is based on fractal analysis yielding a measure of complexity (Mandelbrot, 1967) derived from recording and processing 3D movement and swimming coordinates of an individual. This generates a single measure, the FD. More complex swimming patterns are reflected by higher fractal dimension values and more repetitive, less complex, patterns by lower fractal dimension values. FD can be combined with other behavioural parameters, such as swimming speed and percentage of time spent in the bottom of a tank, in order to assess an animal's welfare, especially after a stressor challenge (Schroeder & Sneddon, 2017).

Final considerations

As highlighted throughout this chapter, the environmental and social setting of the fish can influence their behaviour and significantly impact or confound the results of behavioural tests. Hence experimental design needs to consider the following: (1) housing conditions (number of fish per tank; male/female ratio, age groups, size grading and possible presence of gravel and/or plants in the tanks); (2) the time of the day when the experiment is performed, taking into account the circadian cycle; (3) diet (type and quantity of feed, as well as frequency and presentation); (4) the most suitable strain; and (5) any appropriate control groups required to validate the results. The same conditions must be applied to all groups in the experiment, and all fish should be acclimatized to these conditions before the experiment. Taking all these variables into account will improve the validity and reproducibility of the experiments.

References

Abril-De-Abreu, R., Cruz, J., & Oliveira, R. F. (2015). Social eavesdropping in zebrafish: Tuning of attention to social interactions. *Scientific Reports*, *5*, 12678. https://doi.org/10.1038/srep12678

Agrillo, C., Miletto Petrazzini, M. E., Tagliapietra, C., & Bisazza, A. (2012). Inter-specific differences in numerical abilities among teleost fish. *Frontiers in Psychology*, *3*, 1–9. https://doi.org/10.3389/fpsyg.2012.00483

Ahmad, F., & Richardson, M. K. (2013). Exploratory behaviour in the open field test adapted for larval zebrafish: Impact of environmental complexity. *Behavioural Processes*, *92*, 88–98. https://doi.org/10.1016/j.beproc.2012.10.014

Al-Imari, L., & Gerlai, R. (2008). Sight of conspecifics as reward in associative learning in zebrafish (Danio rerio). *Behavioural Brain Research, 189*(1), 216–219. https://doi.org/10.1016/j.bbr.2007.12.007

Arthur, D., & Levin, E. D. (2001). Spatial and non-spatial visual discrimination learning in zebrafish (Danio rerio). *Animal Cognition, 4*(2), 125–131. https://doi.org/10.1007/s100710100111

Arunachalam, M., Raja, M., Vijayakumar, C., Malaiammal, P., & Mayden, R. L. (2013). Natural history of zebrafish (Danio rerio) in India. *Zebrafish, 10*(1), 1–14. https://doi.org/10.1089/zeb.2012.0803

Audira, G., Siregar, P., Strungaru, S. A., Huang, J. C., & Hsiao, C. D. (2020). Which zebrafish strains are more suitable to perform behavioral studies? A comprehensive comparison by phenomic approach. *Biology, 9*(8), 1–21. https://doi.org/10.3390/biology9080200

Avdesh, A., Chen, M., Martin-Iverson, M., Verdile, G., Mondal, A., & Martins, R. (2010). Natural colour preference in the zebrafish (Danio rerio). *Proceedings of the Measuring Behavior, 8*, 155–157.

Bass, S. L. S., & Gerlai, R. (2008). Zebrafish (Danio rerio) responds differentially to stimulus fish: The effects of sympatric and allopatric predators and harmless fish. *Behavioural Brain Research, 186*(1), 107–117. https://doi.org/10.1016/j.bbr.2007.07.037

Best, J. D., Berghmans, S., Hunt, J. J., Clarke, S. C., Fleming, A., Goldsmith, P., & Roach, A. G. (2008). Non-associative learning in larval zebrafish. *Neuropsychopharmacology, 33*(5), 1206–1215. https://doi.org/10.1038/sj.npp.1301489

Bondi, C. O., Rodriguez, G., Gould, G. G., Frazer, A., & Morilak, D. A. (2008). Chronic unpredictable stress induces a cognitive deficit and anxiety-like behavior in rats that is prevented by chronic antidepressant drug treatment. *Neuropsychopharmacology, 33*(2), 320–331. https://doi.org/10.1038/sj.npp.1301410

Borrelli, L., Aceto, S., Agnisola, C., De Paolo, S. D., Dipineto, L., Stilling, R. M., Dinan, T. G., Cryan, J. F., Menna, L. F., & Fioretti, A. (2016). Probiotic modulation of the microbiota-gut-brain axis and behaviour in zebrafish. *Scientific Reports, 6*, 30046. https://doi.org/10.1038/srep30046

Braubach, O. R., Wood, H. D., Gadbois, S., Fine, A., & Croll, R. P. (2009). Olfactory conditioning in the zebrafish (Danio rerio). *Behavioural Brain Research, 198*(1), 190–198. https://doi.org/10.1016/j.bbr.2008.10.044

Brydges, N. M., Leach, M., Nicol, K., Wright, R., & Bateson, M. (2011). Environmental enrichment induces optimistic cognitive bias in rats. *Animal Behaviour, 81*(1), 169–175. https://doi.org/10.1016/j.anbehav.2010.09.030

Budick, S. A., & O'Malley, D. M. (2000). Locomotor repertoire of the larval zebrafish: Swimming, turning and prey capture. *Journal of Experimental Biology, 203*(17), 2565–2579. https://doi.org/10.1242/jeb.203.17.2565

Buss, R. R., & Drapeau, P. (2001). Synaptic drive to motoneurons during fictive swimming in the developing zebrafish. *Journal of Neurophysiology, 86*(1), 197–210. https://doi.org/10.1152/jn.2001.86.1.197

Cachat, J., Stewart, A., Grossman, L., Gaikwad, S., Kadri, F., Chung, K., Wu, N., Wong, K., Roy, S., Suciu, C., Goodspeed, J., Elegant, M., Bartels, B., Elkhayat, S., Tien, D., Tan, J., Denmark, A., Gilder, T., Kyzar, E., Dileo, J., Frank, K., Chang, K., Utterback, E., Hart, P., & Kalueff, A. (2010). Measuring behavioral and endocrine responses to novelty stress in adult zebrafish. *Nature Protocols* 5, 1786–1799.

Carreño Gutiérrez, H., Vacca, I., Pons, A. I., & Norton, W. H. J. (2018). Automatic quantification of juvenile zebrafish aggression. *Journal of Neuroscience Methods, 296*, 23–31. https://doi.org/10.1016/j.jneumeth.2017.12.012

Chakravarty, S., Reddy, B. R., Sudhakar, S. R., Saxena, S., Das, T., Meghah, V., Brahmendra Swamy, C. V., Kumar, A., & Idris, M. M. (2013). Chronic unpredictable stress (CUS)-induced anxiety and related mood disorders in a zebrafish model: Altered brain proteome profile implicates mitochondrial dysfunction. PloS One, 8(5), e63302. https://doi.org/10.1371/journal.pone.0063302

Champagne, D. L., Hoefnagels, C. C., de Kloet, R. E., & Richardson, M. K. (2010). Translating rodent behavioral repertoire to zebrafish (Danio rerio): Relevance for stress research. *Behavioural Brain Research, 214*(2), 332–342. https://doi.org/10.1016/j.bbr.2010.06.001

Cheng, R. K., Jesuthasan, S., & Penney, T. B. (2011). Time for zebrafish. *Frontiers in Integrative Neuroscience, 5*, 40. https://doi.org/10.3389/fnint.2011.00040

Clark, D. T. (1981). *Visual responses in the developing zebrafish (Brachydanio rerio)*. University of Oregon.

Cleal, M., Fontana, B. D., & Parker, M. O. (2021). The cognitive and behavioral effects of D-amphetamine and nicotine sensitization in adult zebrafish. *Psychopharmacology, 238*(8), 2191–2200. https://doi.org/10.1007/s00213-021-05844-5

Collier, A. D., Kalueff, A. V., & Echevarria, D. J. (2017). Zebrafish models of anxiety-like behaviors. In A. V. Kalueff (Ed.), *Rights and Wrongs of Zebrafish: Behavioral phenotyping of zebrafish* (pp. 45–72). Springer International Publishing.

Colwill, R. M., & Creton, R. (2011). Imaging escape and avoidance behavior in zebrafish larvae. *Reviews in the Neurosciences*, *22*(1), 63–73. https://doi.org/10.1515/RNS.2011.008

Conradsen, C., & McGuigan, K. (2015). Sexually dimorphic morphology and swimming performance relationships in wild-type zebrafish Danio rerio. *Journal of Fish Biology*, *87*(5), 1219–1233. https://doi.org/10.1111/jfb.12784

Crim, M. J. (2020). Viral diseases. In S. C. Cartner et al. (Eds.), *The zebrafish in biomedical research: Biology, husbandry, diseases and research applications* (pp. 509–526). Elsevier.

Darrow, K. O., & Harris, W. A. (2004). Characterization and development of courtship in zebrafish, Danio rerio. *Zebrafish*, *1*(1), 40–45. https://doi.org/10.1089/154585404774101662

Davis, D. J., Bryda, E. C., Gillespie, C. H., & Ericsson, A. C. (2016). Microbial modulation of behavior and stress responses in zebrafish larvae. *Behavioural Brain Research*, *311*, 219–227. https://doi.org/10.1016/j.bbr.2016.05.040

Deakin, A. G., Buckley, J., AlZu'bi, H., Cossins, A. R., Spencer, J. W., Al'Nuimy, W., Young, I. S., Thompson, J. S., & Sneddon, L. U. (2019). Automated monitoring of behaviour in zebrafish after invasive procedures. *Scientific Reports*, *9*, 1.

Deakin, A. G., Spencer, J. W., Cossins, A. R., Young, I. S., & Sneddon, L. U. (2019). Welfare challenges influence the complexity of movement: Fractal analysis of behaviour in zebrafish. *Fishes*, *4*, 1–16.

DePasquale, C., Neuberger, T., Hirrlinger, A. M., & Braithwaite, V. A. (2016). The influence of complex and threatening environments in early life on brain size and behaviour. *Proceedings. Biological Sciences*, *283*(1823), 1–8. https://doi.org/10.1098/rspb.2015.2564

DePasquale, C., & Leri, J. (2018). The influence of exercise on anxiety-like behavior in zebrafish (Danio rerio). *Behavioural Processes*, *157*, 638–644. https://doi.org/10.1016/j.beproc.2018.04.006

do Nascimento, G. S., Walsh-Monteiro, A., & Gouveia, A. J. (2016). A reliable depression-like model in zebrafish (Danio rerio): Learned helplessness. *Psychology and Neuroscience*, *9*(3), 390–397. https://doi.org/10.1037/pne0000057

Engeszer, R. E., Barbiano, L. A., Ryan, M. J., & Parichy, D. M. (2007). Timing and plasticity of shoaling behaviour in the zebrafish, Danio rerio. *Animal Behaviour, 74*(5), 1269–1275. https://doi.org/10.1016/j.anbehav.2007.01.032

Engeszer, R. E., Ryan, M. J., & Parichy, D. M. (2004). Learned social preference in zebrafish. *Current Biology, 14*(10), 881–884. https://doi.org/10.1016/j.cub.2004.04.042

Espigares, F., Martins, R. R., & Oliveira, R. F. (2022) A behavioural assay to investigate judgment bias in zebrafish. *Bio Protoc., 12*(4):e4327. 10.21769/BioProtoc.4327.

Espigares, F., Abad-Tortosa, D., Varela, S. A. M., Ferreira, M. G., & Oliveira, R. F. (2021). Short telomeres drive pessimistic judgement bias in zebrafish. *Biology Letters, 17*(3), 20200745. https://doi.org/10.1098/rsbl.2020.0745

Faustino, A. I., Tacão-Monteiro, A., & Oliveira, R. F. (2017). Mechanisms of social buffering of fear in zebrafish. *Scientific Reports, 7*, 44329. https://doi.org/10.1038/srep44329

Félix, A. S., Faustino, A. I., Cabral, E. M., & Oliveira, R. F. (2013). Noninvasive measurement of steroid hormones in zebrafish Holding-Water. *Zebrafish, 10*(1), 110–115. https://doi.org/10.1089/zeb.2012.0792

Filosa, A., Barker, A. J., Dal Maschio, M. D., & Baier, H. (2016). Feeding state modulates behavioral choice and processing of prey stimuli in the zebrafish tectum. *Neuron, 90*(3), 596–608. https://doi.org/10.1016/j.neuron.2016.03.014

Fulcher, N., Tran, S., Shams, S., Chatterjee, D., & Gerlai, R. (2017). Neurochemical and behavioral responses to unpredictable chronic mild stress following developmental isolation: The zebrafish as a model for major depression. *Zebrafish, 14*(1), 23–34. https://doi.org/10.1089/zeb.2016.1295

Gaikwad, S., Stewart, A., Hart, P., Wong, K., Piet, V., Cachat, J., & Kalueff, A. V. (2011). Acute stress disrupts performance of zebrafish in the cued and spatial memory tests: The utility of fish models to study stress-memory interplay. *Behavioural Processes, 87*(2), 224–230. https://doi.org/10.1016/j.beproc.2011.04.004

Galhardo, L., Vital, J., & Oliveira, R. F. (2011). The role of predictability in the stress response of a cichlid fish. *Physiology and Behavior, 102*(3–4), 367–372. https://doi.org/10.1016/j.physbeh.2010.11.035

Genario, R., de Abreu, M. S., Giacomini, A. C. V. V., Demin, K. A., & Kalueff, A. V. (2020). Sex differences in behavior and neuropharmacology of zebrafish. *European Journal of Neuroscience, 52*(1), 2586–2603. https://doi.org/10.1111/ejn.14438

Gerlach, G., Hodgins-Davis, A., Avolio, C., & Schunter, C. (2008). Kin recognition in zebrafish: A 24-hour window for olfactory imprinting. *Proceedings. Biological Sciences, 275*(1647), 2165–2170. https://doi.org/10.1098/rspb.2008.0647

Gerlai, J., & Gerlai, R. (2003). Autism: A large unmet medical need and a complex research problem. *Physiology and Behavior, 79*(3), 461–470. https://doi.org/10.1016/s0031-9384(03)00165-3

Gerlai, R., Lahav, M., Guo, S., & Rosenthal, A. (2000). Drinks like a fish: Zebra fish (Danio rerio) as a behavior genetic model to study alcohol effects. *Pharmacology, Biochemistry, and Behavior, 67*(4), 773–782. https://doi.org/10.1016/s0091-3057(00)00422-6

Gerlai, R., Fernandes, Y., & Pereira, T. (2009). Zebrafish (Danio rerio) responds to the animated image of a predator: Towards the development of an automated aversive task. *Behavioural Brain Research, 201*(2), 318–324. https://doi.org/10.1016/j.bbr.2009.03.003

Ghisleni, G., Capiotti, K. M., Da Silva, R. S., Oses, J. P., Piato, Â. L., Soares, V., Bogo, M. R., & Bonan, C. D. (2012). The role of CRH in behavioral responses to acute restraint stress in zebrafish. *Progress in Neuro-Psychopharmacology and Biological Psychiatry, 36*(1), 176–182. https://doi.org/10.1016/j.pnpbp.2011.08.016

Giacomini, A., de Abreu, M., Koakosk,i G., Idalencio, R., Kalichak, F., Oliveira, T., da Rosa, J., Gusso, D., Piato, A., & Barcellos, L. (2015). My stress, our stress: Blunted cortisol response to stress in isolated housed zebrafish. *Physiology and Behavior,* 139, 182–187.

Giacomotto, J., Carroll, A. P., Rinkwitz, S., Mowry, B., Cairns, M. J., & Becker, T. S. (2016). Developmental suppression of schizophrenia-associated miR-137 alters sensorimotor function in zebrafish. *Translational Psychiatry, 6*(5), e818. https://doi.org/10.1038/tp.2016.88

Gilbert, M. J. H., Zerulla, T. C., & Tierney, K. B. (2014). Zebrafish (Danio rerio) as a model for the study of aging and exercise: Physical ability and trainability decrease with age. *Experimental Gerontology, 50*, 106–113. https://doi.org/10.1016/j.exger.2013.11.013

Gleason, P. E., Weber, P. G., & Weber, S. P. (1977). Effect of group size on avoidance learning in zebra fish, Brachydanio rerio (Pisces: Cyprinidae). *Animal Learning and Behavior, 5*(2), 213–216. https://doi.org/10.3758/BF03214081

Golla, A., Østby, H., & Kermen, F. (2020). Chronic unpredictable stress induces anxiety-like behaviors in young zebrafish. *Scientific Reports, 10*(1), 10339. https://doi.org/10.1038/s41598-020-67182-4

Hall, D., & Suboski, M. D. (1995). Visual and olfactory stimuli in learned release of alarm reactions by zebra danio fish (Brachydanio rerio). *Neurobiology of Learning and Memory, 63*(3), 229–240. https://doi.org/10.1006/nlme.1995.1027

Hamilton, T. J., Morril, A., Lucas, K., Gallup, J., Harris, M., Taylor, P., Schalomon, M., Digweed, S., & Tresguerres, M. (2017). Establishing zebrafish as a model to study the anxiolytic effects of scopolamine. *Scientific Reports, 7*, 1.

Harding, E. J., Paul, E. S., & Mendl, M. (2004). Animal behaviour: Cognitive bias and affective state. *Nature, 427*(6972), 312. https://doi.org/10.1038/427312a

Hinz, F. I., Aizenberg, M., Tushev, G., & Schuman, E. M. (2013). Protein synthesis-dependent associative long-term memory in larval zebrafish. *Journal of Neuroscience, 33*(39), 15382–15387. https://doi.org/10.1523/JNEUROSCI.0560-13.2013

Holley, E. D., Ludwig, K. J., McEwen, D. C., & Wisenden, B. D. (2014). Predictability of food supply, but not ration, increases exploratory behaviour in zebrafish Danio rerio. *Journal of Fish Biology, 85*(3), 597–604. https://doi.org/10.1111/jfb.12441

Hurd, M. W., Debruyne, J., Straume, M., & Cahill, G. M. (1998). Circadian rhythms of locomotor activity in zebrafish. *Physiology and Behavior, 65*(3), 465–472. https://doi.org/10.1016/s0031-9384(98)00183-8

Jones, L. J., & Norton, W. H. J. (2015). Using zebrafish to uncover the genetic and neural basis of aggression, a frequent comorbid symptom of psychiatric disorders. *Behavioural Brain Research, 276*, 171–180. https://doi.org/10.1016/j.bbr.2014.05.055

Kalueff, A. V., Gebhardt, M., Stewart, A. M., Cachat, J. M., Brimmer, M., Chawla, J. S., Craddock, C., Kyzar, E. J., Roth, A., Landsman, S., Gaikwad, S., Robinson, K., Baatrup, E., Tierney, K., Shamchuk, A., Norton, W., Miller, N., Nicolson, T., Braubach, O. (2013). Towards a comprehensive catalog of zebrafish Behavior 1.0 and beyond. *Zebrafish, 10*(1), 70–86. https://doi.org/10.1089/zeb.2012.0861

Kalueff, A. V., & Cachat, J. (2011). *Zebrafish models in neurobehavioral research* (1st ed). Springer International Publishing.

Kalueff, A. V. (2017). Illustrated zebrafish neurobehavioral glossary. In A. V. Kalueff (Ed.), *The rights and wrongs of zebrafish: Behavioral phenotyping of zebrafish* (pp. 291-317). Springer International Publishing.

Kalueff, A. V., Stewart, A. M., & Gerlai, R. (2014). Zebrafish as an emerging model for studying complex brain disorders. *Trends in Pharmacological Sciences, 35*(2), 63–75. https://doi.org/10.1016/j.tips.2013.12.002

Karmakar, A. K., Das, A., & Banerjee, P. K. (2008). Records of the zoological survey of India zoological survey of India: Fish fauna of Subarnarekha River. Director (Ed.), Zoological Survey of India.

Karnik, I., & Gerlai, R. (2012). Can zebrafish learn spatial tasks? An empirical analysis of place and single CS-US associative learning. *Behavioural Brain Research, 233*(2), 415–421. https://doi.org/10.1016/j.bbr.2012.05.024

Katz, R. J., Roth, K. A., & Carroll, B. J. (1981). Acute and chronic stress effects on open field activity in the rat: Implications for a model of depression. *Neuroscience and Biobehavioral Reviews, 5*(2), 247–251. https://doi.org/10.1016/0149-7634(81)90005-1

Lange, M., Neuzeret, F., Fabreges, B., Froc, C., Bedu, S., Bally-Cuif, L., & Norton, W. H. J. (2013). Inter-individual and inter-strain variations in zebrafish locomotor ontogeny. *PloS One*. Public Library of Science ONE, 8(8), e70172. https://doi.org/10.1371/journal.pone.0070172

Lange, M., Froc, C., Grunwald, H., Norton, W. H. J., & Bally-Cuif, L. (2018). Pharmacological analysis of zebrafish lphn3.1 morphant larvae suggests that saturated dopaminergic signaling could underlie the ADHD-like locomotor hyperactivity. *Progress in Neuro-Psychopharmacology and Biological Psychiatry, 84*(A), 181–189. https://doi.org/10.1016/j.pnpbp.2018.02.010

Larsch, J., & Baier, H. (2018). Biological motion as an innate perceptual mechanism driving social affiliation. *Current Biology, 28*(22), 3523–3532.e4. https://doi.org/10.1016/j.cub.2018.09.014

Larson, E. T., Malley, D. M., Melloni, R. H., Hall, M., & Ave, H. (2006). Aggression and vasotocin are associated with dominant – Subordinate relationships in zebrafish. *Behavioural Brain Research, 167*(1), 94–102. https://doi.org/10.1016/j.bbr.2005.08.020

Lawrence, C. (2007). The husbandry of zebrafish (Danio rerio): A review. *Aquaculture, 269*(1–4), 1–20. https://doi.org/10.1016/j.aquaculture.2007.04.077

Luca, R. M., & Gerlai, R. (2012a). Animated bird silhouette above the tank: Acute alcohol diminishes fear responses in zebrafish. *Behavioural*

Brain Research, *229*(1), 194–201. https://doi.org/10.1016/j.bbr.2012.01.021

Luca, R. M., & Gerlai, R. (2012b). In search of optimal fear inducing stimuli: Differential behavioral responses to computer animated images in zebrafish. *Behavioural Brain Research*, *226*(1), 66–76. https://doi.org/10.1016/j.bbr.2011.09.001

Lucon-Xiccato, T., & Dadda, M. (2014). Assessing memory in zebrafish using the one-trial test. *Behavioural Processes*, *106*, 1–4. https://doi.org/10.1016/j.beproc.2014.03.010

Madeira, N., & Oliveira, R. F. (2017). Long-term social recognition memory in zebrafish. *Zebrafish*, *14*(4), 305–310. https://doi.org/10.1089/zeb.2017.1430

Magurran, A. E., & Maciás Garcia, C. (2005). Sex differences in behaviour as an indirect consequence of mating system. *Journal of Fish Biology*, *57*, 839–857.

Mandelbrot, B. (1967). How long is the coast of Britain? Statistical self-similarity and fractional dimension. *Science*, *156*(3775), 636–638. https://doi.org/10.1126/science.156.3775.636

Manuel, R., Gorissen, M., Zethof, J., Ebbesson, L. O., van de Vis, H., Flik, G., & van den Bos, R. (2014). Unpredictable chronic stress decreases inhibitory avoidance learning in Tübingen long-fin zebrafish: Stronger effects in the resting phase than in the active phase. *Journal of Experimental Biology*, *217*(21), 3919–3928. https://doi.org/10.1242/jeb.109736

Marcon, M., Herrmann, A. P., Mocelin, R., Rambo, C. L., Koakoski, G., Abreu, M. S., Conterato, G. M., Kist, L. W., Bogo, M. R., Zanatta, L., Barcellos, L. J., & Piato, A. L. (2016). Prevention of unpredictable chronic stress-related phenomena in zebrafish exposed to Bromazepam, fluoxetine and nortriptyline. *Psychopharmacology*, *233*(21–22), 3815–3824. https://doi.org/10.1007/s00213-016-4408-5

Marcon, M., Mocelin, R., Benvenutti, R., Costa, T., Herrmann, A. P., de Oliveira, D. L., Koakoski, G., Barcellos, L. J. G., & Piato, A. (2018). Environmental enrichment modulates the response to chronic stress in zebrafish. *Journal of Experimental Biology*, *221*(4). https://doi.org/10.1242/jeb.176735

Martins, R. (2020). *Tert-dependent senescence and inflammation in ageing, using zebrafish as a model* [PhD Thesis]. University of Sheffield.

Matthews, M., Trevarrow, B., & Matthews, J. (2002). A virtual tour of the Guide for zebrafish users. *Lab Animal*, *31*(3), 34–40. https://doi.org/10.1038/5000140

Maximino, C., de Brito, T. M., Colmanetti, R., Pontes, A. A., de Castro, H. M., de Lacerda, R. I., Morato, S., & Gouveia, A. (2010a). Parametric analyses of anxiety in zebrafish scototaxis. *Behavioural Brain Research, 210*(1), 1–7. https://doi.org/10.1016/j.bbr.2010.01.031

Maximino, C., Marques de Brito, T., Dias, C. A., Gouveia, A., & Morato, S. (2010b). Scototaxis as anxiety-like behavior in fish. *Nature Protocols, 5*(2), 209–216. https://doi.org/10.1038/nprot.2009.225

Maximino, C. (2011). Modulation of nociceptive-like behavior in zebrafish (Danio rerio) by environmental stressors. *Psychology and Neuroscience, 4*(1), 149–155. https://doi.org/10.3922/j.psns.2011.1.017

Maximino, C., da Silva, A. W., Gouveia, A., & Herculano, A. M. (2011). Pharmacological analysis of zebrafish (Danio rerio) scototaxis. *Progress in Neuro-Psychopharmacology and Biological Psychiatry, 35*(2), 624–631. https://doi.org/10.1016/j.pnpbp.2011.01.006

May, Z., Morrill, A., Holcombe, A., Johnston, T., Gallup, J., Fouad, K., Schalomon, M., & Hamilton, T. J. (2016). Object recognition memory in zebrafish. *Behavioural Brain Research, 296*, 199–210. https://doi.org/10.1016/j.bbr.2015.09.016

Mcaroe, C. L., Craig, C. M., & Holland, R. A. (2017). Shoaling promotes place over response learning but does not facilitate individual learning of that strategy in zebrafish (Danio rerio). *BMC Zoology, 2*, 1–8.

McCann, L. I., & Matthews, J. J. (1974). The effects of lifelong isolation on species identification in zebra fish (Brachydanio rerio). *Developmental Psychobiology, 7*(2), 159–163. https://doi.org/10.1002/dev.420070209

McClure, M. M., McIntyre, P. B., & McCune, A. R. (2006). Notes on the natural diet and habitat of eight danionin fishes, including the zebrafish Danio rerio. *Journal of Fish Biology, 69*(2), 553–570. https://doi.org/10.1111/j.1095-8649.2006.01125.x

Mendl, M., Burman, O. H. P., Parker, R. M. A., & Paul, E. S. (2009). Cognitive bias as an indicator of animal emotion and welfare: Emerging evidence and underlying mechanisms. *Applied Animal Behaviour Science, 118*(3–4), 161–181. https://doi.org/10.1016/j.applanim.2009.02.023

Mendl, M., Brooks, J., Basse, C., Burman, O., Paul, E., Blackwell, E., & Casey, R. (2010). Dogs showing separation-related behaviour exhibit a 'pessimistic' cognitive bias. Current Biology, 20(19), R839–R840. https://doi.org/10.1016/j.cub.2010.08.030

Midttun, H. L. E., Vindas, M. A., Nadler, L. E., Øverli, Ø., & Johansen, I. B. (2020). Behavioural effects of the common brain-infecting parasite

Pseudoloma neurophilia in laboratory zebrafish (Danio rerio). *Scientific Reports, 10*(1), 8083. https://doi.org/10.1038/s41598-020-64948-8
Miller, N., & Gerlai, R. (2007). Quantification of shoaling behaviour in zebrafish (Danio rerio). *Behavioural Brain Research, 184*(2), 157–166. https://doi.org/10.1016/j.bbr.2007.07.007
Miller, N. Y., & Gerlai, R. (2008). Oscillations in shoal cohesion in zebrafish (Danio rerio). *Behavioural Brain Research, 193*(1), 148–151. https://doi.org/10.1016/j.bbr.2008.05.004
Moretz, J. A., Martins, E. P., & Robison, B. D. (2007). Behavioral syndromes and the evolution of correlated behavior in zebrafish. *Behavioral Ecology, 18*(3), 556–562. https://doi.org/10.1093/beheco/arm011
Newby, N. C., Wilkie, M. P., & Stevens, E. D. (2009). Morphine uptake, disposition, and analgesic efficacy in the common goldfish (Carassius auratus). *Canadian Journal of Zoology, 87*(5), 388–399. https://doi.org/10.1139/Z09-023
Ninkovic, J., & Bally-Cuif, L. (2006). The zebrafish as a model system for assessing the reinforcing properties of drugs of abuse. *Methods, 39*(3), 262–274. https://doi.org/10.1016/j.ymeth.2005.12.007
Nunes, A. R., Carreira, L., Anbalagan, S., Blechman, J., Levkowitz, G., & Oliveira, R. F. (2020). Perceptual mechanisms of social affiliation in zebrafish. *Scientific Reports, 10*(1), 3642. https://doi.org/10.1038/s41598-020-60154-8
Oliveira, T. A., Idalencio, R., Kalichak, F., dos Santos Rosa, J. G., Koakoski, G., de Abreu, M. S., Giacomini, A. C. V., Gusso, D., Rosemberg, D. B., Barreto, R. E., & Barcellos, L. J. G. (2017). Stress responses to conspecific visual cues of predation risk in zebrafish. *PeerJ, 5*, e3739. https://doi.org/10.7717/peerj.3739
Oswald, M. E., Drew, R. E., Racine, M., Murdoch, G. K., & Robison, B. D. (2012). Is behavioral variation along the bold-shy continuum associated with variation in the stress axis in zebrafish? *Physiological and Biochemical Zoology, 85*(6), 718–728. https://doi.org/10.1086/668203
Oswald, M., & Robison, B. D. (2008). Strain-specific alteration of zebrafish feeding behavior in response to aversive stimuli. *Canadian Journal of Zoology, 86*(10), 1085–1094. https://doi.org/10.1139/Z08-085
Park, J. S., Ryu, J. H., Choi, T. I., Bae, Y. K., Lee, S., Kang, H. J., & Kim, C. H. (2016). Innate color preference of zebrafish and its use in behavioral analyses. *Molecules and Cells, 39*(10), 750–755. https://doi.org/10.14348/molcells.2016.0173

Parker, M. O., Ife, D., Ma, J., Pancholi, M., Smeraldi, F., Straw, C., & Brennan, C. H. (2013). Development and automation of a test of impulse control in zebrafish. *Frontiers in Systems Neuroscience, 7*, 65. https://doi.org/10.3389/fnsys.2013.00065

Pather, S., & Gerlai, R. (2009). Shuttle box learning in zebrafish (Danio rerio). *Behavioural Brain Research, 196*(2), 323–327. https://doi.org/10.1016/j.bbr.2008.09.013

Pavlidis, M., Sundvik, M., Chen, Y. C., & Panula, P. (2011). Adaptive changes in zebrafish brain in dominant-subordinate behavioral context. *Behavioural Brain Research, 225*(2), 529–537. https://doi.org/10.1016/j.bbr.2011.08.022

Pavlov, I. P. (1927). Conditioned reflexes: An investigation of the physiological activity of the cerebral cortex. *Annals of Neurosciences, 17*(3), 136–141. https://doi.org/10.5214/ans.0972-7531.1017309

Pedroso, G. L., Hammes, T. O., Escobar, T. D., Fracasso, L. B., Forgiarini, L. F., & da Silveira, T. R. (2012). Blood collection for biochemical analysis in adult zebrafish. *Journal of Visualized Experiments: JoVE*, (63), e3865. https://doi.org/10.3791/3865

Pérez-Escudero, A., Vicente-Page, J., Hinz, R. C., Arganda, S., & De Polavieja, G. G. (2014). IdTracker: Tracking individuals in a group by automatic identification of unmarked animals. *Nature Methods, 11*(7), 743–748. https://doi.org/10.1038/nmeth.2994

Pfeiffer, W. (1977). The distribution of fright reaction and alarm substance cells in fishes. *Copeia*, 1977(4), 653–665. https://doi.org/10.2307/1443164

Pham, M., Raymond, J., Hester, J., Kyzar, E., Gaikwad, S., Bruce, I., Fryar, C., Chanin, S., Enriquez, J., Bagawandoss, S., Zapolsky, I., Green, J., Stewart, A. M., Robison, B. D., & Kalueff, A. V. (2012). Assessing social behavior phenotypes in adult zebrafish: Shoaling, social preference, and mirror biting tests. In *Zebrafish protocols for neurobehavioral research* (pp. 231–246). Springer International Publishing. https://doi.org/10.1007/978-1-61779-597-8_17

Piato, Â. L., Capiotti, K. M., Tamborski, A. R., Oses, J. P., Barcellos, L. J., Bogo, M. R., Lara, D. R., Vianna, M. R., & Bonan, C. D. (2011). Unpredictable chronic stress model in zebrafish (Danio rerio): Behavioral and physiological responses. *Progress in Neuro-Psychopharmacology and Biological Psychiatry, 35*(2), 561–567. https://doi.org/10.1016/j.pnpbp.2010.12.018

Pinho, J. S., Castilho, M., Sollari, J. S., & Oliveira, R. F. (2020). Innate chemical, but not visual, threat cues have been co-opted as unconditioned stimulus for social fear learning in zebrafish. *Genes, Brain and Behavior, 19*(8). https://doi.org/10.1111/gbb.12688

Pittman, J., & Piato, A. (2017). Developing zebrafish depression-related models. In A. V. Kalueff (Ed.), *The rights and wrongs of zebrafish: Behavioral phenotyping of zebrafish* (pp. 33–44). Springer International Publishing.

Pritchett, D., & Brennan, C. H. (2020). Classical and operant conditioning in larval zebrafish. In R. T. Gerlai (Ed.), *Behavioral and neural genetics of Zebrafish* (pp. 107–122). Academic Press.

Prut, L., & Belzung, C. (2003). The open field as a paradigm to measure the effects of drugs on anxiety-like behaviors: A review. *European Journal of Pharmacology, 463*(1–3), 3–33. https://doi.org/10.1016/s0014-2999(03)01272-x

Ramsay, J. M., Feist, G. W., Varga, Z. M., Westerfield, M., Kent, M. L., & Schreck, C. B. (2006). Whole-body cortisol is an indicator of crowding stress in adult zebrafish, Danio rerio. *Aquaculture, 258*(1–4), 565–574. https://doi.org/10.1016/j.aquaculture.2006.04.020

Ramsay, J. M., Feist, G. W., Varga, Z. M., Westerfield, M., Kent, M. L., & Schreck, C. B. (2009). Whole-body cortisol response of zebrafish to acute net handling stress. *Aquaculture, 297*(1–4), 157–162. https://doi.org/10.1016/j.aquaculture.2009.08.035

Reilly, S. C., Quinn, J. P., Cossins, A. R., & Sneddon, L. U. (2008). Behavioural analysis of a nociceptive event in fish: Comparisons between three species demonstrate specific responses. *Applied Animal Behaviour Science, 114*(1–2), 248–259. https://doi.org/10.1016/j.applanim.2008.01.016

Reolon, G. K., de Melo, G. M., da Rosa, J. G. D. S., Barcellos, L. J. G., & Bonan, C. D. (2018). Sex and the housing: Effects on behavior, cortisol levels and weight in zebrafish. *Behavioural Brain Research, 336*, 85–92. https://doi.org/10.1016/j.bbr.2017.08.006

Ribas, L., Valdivieso, A., Díaz, N., & Piferrer, F. (2017). Appropriate rearing density in domesticated zebrafish to avoid masculinization: Links with the stress response. Journal of Experimental Biology, 220(6), 1056–1064. https://doi.org/10.1242/jeb.144980

Roberts, A. C., Bill, B. R., & Glanzman, D. L. (2013). Learning and memory in zebrafish larvae. *Frontiers in Neural Circuits, 7*, 126. https://doi.org/10.3389/fncir.2013.00126

Romero-Ferrero, F., Bergomi, M. G., Hinz, R. C., Heras, F. J. H., & de Polavieja, G. G. (2019). idtracker.ai: Tracking all individuals in small or large collectives of unmarked animals. *Nature Methods, 16*(2), 179–182. https://doi.org/10.1038/s41592-018-0295-5

Roques, J. A., Abbink, W., Geurds, F., van de Vis, H., & Flik, G. (2010). Tailfin clipping, a painful procedure: Studies on Nile tilapia and common carp. *Physiology and Behavior, 101*(4), 533–540. https://doi.org/10.1016/j.physbeh.2010.08.001

Rose, J. D. (2003). A critique of the paper "Do fish have nociceptors: evidence for the evolution of a vertebrate sensory system" by Sneddon et al. (2003). In H. S. Erickson (Ed.), *Information resources on fish welfare* (13th ed) (pp. 49–51). Animal Welfare Information Center (AWIC) Resources Series No. 20. AWIC. National Agricultural Library, United States Department of Agriculture.

Ruhl, N., & McRobert, S. P. (2005). The effect of sex and shoal size on shoaling behaviour in Danio rerio. *Journal of Fish Biology, 67*(5), 1318–1326. https://doi.org/10.1111/j.0022-1112.2005.00826.x

Saverino, C., & Gerlai, R. (2008). The social zebrafish: Behavioral responses to conspecific, heterospecific, and computer animated fish. *Behavioural Brain Research, 191*(1), 77–87. https://doi.org/10.1016/j.bbr.2008.03.013

Schnörr, S. J., Steenbergen, P. J., Richardson, M. K., & Champagne, D. L. (2012). Measuring thigmotaxis in larval zebrafish. *Behavioural Brain Research, 228*(2), 367–374. https://doi.org/10.1016/j.bbr.2011.12.016

Schroeder, P., Jones, S., Young, I. S., & Sneddon, L. U. (2014). What do zebrafish want? Impact of social grouping dominance and gender on preference for enrichment. *Laboratory Animals, 48*(4), 328–337. https://doi.org/10.1177/0023677214538239

Schroeder, P. G., & Sneddon, L. U. (2017). Exploring the efficacy of immersion analgesics in zebrafish using an integrative approach. *Applied Animal Behaviour Science, 187*, 93–102. https://doi.org/10.1016/j.applanim.2016.12.003

Serra, E. L., Medalha, C. C., & Mattioli, R. (1999). Natural preference of zebrafish (Danio rerio) for a dark environment. *Brazilian Journal of Medical and Biological Research, 32*(12), 1551–1553. https://doi.org/10.1590/s0100-879x1999001200016

Sireeni, J., Bakker, N., Jaikumar, G., Obdam, D., Slabbekoorn, H., Tudorache, C., & Schaaf, M. (2020). Profound effects of glucocorticoid resistance on anxiety-related behavior in zebrafish adults but not in larvae. *General and Comparative Endocrinology, 292*, 113461. https://doi.org/10.1016/j.ygcen.2020.113461

Sison, M., & Gerlai, R. (2010). Associative learning in zebrafish (Danio rerio) in the plus maze. *Behavioural Brain Research, 207*(1), 99–104. https://doi.org/10.1016/j.bbr.2009.09.043

Song, C., Liu, B. P., Zhang, Y. P., Peng, Z., Wang, J., Collier, A. D., Echevarria, D. J., Savelieva, K. V., Lawrence, R. F., Rex, C. S., Meshalkina, D. A., & Kalueff, A. V. (2018). Modeling consequences of prolonged strong unpredictable stress in zebrafish: Complex effects on behavior and physiology. *Progress in Neuro-Psychopharmacology and Biological Psychiatry, 81*, 384–394. https://doi.org/10.1016/j.pnpbp.2017.08.021

Spagnoli, S., Sanders, J., & Kent, M. L. (2017). The common neural parasite Pseudoloma neurophilia causes altered shoaling behavior in adult laboratory zebrafish (Danio rerio) and its implications for neurobehavioral research. *Journal of Fish Diseases, 40*(3), 443–446. https://doi.org/10.1111/jfd.12512

Spagnoli, S., Xue, L., & Kent, M. L. (2015). The common neural parasite Pseudoloma neurophilia is associated with altered startle response habituation in adult zebrafish (Danio rerio): Implications for the zebrafish as a model organism. *Behavioural Brain Research, 291*, 351–360. https://doi.org/10.1016/j.bbr.2015.05.046

Speedie, N., & Gerlai, R. (2008). Alarm substance induced behavioral responses in zebrafish (Danio rerio). *Behavioural Brain Research, 188*(1), 168–177. https://doi.org/10.1016/j.bbr.2007.10.031

Spence, R., Fatema, M. K., Reichard, M., Huq, K. A., Wahab, M. A., Ahmed, Z. F., & Smith, C. (2006). The distribution and habitat preferences of the zebrafish in Bangladesh. *Journal of Fish Biology, 69*(5), 1435–1448. https://doi.org/10.1111/j.1095-8649.2006.01206.x

Spence, R., Gerlach, G., Lawrence, C., & Smith, C. (2008). The behaviour and ecology of the zebrafish, Danio rerio. *Biological Reviews of the Cambridge Philosophical Society, 83*(1), 13–34. https://doi.org/10.1111/j.1469-185X.2007.00030.x

Spence, R., Magurran, A. E., & Smith, C. (2011). Spatial cognition in zebrafish: The role of strain and rearing environment. *Animal Cognition, 14*(4), 607–612. https://doi.org/10.1007/s10071-011-0391-8

Spence, R., & Smith, C. (2005). Male territoriality mediates density and sex ratio effects on oviposition in the zebrafish, Danio rerio. *Animal Behaviour, 69*(6), 1317–1323. https://doi.org/10.1016/j.anbehav.2004.10.010

Spence, R., & Smith, C. (2008). Innate and learned colour preference in the zebrafish, Danio rerio. *Ethology, 114*(6), 582–588. https://doi.org/10.1111/j.1439-0310.2008.01515.x

Steenbergen, P. J., Richardson, M. K., & Champagne, D. L. (2011). Patterns of avoidance behaviours in the light/dark preference test in young juvenile zebrafish: A pharmacological study. *Behavioural Brain Research*, *222*(1), 15–25. https://doi.org/10.1016/j.bbr.2011.03.025

Sterba, G. (1962). *Freshwater fishes of the world* (6th ed). Vista Books.

Stewart, A., Cachat, J., Wong, K., Gaikwad, S., Gilder, T., Dileo, J., Chang, K., Utterback, E., & Kalueff, A. V. (2010). Homebase behavior of zebrafish in novelty-based paradigms. *Behavioural Processes*, *85*(2), 198–203. https://doi.org/10.1016/j.beproc.2010.07.009

Stewart, A., Wu, N., Cachat, J., Hart, P., Gaikwad, S., Wong, K., Utterback, E., Gilder, T., Kyzar, E., Newman, A., Carlos, D., Chang, K., Hook, M., Rhymes, C., Caffery, M., Greenberg, M., Zadina, J., & Kalueff, A. V. (2011). Pharmacological modulation of anxiety-like phenotypes in adult zebrafish behavioral models. *Progress in Neuro-Psychopharmacology and Biological Psychiatry*, *35*(6), 1421–1431. https://doi.org/10.1016/j.pnpbp.2010.11.035

Stewart, A. M., Gaikwad, S., Kyzar, E., & Kalueff, A. V. (2012). Understanding spatio-temporal strategies of adult zebrafish exploration in the open field test. *Brain Research*, *1451*, 44–52. https://doi.org/10.1016/j.brainres.2012.02.064

Suriyampola, P. S., Shelton, D. S., Shukla, R., Roy, T., Bhat, A., & Martins, E. P. (2016). Zebrafish social behavior in the wild. *Zebrafish*, *13*(1), 1–8. https://doi.org/10.1089/zeb.2015.1159

Teles, M. C., Cardoso, S. D., & Oliveira, R. F. (2016). Social plasticity relies on different neuroplasticity mechanisms across the brain social decision-making network in zebrafish. *Frontiers in Behavioral Neuroscience*, *10*, 16. https://doi.org/10.3389/fnbeh.2016.00016

Thompson, R. F., & Spencer, W. A. (1966). Habituation: A model phenomenon for the study of neuronal substrates of behavior. *Psychological Review*, *73*(1), 16–43. https://doi.org/10.1037/h0022681

Thörnqvist, P. O., McCarrick, S., Ericsson, M., Roman, E., & Winberg, S. (2019). Bold zebrafish (Danio rerio) express higher levels of delta opioid and dopamine D2 receptors in the brain compared to shy fish. *Behavioural Brain Research*, *359*, 927–934. https://doi.org/10.1016/j.bbr.2018.06.017

Tracey, W. D. (2017). Nociception. *Current Biology*, *27*(4), R129–R133. https://doi.org/10.1016/j.cub.2017.01.037

Valente, A., Huang, K. H., Portugues, R., & Engert, F. (2012). Ontogeny of classical and operant learning behaviors in zebrafish. *Learning and Memory, 19*(4), 170–177. https://doi.org/10.1101/lm.025668.112

vom Berg-Maurer, C. M., Trivedi, C. A., Bollmann, J. H., De Marco, R. J., & Ryu, S. (2016). The severity of acute stress is represented by increased synchronous activity and recruitment of hypothalamic CRH neurons. *Journal of Neuroscience, 36*(11), 3350–3362. https://doi.org/10.1523/JNEUROSCI.3390-15.2016

von Krogh, K., Sørensen, C., Nilsson, G. E., & Øverli, Ø. (2010). Forebrain cell proliferation, behavior, and physiology of zebrafish, Danio rerio, kept in enriched or barren environments. *Physiology and Behavior, 101*(1), 32–39. https://doi.org/10.1016/j.physbeh.2010.04.003

Wafer, L. N., Jensen, V. B., Whitney, J. C., Gomez, T. H., Flores, R., & Goodwin, B. S. (2016). Effects of environmental enrichment on the fertility and fecundity of zebrafish (Danio rerio). *Journal of the American Association for Laboratory Animal Science, 55*(3), 291–294.

Waldman, B. (1982). Quantitative and developmental analyses of the alarm reaction in the zebra danio, Brachydanio rerio. *Copeia*. American Society of Ichthyologists and Herpetologists, *1982*(1), 1–9. https://doi.org/10.2307/1444261

Williams, F. E., White, D., & Messer, W. S. (2002). A simple spatial alternation task for assessing memory function in zebrafish. *Behavioural Processes, 58*(3), 125–132. https://doi.org/10.1016/s0376-6357(02)00025-6

Wilson, C. A., High, S. K., McCluskey, B. M., Amores, A., Yan, Y. L., Titus, T. A., Anderson, J. L., Batzel, P., Carvan, M. J., Schartl, M., & Postlethwait, J. H. (2014). Wild sex in zebrafish: Loss of the natural sex determinant in domesticated strains. *Genetics, 198*(3), 1291–1308. https://doi.org/10.1534/genetics.114.169284

Wolman, M., & Granato, M. (2012). Behavioral genetics in larval zebrafish: Learning from the young. *Developmental Neurobiology, 72*(3), 366–372. https://doi.org/10.1002/dneu.20872

Wong, K., Elegante, M., Bartels, B., Elkhayat, S., Tien, D., Roy, S., Goodspeed, J., Suciu, C., Tan, J., Grimes, C., Chung, A., Rosenberg, M., Gaikwad, S., Denmark, A., Jackson, A., Kadri, F., Chung, K. M., Stewart, A., Gilder, T., . . . Kalueff, A. V. (2010). Analyzing habituation responses to novelty in zebrafish (Danio rerio). *Behavioural Brain Research, 208*(2), 450–457. https://doi.org/10.1016/j.bbr.2009.12.023

Yeh, C. M., Glöck, M., & Ryu, S. (2013). An optimized whole-body cortisol quantification method for assessing stress levels in larval zebrafish. *PloS One*. Public Library of Science ONE, *8*(11), e79406. https://doi.org/10.1371/journal.pone.0079406

Yu, L., Tucci, V., Kishi, S., & Zhdanova, I. V. (2006). Cognitive aging in zebrafish. *PloS One*. Public Library of Science ONE, *1*(1), e14. https://doi.org/10.1371/journal.pone.0000014

Zimmermann, F. F., Altenhofen, S., Kist, L. W., Leite, C. E., Bogo, M. R., Cognato, G. P., & Bonan, C. D. (2016). Unpredictable chronic stress alters adenosine metabolism in zebrafish brain. *Molecular Neurobiology*, *53*(4), 2518–2528. https://doi.org/10.1007/s12035-015-9270-7

Chapter 13

Anaesthesia, analgesia and euthanasia of zebrafish

Protocol evaluation

Jean-Philippe Mocho

Abstract

The vast majority of laboratory *Danio rerio* are anaesthetized at least once in their life. The purpose and depth of anaesthesia vary from immobilization and sedation to surgical anaesthesia and euthanasia. The length of the scientific procedure and its potential to induce perioperative pain are essential criteria to consider when tailoring an anaesthesia protocol. Moreover, the choice of the anaesthetic compound is influenced by both national legislation on drug delivery and health and safety restrictions. In consequence, to determine the appropriate methods of anaesthesia, analgesia and euthanasia for a specific experiment, each possible technique can be screened through a set of requirements: Is this protocol a reliable and efficient option to induce sedation or euthanasia? Is this substance likely to procure analgesia perioperatively? Is this compound likely to induce aversion? What is the expected survival rate with this protocol? Are there adverse effects that would affect the health or fertility of the fish in the longer term or the scientific outcome? In the case of a repeat anaesthesia, is there any increased risk of mortality, aversion or other adverse effects? This guidance details how general zebrafish anaesthetics fulfil these requirements and how to determine the appropriate protocols for the main scientific procedures.

Claire Allen and Jean-Philippe Mocho (eds) **Zebrafish**
DOI: 10.1079/9781800629431.0013, © CAB International, 2024

Keywords

zebrafish, *Danio rerio*, fish, anaesthesia, analgesia, euthanasia

Introduction

An accumulation of experimental evidence supports the idea that zebrafish have the ability to feel animal pain (Sneddon et al., 2014). They are equipped with the required anatomical and functional features and receptors (nociceptors) to perceive pain-causing stimuli: i.e. zebrafish are capable of nociception (Sneddon, 2003; Sneddon et al., 2003). Following a direct or delayed noxious stimulus, *D. rerio* displays whole-animal responses and long-term motivational and behavioural changes characteristic of pain (Magalhaes et al., 2017; Reilly et al., 2008; Sneddon, 2015; Taylor et al., 2017). In consequence, zebrafish may feel pain during or after experimental procedures. Analgesia is the reduction of pain, and painkillers (i.e. analgesics) should be used to alleviate procedural adverse effects (Sloman et al., 2019). For that purpose, some anaesthetics can be used as analgesics perioperatively (i.e. before, during and after the procedure) as described in Table 13.1 (Deakin et al., 2019a,b; Schroeder & Sneddon, 2017). Anaesthetics are compounds used to perform anaesthesia. Anaesthesia is a state of muscle relaxation, analgesia and loss of consciousness. This is relevant since the vast majority of laboratory *D. rerio* are anaesthetized at least once in their life, e.g. for painful procedures with recovery such as fin-clipping or for a terminal purpose such as euthanasia.

Euthanasia is the killing of an animal with minimal distress. The purpose, length and depth of anaesthesia vary in parallel with the two main objectives leading the design of the anaesthesia protocol: optimization of animal welfare during and after the procedure, and achievement of the scientific endpoints. For example, when performing euthanasia, an overdose of anaesthesia may be used. The minutes of anaesthesia leading to the death of the fish should not induce avoidable distress to the animal, and a balance must be reached between a rapid and potentially unpleasant experience and a slow

Table 13.1 Steps of analgesia, anaesthesia and euthanasia.

Step	Aim	Examples
Pre-emptive analgesia	When painkillers (i.e. analgesics) are administered before the procedure, the postoperative analgesia (i.e. painkilling) is more efficient.	Lidocaine hydrochloride (2–5 mg/l)
Anaesthesia and procedure	Under balanced anaesthesia, the animal loses consciousness, mobility and pain perception, and we can thus perform a procedure.	Buffered tricaine
Postoperative analgesia	As the drugs used for anaesthesia lose their efficacy, the animal recovers from the anaesthesia and the procedure. The animal can now feel the pain induced by the procedure. Therefore, painkillers (i.e. analgesics) must be administered to reduce this postoperative pain. Painkillers can be administered by immersion over several hours. This method of analgesic administration is efficient providing the fish is immersed in the analgesic-medicated water. Painkillers can also be injected at the time of the anaesthesia and procedure, but we lack data on the length of efficacy following injection.	Lidocaine hydrochloride (2–5 mg/l) Morphine
Euthanasia	Fish can be killed by an overdose of anaesthesia.	Lidocaine hydrochloride (1,000 mg/l) with sodium bicarbonate (2,000 mg/l)

and comfortable loss of consciousness. Nevertheless, death must be complete before a sample is taken, and the sample must remain intact for the scientific purpose during the euthanasia procedure.

The effects of anaesthetics depend on their mechanisms of action (e.g. which receptors they interact with) and the dose, route and length of exposure (Ross & Ross, 2008). For example, not all anaesthetics provide analgesia and painkilling potency may be reached at lower doses than concentrations used to induce anaesthesia (de Abreu et al., 2019; Deakin et al., 2019a,b; Schroeder & Sneddon, 2017). The depth of anaesthesia is defined as light for sedation or increasingly deeper levels for immobilization, surgical anaesthesia and, finally, overdose (corresponding to a depth not allowing animal recovery if administration is prolonged). Table 13.2 shows that, as the dose of and exposure to an anaesthetic increase, reflexes are faint and anaesthesia deepens. In consequence, performing anaesthesia implies dosing the toxic effects of a poison – the anaesthetic. In order to reduce adverse effects and to achieve the required anaesthesia depth, the compound's exposure must be adapted to chronobiology (Sánchez-Vázquez et al., 2011), local husbandry and water parameters (von Krogh et al., 2021), fish strain and size (Martins et al., 2016) and developmental stages (Mocho et al., 2022; Owen & Kelsh, 2021). Indeed, for the first few days following fertilization, the main absorption route will be the skin; gills will be fully functional only a few days later and will then become the main route of entry for anaesthetics (Macova et al., 2008; Rombough, 2007). The toxicity of an anaesthetic or a dose rate may be detected rapidly or suspected during or after a procedure (e.g. difficult recovery, behaviour revealing averseness) or it may have a longer impact on physiologic parameters in the weeks following the procedure for the anaesthetized individual, the environment or the operator. The choice of an anaesthesia protocol should therefore rely on a multifactorial evaluation of each candidate compound to achieve the required anaesthesia depth with ease and reliability, and with limited undesired effects.

Such an assessment system for anaesthesia and euthanasia compounds is proposed below. First, further general considerations on anaesthesia and other immobilization methods are discussed and the main anaesthetics used in zebrafish are introduced. The concluding part of this chapter reviews options for the administration of analgesics to zebrafish.

Immobilization and euthanasia using non-anaesthetic methods

For some scientific aims, immobilization rather than anaesthesia is required. Fish can be immobilized by either physical methods or

Table 13.2 Main stages of zebrafish anaesthesia. Anaesthesia starts with induction, leading to sedation, immobilization, surgical anaesthesia and up to overdose, depending on the amount of anaesthetic administered. In some cases (e.g. short procedures such as fin-clipping), anaesthesia stops after induction, the animal is dealt with and recovers. In other cases (e.g. for the purposes of imaging), anaesthesia is prolonged with a maintenance phase. The animal then either recovers or it is maintained under an overdose of anaesthesia to achieve euthanasia. Depth of anaesthesia is indicated by alteration in or preservation or absence of clinical signs: righting reflex: ability of the fish to right its body to a ventral position; mobility: ability to swim; reaction to stimulus: noise, vibration or touch triggering movement. Opercular beat rate indicates respiratory rate.

Stage of anaesthesia	Clinical signs indicative of anaesthesia depth				Dose required to achieve stage	
	Righting reflex	Mobility	Reaction to stimulus	Opercular beat rate	Induction	Maintenance
Sedation	Preserved	Reduced	Reduced	May be affected	Low	Low
Immobilization	Absent	Absent	Reduced	Reduced	Medium	Medium
Surgical anaesthesia	Absent	Absent	Absent	Reduced	High	Medium
Overdose	Absent	Absent	Absent	Absent	High	High

neuromuscular blocking agents (NMBA; Aoki et al., 2013). NMBA do not induce a loss of consciousness or analgesia, nor do they induce anaesthesia. They prevent animals from showing signs of discomfort by blocking the signal from the brain to the muscles to trigger movement: animals appear anaesthetized when they are fully conscious. The use of such non-anaesthetic methods must be ethically justified, considering welfare impairments versus scientific benefits. Alternatively, the use of physical methods presents the advantage of avoiding any environmental pollution due to medication residue in waste waters. However, physical methods such as hypothermic shock (Chen et al., 2014), gradual cooling (Collymore et al., 2014) or electrical stunning (Schroeder et al., 2021) have their limits regarding welfare and scientific objectives. For example, fish being immobilized with recovery using chilled water might recover with significant harm while warming back to housing temperature. There is currently no understanding of fish experience during hypothermic events or electrical stunning, and there is no provision of analgesia – thus physical methods do not qualify as anaesthesia. Nonetheless, these methods may be suitable for euthanasia, with the advantage of a fast speed of action. It requires only 30 s to euthanize an adult zebrafish kept above 26 °C and dropped in chilled water (0–4 °C), due to the abruptly induced hypothermic shock (Wallace et al., 2018). This may seem a significant improvement, when very fresh and non-medicated samples are needed, compared with an overdose of traditional anaesthetic such as tricaine, for which exposure to overdose concentration for several minutes is required to achieve death (Leary et al., 2013). However, overdose with the anaesthetic lidocaine allows a very fast euthanasia over a comparable duration and with less aversion (von Krogh et al., 2021). The speed advantage of hypothermic shock is lost in *D. rerio* below 2 weeks post fertilization (Mocho et al., 2022; Strykowski et al., 2015; Wallace et al., 2018), for which overnight exposure in chilled water (0–4 °C) might be necessary to complete euthanasia, potentially inducing aversion for a prolonged time. Indeed, a significant proportion of fish undergoing hypothermic shock displays signs of discomfort (Mocho et al., 2022; Wallace et al., 2018; Wilson et al., 2009). Moreover, there are no data on fish perception and brain activity while hypothermic conditions are induced and maintained. Finally, sample quality might be compromised by physical methods. The water used to supply ice for hypothermic shock is usually different from system or reverse osmosis water, and it therefore constitutes a potential source of pollutants. Electrical stunning seems the method of choice for rapid euthanasia of larvae, although the impact of a stunning electrical current on biological samples is unknown (Mocho et al., 2022).

General considerations for zebrafish anaesthesia

Zebrafish typically undergo two types of anaesthesia: with or without maintenance. Maintenance of anaesthesia is required when animals are under procedure for a longer time than the period between induction and recovery. Induction is the passage from the alert state to the sedative or unconscious state (Table 13.2); recovery is the reverse passage, from the anaesthetized state to the alert state of full consciousness. Induction and recovery are the two steps of fish anaesthesia with the highest risk for health, and mortality is more likely to occur when induction and recovery are repeated in a short timeframe (Dang et al., 2016). Maintenance of the anaesthetized state should represent the low-risk step once the set-up and effects of the anaesthetics are well controlled. There is little limit to the maintenance, providing physiological requirements are fulfilled (e.g. provision of oxygen and nutrients and elimination of waste). Note that maintenance is required only for procedures involving immobilization for more than a few minutes, such as prolonged imaging, and the choice of anaesthetics may be key to the scientific outcome; this will not be detailed in length here. Such procedures usually focus on one fish at a time and allow individual adjustments (Lockwood et al., 2017; Olt et al., 2016; Wynd et al., 2017). In this chapter we will focus on the more common anaesthesia for short procedures such as fin-clipping, gamete recovery or compound administration. These procedures are often performed on a whole tank at a time, with several fish being sedated simultaneously and/or being anaesthetized in quick succession, in contradiction with many good practice guidelines usually adapted to terrestrial animals. Indeed, one caveat of the group procedure is stress management. When *D. rerio* is touched by a net during capture it releases an alarm substance alerting and raising the stress levels of other fish in the tank or in the anaesthetic bath (Carter et al., 2011; Speedie & Gerlai, 2008). Although it would be advisable to have an anaesthetic bath for each fish, practicalities in the busy and space-limited laboratory rarely allow for such refinement. The consequences of fish-handling stress are increases in aversive behaviour and cortisol levels and physiological disturbance to the formulation of red and white blood cells, the latter lasting for a few days or even weeks, depending on the anaesthetic's own adverse effects (Deebani et al., 2019; Kristan et al., 2012; Philippe et al., 2023; Shishis et al., 2023; Witeska et al., 2015). In practice, anaesthesia is induced by bathing the fish in an anaesthetic solution (i.e. by immersion or inhalation); there is no maintenance step and the procedure is performed, before the fish recovers in a tank containing system water and no anaesthetic solution. The physiology of zebrafish and their resistance to lack of oxygen compared with other fish species often makes the provision of an air stone during or after anaesthesia unnecessary in practice. For example, following a short procedure such as fin-clipping, fish are routinely kept for 48 h out of the recirculating system, at a density of five fish per litre, with reduced daily feed provision, without water renewal or air stone supply and immersed in low doses of lidocaine hydrochloride (2–5 mg/l) to provide analgesia whenever possible (Schroeder et al., 2021). Finally, one disadvantage of the use of anaesthetics is the environmental pollution induced by the release into the wild of medicated waste waters, although this can be reduced by carbon filtration before disposal (Howe et al., 1990).

Main anaesthetics used in zebrafish

The family of molecules licensed for immersive anaesthesia and/or euthanasia of fish are now introduced. Table 13.3 summarizes modes of action, formulations and efficacious doses for sedation, deep anaesthesia and euthanasia according to author's experience and published data for adult *D. rerio* (Attili & Hughes, 2014;

Table 13.3 Immersion anaesthetics for adult *D. rerio*. This table summarizes the main effects on nerve sensitivity, preparation, classification and dose rates for three levels of immersion anaesthesia in adult zebrafish (sedation, deep anaesthesia and euthanasia by overdose of anaesthesia) based on the author's experience, and on the following; Attili and Hughes (2014), Baldisserotto et al. (2018), Collymore et al. (2014), Forman (2011). Herring et al. (2011), Olt et al. (2016), Sánchez-Vázquez et al. (2011), Sloman et al. (2019), Sneddon et al. (2012), von Krogh et al. (2021) and Westerfield (1994). Deep anaesthesia corresponds to the stages from heavy sedation to surgical anaesthesia, depending on length of immersion. Fish must be maintained under the euthanasia dose to achieve euthanasia. LAP: the use of this compound for this application is not recommended due to its lack of anaesthetic potency. N/A, not applicable because the compound can be purchased in a water-soluble format that does not affect pH significantly. TBD: the dose for this depth of anaesthesia needs to be defined according to precise needs (e.g. short heavy sedation or prolonged surgical anaesthesia).

Anaesthetic	Sedation dose	Deep anaesthesia dose	Euthanasia dose	Buffer or mix in	Drug classification	Main effect on nerve sensitivity
2-Phenoxyethanol	TBD	400 mg/l	3 g/l	N/A	2-Phenoxyethanol	Expansion of neuronal cell membranes
Benzocaine	TBD	100 mg/l	1 g/l	Mix in ethanol 20 g/l	Local anaesthetic	Blockade of voltage-sensitive sodium channels
Lidocaine hydrochloride	180 mg/l	TBD	1 g/l	[NaHCO3]= 2×[Lidocaine]		
Tricaine methanesulfonate (MS222)	40 mg/l	168 mg/l	1 g/l	[NaHCO3]= 2×[tricaine]		
Etomidate	2 mg/l	LAP	LAP	N/A	Non-barbiturate imidazole	Bind GABAA receptors and inhibit neurotransmitter release in neurosecretory cells
Medetomidate	2 mg/l	LAP	LAP	N/A		
Clove oil	45 mg/l	TBD	1 g/l	Mix 1 in 10 ethanol 96%	*Syzygium aromaticum* derivative	Inhibition of calcium, potassium and sodium channels and of NMDA receptors, potentiation of GABAA receptors
Isoeugenol	5 mg/l	20 mg/l	540 mg/l	N/A		

Baldisserotto et al., 2018; Collymore et al., 2014; Forman, 2011; Herring et al., 2011; Olt et al., 2016; Sánchez-Vázquez et al., 2011; Sloman et al., 2019; Sneddon et al., 2012; von Krogh et al., 2021; Westerfield, 1994). Data on anaesthesia of larvae and juveniles are less easily obtained, and dose rates should be adapted to the developmental stage of the fish (Ehrlich et al., 2019; Macova et al., 2008; Mocho et al., 2022; Olt et al., 2016; Rombough, 2007).

Amino amide and amino ester local anaesthetics used for general anaesthesia

So-called local anaesthetics are commonly used to provide general anaesthesia or euthanasia in aquatic animals. The general analgesia effect is also used in mammals and fish. Only three compounds are considered because data are lacking for others. Tricaine methane sulfonate (also called MS222) is licensed in many countries for use in fish and is commonly used in fish laboratories (Mocho & von Krogh, 2022). It is not licensed in humans. Benzocaine is licensed in some countries for use in fish and as a local anaesthetic in humans. Lidocaine (also called xylocaine or lignocaine) is licensed as a local anaesthetic for humans and many other terrestrial species, and it is also used for its general analgesic effects in both human and veterinary medicine. Tricaine and benzocaine are amino esters and have low pKa values. Lidocaine is an amino amide and its pKa value is 7.75. When pH is above the pKa value, these compounds are mainly present under their non-ionized form and this significantly increases their absorption, and therefore their potency. These compounds may reduce pH and therefore buffering to the required pH is recommended (Carrasco et al., 1984; Carter et al., 2011; Mocho et al., 2022; Topic Popovic et al., 2012; von Krogh et al., 2021). Preparing euthanasia solutions with lidocaine hydrochloride and tricaine remains straightforward because these dissolve in water at the required concentrations. Benzocaine solutions are more complex, requiring the addition of a solvent such as ethanol. These three compounds provide analgesia providing the fish is immersed in the medicated solutions. Benzocaine is often considered a safer option despite the lack of supporting published data (Ross & Ross, 2008).

Lidocaine hydrochloride is used at very low doses (2–5 mg/l) to provide analgesia without sedation (Schroeder & Sneddon, 2017; Schroeder et al., 2021), and at very high doses (1 g/l with 2 g/l sodium bicarbonate – and 50 ml/l of ethanol 96% for larvae) to euthanize zebrafish (Collymore et al., 2016; Mocho et al., 2022; von Krogh et al., 2021). The use of lidocaine for anaesthesia is mainly recommended when mixed with propofol, an anaesthetic licensed in human and veterinary medicine (Martins et al., 2018; Valentim et al., 2016). Such a combination has the advantage of mixing an analgesic, lidocaine, with a hypnotic anaesthetic deprived of suitable anti-nociceptive properties postoperatively: propofol, a GABAA receptor agonist (Fassoulaki, 2011; Martins et al., 2019). Tricaine is commonly used in the laboratory for zebrafish anaesthesia (168 mg/l, buffered to system water pH; Attili & Hughes, 2014; Westerfield, 1994). Mortality is usually <1% of anaesthetized animals and can be reduced further with a premedication sedative bath (30–48 mg/l, buffered to system water pH; Matthews et al., 2018). Daily repetition of tricaine anaesthesia induces significant mortality, although it is difficult to determine whether this is just due to the repeat of anaesthesia rather than tricaine repeated exposure (Dang et al., 2016). Nevertheless, it is recommended to leave 2 weeks between administrations (Deebani et al., 2019; Kristan et al., 2012). All three molecules have been classified as aversive at low doses in zebrafish (Readman et al., 2013). However, aversion is more complex to assess at anaesthetic doses for tricaine and lidocaine (Ferreira et al., 2022; Wong et al., 2014). When inducing overdose of anaesthesia, benzocaine induces strong aversive reactions whereas lidocaine induces little aversion (Mocho et al., 2022; von Krogh et al., 2021). Other adverse effects of amino esters (benzocaine and tricaine) are the risk of allergies to exposed staff. To avoid skin contact, transcutaneous absorption or inhalation, personal protective equipment (PPE) and other specific steps should be taken when handling powders and medicated water (including with amino amides like lidocaine). Due to the licensed use of lidocaine in human and veterinary medicine, the environmental toxicity of the compound has been well studied. It is expected that the predicted environmental concentration will remain below the predicted no-effect concentration. Lidocaine is degraded in surface water, does not infiltrate into groundwater and has a low potential to bio-accumulate (Rúa-Gómez & Püttmann, 2012, 2013; Strichartz et al., 1990).

Non-barbiturate imidazole

Etomidate and metomidate are licensed for use in humans and fish in some countries. Both lack analgesic properties (Ross & Ross, 2008) and are available as solutions, dissolved in water with solvents such as propylene glycol; Otherwise, dissolution of the raw powder in water with ethanol is an option. Etomidate and metomidate are mainly used in larger fish species as a sedative. Reaching deep anaesthesia or euthanasia can be challenging with these compounds because the response of individual fish to higher doses is variable (Collymore et al., 2014; von Krogh et al., 2021). The main advantage of these compounds is that zebrafish do not show signs of aversion at low immersion concentrations (Readman et al., 2013). At anaesthetic concentrations, aversion is more complex to assess, with etomidate triggering aversion but not metomidate (Ferreira et al., 2022; Wong et al., 2014). Moreover, these imidazoles induce significant long-lasting changes on complete blood count of fish (Kristan et al., 2012; Witeska et al., 2015). They also suppress corticosteroid synthesis, hence reducing cortisol peak due to handling (Forman, 2011; Ferreira et al., 2022; Herring et al., 2011; Sneddon, 2012). Nonetheless, steroidogenesis is an important pathway of fish metabolism.

No data are available regarding the length of suppression, which may vary among various transgenic lines. Eventually, suppression of corticosteroid synthesis could lead to adrenocortical dysfunction and impact on fish health and fertility in the long term. PPE should be worn when handling medicated water or the compounds as powder.

2-Phenoxyethanol

2-Phenoxyethanol is used in many domestic products, in licensed medicines as a preservative and as a licensed fish anaesthetic in some countries. Its analgesic properties are unclear (Ross & Ross, 2008).

The raw compound is a liquid readily mixed in water when shaken thoroughly. It is used in many species to induce sedation, general anaesthesia or euthanasia, and the dose and depth of anaesthesia are easily adjusted (Owen & Kelsh, 2021). Zebrafish display aversion towards low doses of 2-phenoxyethanol (Readman et al., 2013), and other studies have demonstrated long-lasting adverse effects on complete blood count in fish (Kristan et al., 2012; Witeska et al., 2015). In humans, exposure is sometimes related to various impairments of the central nervous system and digestive or respiratory tract. It is recommended to prepare the anaesthetic solution in a well-ventilated area because 2-phenoxyethanol gas escapes from solution when mixed thoroughly (Ross & Ross, 2008). Appropriate PPE to avoid skin contact should be worn.

Syzygium aromaticum derivative products

Clove oil is used as a food product. Its main active ingredient, eugenol, and a structurally related derivative, isoeugenol, are licensed in human and fish anaesthetics, and in analgesics in some countries (Ross & Ross, 2008). Clove oil formulations vary between suppliers in quantity and quality (Chaieb et al., 2007) because they are composed of eugenol and terpenoids from the leaves, buds and stems of the clove tree (Baldisserotto et al., 2018) and mixed in ethanol. In practice it is recommended to add ethanol to clove oil anaesthetic solution, otherwise dissolution may be an issue (Grush et al., 2004; von Krogh et al., 2021). Some veterinary products contain isoeugenol in solution, and these are readily mixed in water (Schroeder et al., 2021). The main uses of these products are euthanasia with clove oil (Davis et al., 2015) and sedation with isoeugenol (Baldisserotto et al., 2018; Sánchez-Vázquez et al., 2011). To date there is one published experiment with very encouraging results for the use of 20 mg/l isoeugenol as an anaesthetic for fin-clipping of zebrafish (Musk et al., 2020). Trials comparing recovery anaesthesia with *S. aromaticum* derivative products and tricaine revealed a slower recovery for eugenol-related compounds (Ehrlich et al., 2019; Grush et al., 2004; Musk et al., 2020). However, clove oil is deemed less aversive than tricaine, possibly due to the presence of ethanol (Wong et al., 2014).

All these clove-derived compounds appear to induce less toxicity than other licensed fish anaesthetics (Grush et al., 2004; Kristan et al., 2012). One caveat with their use is their strong smell and volatility, leading to potential headaches and other discomfort for staff if the area is not well ventilated (Ross & Ross, 2008). Appropriate PPE should be worn to avoid skin contact.

Multifactorial evaluation of anaesthetics

To choose which anaesthetic to use for a procedure, researchers can perform a theoretical evaluation based on both literature and personal experience. The result should be confirmed in practice by performing the protocol on a small number of animals and assessing key outcomes, including the ability to achieve consistently the required depth of anaesthesia, the uneventful recovery of the animals and rapid euthanasia with minimal distress (Martins et al., 2019; Mocho et al., 2022; Schroeder et al., 2021; von Krogh et al., 2021). Indeed, literature specific to zebrafish is restricted and water and fish conditions vary, sometimes significantly, among facilities. The first step in theoretical evaluation is to establish a set of criteria that define good anaesthesia (Marking & Meyer, 1985; Mocho et al., 2022; Ross & Ross, 2008; Schroeder et al., 2021; von Krogh et al., 2021). For example, preparation of the anaesthetic solution should be simple in order to avoid mistakes. Induction and recovery should be rapid, although this is often only checked through practice. The analgesic properties of the compounds must be taken into account in regard to painful procedures; a combination of two compounds is also an option. Zebrafish aversion towards a compound must be considered to optimize animal welfare during the procedure. Nevertheless, there are three other very important parameters. The anaesthesia should be reliable, meaning that it must be easy to achieve the required depth of anaesthesia for all animals without having to modify concentration or exposure time excessively. The protocol should not lead to increased mortality or long-term adverse effects, even more so when the procedure is to be repeated. Health and safety measures must be able to contain the risk associated with handling of the compound, or with staff and environmental exposure to medicated water. To perform a comparison, each compound can be given a score between 1 and 4, 4 being the optimal score for one criterion and 1 being the worst. Scores are added and the compound with the highest score is the preferred candidate. It is also possible to weight the evaluation and, for example, to give a coefficient of 2 for reliability, mortality/toxicity or health and safety.

Selection of an anaesthetic for a short and painful procedure in adult zebrafish

Table 13.4 shows an example of evaluation of candidates for a short and painful procedure performed on an adult *D. rerio* and requiring analgesia and heavy sedation to surgical anaesthesia, for example, for fin-clipping, injection, oral gavage or squeezing for gamete harvesting. This table is not weighted: all parameters have a coefficient of 1 for score calculation. The list of compounds corresponds to what is available in a hypothetical country. Compounds receive a score of 4 if they are recognized analgesics, 2 if this is unclear and 1 if they are not. For the preparation column, the scores relate to the risk of mistakes when preparing the anaesthetic bath: a score of 4 is given when two solutions are mixed, and the score decreases when powders need to be measured and solvents are required. A candidate's reliability is defined as its ability to provide a set depth of anaesthesia for all fish in the anaesthetic solution in a similar timeframe. Averseness receives a binary score: 4 when data do not show aversion in zebrafish, 1 when they do. Toxicity corresponds to the reported adverse effects and mortality in fish following anaesthesia. The hazard due to handling of the candidate and its solution is considered in the health and safety column. For each row, the addition of values in each column constitutes the total score. In this example of theoretical evaluation, isoeugenol is the first-choice candidate to trial, with the highest score of 19 out of a maximum possible score of 24; isoeugenol's score could be further improved by using the product in a fume-capturing safety cabinet. Otherwise, tricaine remains a valuable candidate in this scenario, with a score of 18, even more

Table 13.4 An example of evaluation of candidates for a short and painful procedure requiring general anaesthesia in adult zebrafish. Each anaesthetic candidate is evaluated for a set of criteria adapted to a short and painful procedure requiring analgesia and heavy sedation to surgical anaesthesia, for example, fin-clipping, injection, oral gavage or squeezing for gamete harvesting. A score from 1 to 4 is attributed to each candidate for each criterion; scores for each criterion are added to give a total score for each candidate. The candidate with the highest total score is the first-choice anaesthetic candidate.

Anaesthesia	Analgesia	Preparation	Reliability	Averseness	Toxicity	Health & safety	Total score
2-Phenoxyethanol	2	4	4	1	1	2	14
Benzocaine	4	1	4	1	3	3	16
Isoeugenol	4	4	4	1	4	2	19
Etomidate	1	4	1	1	1	4	12
Lidocaine	4	3	4	1	2	3	17
Tricaine	4	3	4	1	3	3	18

so when the space is too confined and not sufficiently ventilated to avoid staff exposure to the smell of isoeugenol.

Selection of anaesthetic for euthanasia of adult zebrafish

Table 13.5 shows an example of an evaluation of anaesthetic candidates for adult euthanasia. This table is weighted with a coefficient of 2 given to reliability because euthanasia is often performed on a group of fish and aiming for minimal distress implies reducing the time of impaired alertness and ensuring that no fish significantly suffers longer than others. The list of available compounds corresponds to that available in a hypothetical country. For the preparation column, the scores relate to the risk of mistakes when preparing the anaesthetic bath: a score of 4 is given when two solutions are mixed, and this decreases when powders need to be measured and solvents are required. A candidate's reliability is defined as its ability to induce death in all fish in a restricted timeframe, death here being the permanent loss of righting reflex, mobility, operculum movement and other reflexes even when stimulated. Averseness receives a binary score: 4 when data do not show aversion in zebrafish, 1 when

Table 13.5 An example of evaluation of candidates for adult zebrafish euthanasia by overdose of anaesthesia, with a weighted coefficient of 2 for reliability. Each anaesthetic candidate is evaluated for a set of criteria adapted to a scenario of euthanasia by overdose of anaesthesia. A score from 1 to 4 is attributed to each candidate for each of these criteria: preparation, reliability, averseness, health and safety. Scores for each criterion are added, with the score for reliability added twice. This gives a total score for each candidate, with a weighted coefficient of 2 for reliability. The candidate with the highest total score is the first-choice anaesthetic.

Anaesthetic	Preparation	Reliability	Averseness	Health & safety	Total score
2-Phenoxyethanol	4	3	1	2	13
Benzocaine	1	4	1	3	13
Isoeugenol	4	3	1	2	13
Etomidate	4	1	1	4	11
Lidocaine	3	4	2	3	16
Tricaine	3	3	1	3	13

they do. The hazard due to handling of the candidate and its solution is considered in the health and safety column. For each row, the addition of values in each column constitutes the total score, with the score for reliability counted twice. In this example, lidocaine is the first-choice candidate with the highest score of 16 out of a maximum possible score of 20. Etomidate is the last choice due to its demonstrated averseness at concentration doses and to its absence of reliability in induction of euthanasia of all fish in a short timeframe. Moreover, when trialling and designing the euthanasia protocol it is also relevant to consider completion of death, i.e. how to confirm that the animal is dead following overdose of anaesthesia and that it will not recover if removed from immersion. Good practice for adult zebrafish is to prolong immersion for at least 10 min following permanent cessation of all reflexes, including stimulation by touch, pinch and startle (von Krogh et al., 2021). Either maceration or severing the head, the gills or the caudal vein will then ensure that death is complete.

General considerations for zebrafish analgesia

Analgesia should be used perioperatively when performing painful procedures, unless justified and ethically agreed otherwise. As explained in Table 13.1, this means that painkillers should be provided before any painful stimulus in order to apply pre-emptive analgesia and optimize analgesic efficacy. Analgesia should be administered during and after the painful procedure, while the animal is recovering and for as long as pain is deemed to last. To achieve anti-nociceptive cover without inducing unnecessary sedation, the choice of analgesic compound can be differentiated from that of the anaesthetic, and a combination of drugs administered perioperatively. Painkillers may be added to the anaesthetic solution or they may be given independently before anaesthesia induction, during recovery and postoperatively. Whereas anti-nociception in larval *D. rerio* has been studied to establish screening models for human analgesic candidates (Chatigny et al., 2018; Curtright et al., 2015; Lopez-Luna et al., 2017a,b; Steenbergen & Bardine, 2014), experiments on adult zebrafish have demonstrated analgesic properties for three families of compounds routinely used in veterinary medicine, allowing refinements of experimental procedures (data are summarized in Table 13.6). Lidocaine, from the local anaesthetics' family, has a local and general effect by blocking voltage-sensitive sodium channels (Audira et al., 2022; Deakin et al., 2019a,b; Schroeder & Sneddon, 2017; Schroeder et al., 2021). Morphine, an opioid, is an agonist of the μ receptor in the central nervous system (Correia et al., 2011; Costa et al., 2019; Deakin et al., 2019a; Magalhaes et al., 2017, 2018; Taylor et al., 2017). Non-steroidal anti-inflammatory drugs (NSAIDs) inhibit cyclooxygenase enzymes (COX-1 and/or COX-2; Costa et al., 2019; Deakin et al., 2019a; Magalhaes et al., 2017; Schroeder & Sneddon, 2017). Each family of analgesics interferes with a different pathway and more than one pathway can be compromised concurrently. Therefore, compounds from different families can be used concurrently to interfere with as many pathways as possible and provide a so-called multi-modal analgesia and optimize perioperative care. Despite data building up on the anti-nociceptive properties of compounds, there are still a few key considerations that have not been investigated. There are no data evaluating the exposure time required for the pre-emptive analgesia to take effect. This may vary with water parameters (e.g. pH and lidocaine) and the procedure's severity, just as the efficacy of any medication depends on husbandry, genotype, microbiome, chronobiology, fish health, strain, developmental stage, body condition, size and, eventually, sex. Another unanswered question is this: for how long should analgesia be provided following procedures? Zebrafish display signs of pain for at least 6 h post fin-clipping (Audira et al., 2022; Deakin et al., 2019a,b), but there are no data suggesting the length of time required for the fish not to require painkillers. Neither are there any established means to monitor analgesic efficacy at the tank side by direct observation, even though score sheets have been proposed (Martins et al., 2016; see Chapter 14 for more information). This impairs our ability to check on protocol efficacy or fish needs for repeat, prolonged or supplemented administration.

Table 13.6 Effective dose and route of analgesics reported in the literature for adult *D. rerio*. The table summarizes published data on efficacy of trialled analgesics in adult zebrafish. For each compound, only dose ranges with demonstrated anti-nociceptive properties are indicated, plus route of administration, reported adverse effects, indication of compound classification and literature reference(s) supporting data (Audira et al., 2022; Correia et al., 2011; Costa et al., 2019; Deakin et al., 2019a,b; Magalhaes et al., 2017, 2018; Schroeder & Sneddon, 2017; Taylor et al., 2017).

Analgesic	Reported efficacious dose	Route	Adverse effects	Drug classification	Reference(s)
Lidocaine	2–5 mg/l	Immersion	None observed	Local anaesthetics	Schroeder & Sneddon, 2017; Deakin et al., 2019a; Deakin et al., 2019b; Audira et al., 2022
Morphine	3–13 mg/kg	Intra-coelomic	None observed	Opioids	Magalhaes et al., 2017
	8–16 mg/kg	Intra-coelomic	None observed		Magalhaes et al., 2018
	2.5 mg/kg	Intra-coelomic	None observed		Costa et al., 2019
	3–6 mg/kg	Intramuscular	None observed		Correia et al., 2011
	2.5-5 mg/kg	Intramuscular	None observed		Taylor et al., 2017
	48 mg/l	Immersion	None observed		Deakin et al., 2019a
Diclofenac	40 mg/kg	Intra-coelomic	None observed	NSAIDs	Costa et al., 2019
Indomethacin	3–13 mg/kg	Intra-coelomic	None observed		Magalhaes et al., 2017
Aspirin	2.5 mg/l	Immersion	None observed		Schroeder & Sneddon, 2017
Flunixin	8 mg/l	Immersion	None observed		Deakin et al., 2019a

Selection of an analgesic and suitable protocol

Data on NSAIDs used as painkillers in zebrafish remain scant. Experiments demonstrating efficacy often suggest a higher analgesic potency with lidocaine or morphine (Audira et al., 2022; Costa et al., 2019; Deakin et al., 2019a; Magalhaes et al., 2017; Schroeder & Sneddon, 2017). As a consequence, although NSAIDs are not the first-choice analgesics in zebrafish, they could be used in combination with compounds of other families to achieve multimodal analgesia for moderately to severely painful procedures. Opioids such as morphine could be recommended to alleviate such nociception, with dose adapted to the severity. Indeed, 4 mg/kg of morphine failed to alleviate corneal nociception (Magalhaes et al., 2018) whereas such a dose range is efficient in other models such as complete Freund's adjuvant, formalin or acetic acid injection (Correia et al., 2011; Costa et al., 2019; Magalhaes et al., 2017; Taylor et al., 2017). For severe procedures such as cardiac cryo-injury, higher doses of morphine should therefore be used, e.g. at least 8–16 mg/kg. To date, publications on regeneration models using opioids have not revealed any experimental interference regarding the use of morphine immersion or buprenorphine injection (Hu et al., 2022; Murugan et al., 2022). The means of administration is also relevant when choosing an analgesic protocol. For example, the benefits of injecting fish with painkillers when fin-clipping would be outweighed by the increased welfare impairment due to the prolonged time out of water and the creation of a second wound (the injection site). However, analgesic injection could be considered a valuable approach for more severe procedures. On the other hand, immersion is a practical means of administration when using a large number of animals in a single session. This method's main limitations are (1) the absence of data regarding the ability of the published immersion protocols to alleviate severe pain and (2) care should be taken when handling medicated water to avoid staff exposure (e.g. highly concentrated morphine solution; Deakin et al., 2019a). Lidocaine is the immersion analgesic of choice for mild to moderate procedures such as fin-clipping (Deakin et al.,

2019a,b; Schroeder & Sneddon, 2017). At the recommended range (2–5 mg/l), adverse effects are not detected. Increasing the dose to 10 mg/l would induce some sedation (de Abreu et al., 2019). Similarly, eugenol is not proposed as an analgesic because no analgesia has been demonstrated at low doses (5 mg/l) and sedation commences from 10 mg/l, limiting the exploration of the analgesic potency of the compound and preventing its use post-operatively during the recovery phase (Baldisserotto et al., 2018).

Administration of analgesics in adult zebrafish

Table 13.6 lists routes for analgesic administration that have been used with success in adult zebrafish. Intra-coelomic and intramuscular injections can be used only on an anaesthetized animal. The fish is placed on a sponge and a suitably narrow needle is used. The sponge is soaked with system water and autoclaved between sessions. The fish should not be anaesthetized simply for the purpose of receiving painkillers since this would most likely be counterproductive for its welfare. Therefore, the animal should be anaesthetized for the purpose of the procedure requiring analgesic cover, and the analgesic injection should be simply an additional step performed sufficiently rapidly not to increase significantly the time spent out of water, unless the fish is maintained under general anaesthesia by continuous watering of the gills. This eliminates injection as a route for either pre-emptive analgesia or repeat analgesic administration. Nonetheless, the half-life of many compounds is longer in fish than in mammals and this can be an advantage when injecting morphine (Newby et al., 2008; Sloman et al., 2019). Another option is to mix analgesics in the syringe containing other substance(s) injected for the purpose of the experimental procedure. This should be considered with great caution depending on the compounds and their excipients' chemical properties, since it may interfere significantly with the pharmacokinetics of either the analgesic or the tested compounds.

As an alternative to injection, immersion can be used to administer analgesia to an alert population of fish in their holding tank before the procedure, or in recovery tanks, isolated or on a recirculating system. In practice, to accelerate administration of lidocaine when fin-clipping, several options are available. A stock solution with 530 mg/l lidocaine hydrochloride (or higher concentrations to medicate >1-l recovery tanks) can be prepared; 3-ml pipettes are used to administer 12 ml (four pipettes) to 3-l holding tanks pre-emptively. Each recovery tank, containing one or two fin-clipped and identified zebrafish in 400–800 ml of system water, is medicated with one 3-ml pipette. Fish remain for up to 48 h in recovery tanks, fed once daily at reduced quantity, before water is changed for clean, non-medicated system water. Such a protocol allows the rapid medication of tanks and fish with 2–5 mg/l lidocaine hydrochloride (Schroeder et al., 2021). Alternatively, an entire recirculating system can be medicated and a carbon filtration used to adsorb the lidocaine treatment and control exposure to the painkiller. One caveat is the exposure of non-procedural fish to the analgesic compound; this could be negligible in the absence of toxicity or in regard to interference with research models. There are two options when treating a whole system: (1) the carbon filter can be kept switched off for the duration of the treatment and switched back on when analgesic provision is no longer required; however, adsorption is not instantaneous; and (2) controlling the speed of adsorption. As the whole system is medicated with 5 mg/l lidocaine hydrochloride, a water pump creates a flow through a carbon filter so that the lidocaine is slowly adsorbed, ensuring that painkiller concentration remains >2 mg/l for the required number of hours. As an analogy, the concentration of lidocaine in the system water, i.e. the dose of painkiller administered by immersion, decreases with time and reproduces the pharmacokinetic curve of a compound administered by intravenous bolus and eliminated by kidney filtration. The set-up of the carbon filtration unit requires some fine-tuning because flow rate must be adjusted to the required 'clearance'. To do so, the lidocaine concentration in the system water is measured by high-performance liquid chromatography. Figure 13.1 shows data obtained from a zebrafish-genotyping recirculating rack with a carbon filtration unit developed in house while facilitating perfect control of lidocaine immersion at 2–5 mg/l for 5 h.

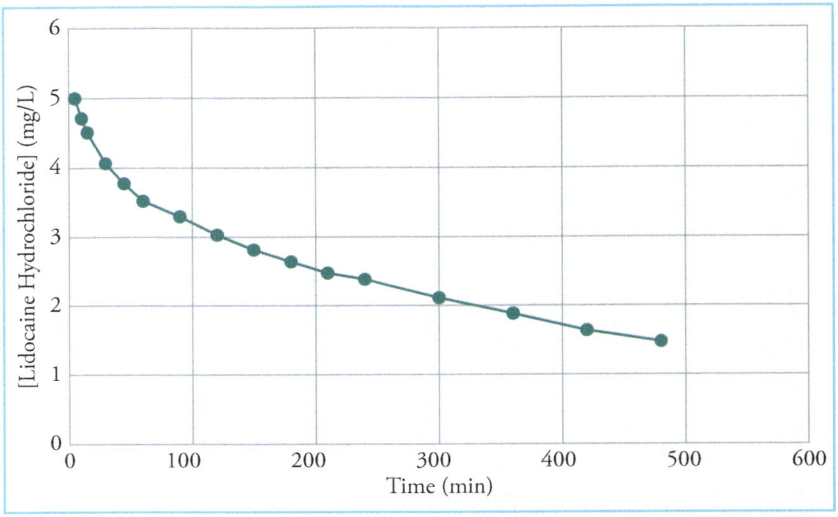

Figure 13.1 Concentration of lidocaine controlled by carbon filtration in a zebrafish-genotyping recirculating system following whole-system treatment. System water is treated at 0 min with 5 mg/l lidocaine hydrochloride. Water samples are taken and lidocaine concentration is measured by high-performance liquid chromatography. The carbon filtration unit was adjusted to maintain at least 2 mg/l of lidocaine for 300 min.

Conclusion

The choice of analgesics and the proposed evaluation system for anaesthetic candidates are limited by the small range of licensed products at the national level. Local requirements regarding the species-exclusive use of certain products or access to off-licence medicines can also be a hurdle, although such drug evaluation systems have been used as a tool to justify imports of products licensed in other countries. The market for, and availability of, licensed fish, veterinary and human anaesthetics are in permanent transformation, and surveys of available candidates must be performed regularly. This might affect the cost of the products because there could be significant variation between formulations containing the same candidate. Similarly, evaluations should regularly be adapted with new published knowledge. For example, the use of essential oils for zebrafish anaesthesia, euthanasia or analgesia could be explored further (Martins et al., 2019) and data on mixed anaesthetic solutions could be considered, such as combinations of propofol and lidocaine (Martins et al., 2018; Valentim et al., 2016). Bearing in mind that evaluation is only a theoretical exercise, the choice of candidates should be supported by in-house experience and trials. Moreover, a degree of subjectivity in the scoring is unavoidable. For example, individuals may deem one parameter more relevant than another or may be more confident or interested in assessing some specific adverse effect(s) or welfare parameter(s). There are also knowledge gaps in published data, such as protocols for larval and juvenile anaesthesia and euthanasia. The choice of analgesics remains limited too, even more so for procedures likely to induce severe pain. When data on analgesics are more developed, similar evaluation systems to those for anaesthesia and euthanasia can be proposed for various analgesic scenarios. Evaluation systems are destined to become self-feeding loops. Indeed, as data sharing is increased, evaluations are more educated, new gaps of knowledge are identified, new questions are investigated, evaluation systems are improved and new data are generated and shared.

References

Aoki, T., Kinoshita, M., Aoki, R., Agetsuma, M., Aizawa, H., Yamazaki, M., Takahoko, M., Amo, R., Arata, A., Higashijima, S., Tsuboi, T., & Okamoto, H. (2013). Imaging of neural ensemble for the retrieval of a learned behavioral program. *Neuron*, *78*(5), 881–894. https://doi.org/10.1016/j.neuron.2013.04.009

Attili, S., & Hughes, S. M. (2014). Anaesthetic tricaine acts preferentially on neural voltage-gated sodium channels and fails to block directly evoked muscle contraction. *PloS One*. Public Library of Science ONE, *9*(8), e103751. https://doi.org/10.1371/journal.pone.0103751

Audira, G., Suryanto, M. E., Chen, K. H., Vasquez, R. D., Roldan, M. J. M., Yang, C. C., Hsiao, C. D., & Huang, J. C. (2022). Acute and chronic effects of fin amputation on behavior performance of adult zebrafish in 3D locomotion test assessed with fractal dimension and entropy analyses and their relationship to fin regeneration. *Biology, 11*(7), 969. https://doi.org/10.3390/biology11070969

Baldisserotto, B., Parodi, T. V., & Stevens, E. D. (2018). Lack of postexposure analgesic efficacy of low concentrations of eugenol in zebrafish. *Veterinary Anaesthesia and Analgesia, 45*(1), 48–56. https://doi.org/10.1016/j.vaa.2017.08.009

Carrasco, S., Sumano, H., & Navarro-Fierro, R. (1984). The use of lidocaine-sodium bicarbonate as anaesthetic in fish. *Aquaculture, 41*(4), 395–398. https://doi.org/10.1016/0044-8486(84)90207-2

Carter, K. M., Woodley, C. M., & Brown, R. S. (2011). A review of tricaine methanesulfonate for anesthesia of fish. *Reviews in Fish Biology and Fisheries, 21*(1), 51–59. https://doi.org/10.1007/s11160-010-9188-0

Chaieb, K., Hajlaoui, H., Zmantar, T., Kahla-Nakbi, A. B., Rouabhia, M., Mahdouani, K., & Bakhrouf, A. (2007). The chemical composition and biological activity of clove essential oil, Eugenia caryophyllata (Syzigium aromaticum L. Myrtaceae): A short review. *Phytotherapy Research, 21*(6), 501–506. https://doi.org/10.1002/ptr.2124

Chatigny, F., Creighton, C. M., & Stevens, E. D. (2018). Updated review of fish analgesia. *Journal of the American Association for Laboratory Animal Science, 57*(1), 5–12.

Chen, K., Wang, C. Q., Fan, Y. Q., Xie, Y. S., Yin, Z. F., Xu, Z. J., Zhang, H. L., Cao, J. T., Han, Z. H., Wang, Y., & Song, D. Q. (2014). The evaluation of rapid cooling as an anesthetic method for the zebrafish. *Zebrafish, 11*(1), 71–75. https://doi.org/10.1089/zeb.2012.0858

Collymore, C., Banks, E. K., & Turner, P. V. (2016). Lidocaine hydrochloride compared with MS222 for the euthanasia of zebrafish (Danio rerio). *Journal of the American Association for Laboratory Animal Science, 55*(6), 816–820.

Collymore, C., Tolwani, A., Lieggi, C., & Rasmussen, S. (2014). Efficacy and safety of 5 anesthetics in adult zebrafish (Danio rerio). *Journal of the American Association for Laboratory Animal Science, 53*(2), 198–203.

Correia, A. D., Cunha, S. R., Scholze, M., & Stevens, E. D. (2011). A novel behavioral fish model of nociception for testing analgesics. *Pharmaceuticals, 4*(4), 665–680. https://doi.org/10.3390/ph4040665

Costa, F. V., Rosa, L. V., Quadros, V. A., Santos, A. R. S., Kalueff, A. V., & Rosemberg, D. B. (2019). Understanding nociception-related phenotypes in adult zebrafish: Behavioral and pharmacological characterization using a new acetic acid model. *Behavioural Brain Research, 359*, 570–578. https://doi.org/10.1016/j.bbr.2018.10.009

Curtright, A., Rosser, M., Goh, S., Keown, B., Wagner, E., Sharifi, J., Raible, D. W., & Dhaka, A. (2015). Modeling nociception in zebrafish: A way forward for unbiased analgesic discovery. *PloS One*. Public Library of Science ONE, *10*(1), e0116766. https://doi.org/10.1371/journal.pone.0116766

Dang, M., Henderson, R. E., Garraway, L. A., & Zon, L. I. (2016). Long-term drug administration in the adult zebrafish using oral gavage for cancer preclinical studies. *Disease Models and Mechanisms*, *9*(7), 811–820. https://doi.org/10.1242/dmm.024166

Davis, D. J., Klug, J., Hankins, M., Doerr, H. M., Monticelli, S. R., Song, A., Gillespie, C. H., & Bryda, E. C. (2015). Effects of clove oil as a euthanasia agent on blood collection efficiency and serum cortisol levels in Danio rerio. *Journal of the American Association for Laboratory Animal Science*, *54*(5), 564–567.

de Abreu, M. S., Giacomini, A. C. V. V., dos Santos, B. E., Genario, R., Marchiori, N. I., da Rosa, L. G. D., & Kalueff, A. V. (2019). Effects of lidocaine on adult zebrafish behavior and brain acetylcholinesterase following peripheral and systemic administration. *Neuroscience Letters*, *692*, 181–186. https://doi.org/10.1016/j.neulet.2018.11.004

Deakin, A. G., Buckley, J., AlZu'bi, H. S., Cossins, A. R., Spencer, J. W., Al'Nuaimy, W., Young, I. S., Thomson, J. S., & Sneddon, L. U. (2019a). Automated monitoring of behaviour in zebrafish after invasive procedures. *Scientific Reports*, *9*(1), 9042. https://doi.org/10.1038/s41598-019-45464-w

Deakin, A. G., Spencer, J. W., Cossins, A. R., Young, I. S., & Sneddon, L. U. (2019b). Welfare challenges influence the complexity of movement: Fractal analysis of behaviour in zebrafish. *Fishes*, *4*(1), 8. https://doi.org/10.3390/fishes4010008

Deebani, A., Iyer, N., Raman, R., & Jagadeeswaran, P. (2019). Effect of MS222 on hemostasis in zebrafish. *Journal of the American Association for Laboratory Animal Science*, *58*(3), 390–396. https://doi.org/10.30802/AALAS-JAALAS-18-000069

Ehrlich, O., Karamalakis, A., Krylov, A. J., Dudczig, S., Hassell, K. L., & Jusuf, P. R. (2019). Clove oil and AQUI-S efficacy for zebrafish embryo, larva, and adult anesthesia. *Zebrafish*, *16*(5), 451–459. https://doi.org/10.1089/zeb.2019.1737

Fassoulaki, A. (2011). Is propofol an analgesic? *European Journal of Anaesthesiology*, *28*(7), 481–482. https://doi.org/10.1097/EJA.0b013e32834584a5

Ferreira, J. M., Jorge, S., Félix, L., Morello, G. M., Olsson, I. A. S., & Valentim, A. M. (2022). Behavioural aversion and cortisol level assessment when adult zebrafish are exposed to different anaesthetics. *Biology*, *11*(10), 1433. https://doi.org/10.3390/biology11101433

Forman, S. A. (2011). Clinical and molecular pharmacology of etomidate. *Anesthesiology*, *114*(3), 695–707. https://doi.org/10.1097/ALN.0b013e3181ff72b5

Grush, J., Noakes, D. L., & Moccia, R. D. (2004). The efficacy of clove oil as an anesthetic for the zebrafish, Danio rerio (Hamilton). *Zebrafish*, *1*(1), 46–53. https://doi.org/10.1089/154585404774101671

Herring, B. E., McMillan, K., Pike, C. M., Marks, J., Fox, A. P., & Xie, Z. (2011). Etomidate and propofol inhibit the neurotransmitter release machinery at different sites. *Journal of Physiology*, *589*(5), 1103–1115. https://doi.org/10.1113/jphysiol.2010.200964

Howe, G. E., Bills, T. D., & Marking, L. L. (1990). Communications: Removal of benzocaine from water by filtration with activated carbon. *Progressive Fish-Culturist*, *52*(1), 32–35. https://doi.org/10.1577/1548-8640(1990)052<0032:CROBFW>2.3.CO;2

Hu, B., Lelek, S., Spanjaard, B., El-Sammak, H., Simões, M. G., Mintcheva, J., Aliee, H., Schäfer, R., Meyer, A. M., Theis, F., Stainier, D. Y. R., Panáková, D., & Junker, J. P. (2022). Origin and function of activated fibroblast states during zebrafish heart regeneration. *Nature Genetics*, *54*(8), 1227–1237. https://doi.org/10.1038/s41588-022-01129-5

Kristan, J., Stara, A., Turek, J., Policar, T., & Velisek, J. (2012). Comparison of the effects of four anaesthetics on haematological and blood biochemical profiles in pikeperch (Sander lucioperca L.). *Neuro Endocrinology Letters*, *33*(3) Suppl. 3, 66–71.

Leary, S., Underwood, W., Anthony, R., Cartner, S., Corey, D., Grandin, T., Greenacre, C., Gwaltney-Brant, S., McCrackin, M. A., Meyer, R., Miller, D., Shearer, J., & Yanong, R. (2013). *AVMA guidelines for the euthanasia of animals* (2013 ed) p. 1. American Veterinary Medical Association 2013(0).

Lockwood, N., Parker, J., Wilson, C., & Frankel, P. (2017). Optimal anesthetic regime for motionless three-dimensional image acquisition during longitudinal studies of adult non-pigmented zebrafish. *Zebrafish*, *14*, 2.

Lopez-Luna, J., Al-Jubouri, Q., Al-Nuaimy, W., & Sneddon, L. U. (2017a). Activity reduced by noxious chemical stimulation is ameliorated by immersion in analgesic drugs in zebrafish. *Journal of Experimental Biology*, *220*(8), 1451–1458. https://doi.org/10.1242/jeb.146969

Lopez-Luna, J., Al-Jubouri, Q., Al-Nuaimy, W., & Sneddon, L. U. (2017b). Impact of analgesic drugs on the behavioural responses of larval zebrafish to potentially noxious temperatures. *Applied Animal Behaviour Science*, *188*, 97–105. https://doi.org/10.1016/j.applanim.2017.01.002

Macova, S., Dolezelova, P., Pistekova, V., Svobodova, Z., Bedanova, I., & Voslarova, E. (2008). Comparison of acute toxicity of 2-phenoxyethanol and clove oil to juvenile and embryonic stages of Danio rerio. *Neuro Endocrinology Letters*, *29*(5), 680–684.

Magalhães, F. E. A., Batista, F. L. A., Lima, L. M. G., Abrante, I. A., Batista, F. L. A., Abrante, I. A., de Araújo, J. I. F., Santos, S. A. A. R., de Oliveira, B. A., Raposo, R. D. S., & Campos, A. R. (2018). Adult zebrafish (Danio rerio) as a model for the study of corneal antinociceptive compounds. *Zebrafish*, *15*(6), 566–574. https://doi.org/10.1089/zeb.2018.1633

Magalhães, F. E. A., de Sousa, C. Á. P. B., Santos, S. A. A. R., Menezes, R. B., Batista, F. L. A., Abreu, Â. O., de Oliveira, M. V., Moura, L. F. W. G., Raposo, R. D. S., & Campos, A. R. (2017). Adult zebrafish (Danio rerio): An alternative behavioral model of formalin-induced nociception. *Zebrafish*, *14*(5), 422–429. https://doi.org/10.1089/zeb.2017.1436

Marking, L. L., & Meyer, F. P. (1985). Are better anesthetics needed in fisheries? *Fisheries*, *10*(6), 2–5. https://doi.org/10.1577/1548-8446(1985)010<0002:ABANIF>2.0.CO;2

Martins, T., Diniz, E., Félix, L. M., & Antunes, L. (2018). Evaluation of anaesthetic protocols for laboratory adult zebrafish (Danio rerio). *PloS One*. Public Library of Science ONE, *13*(5), e0197846. https://doi.org/10.1371/journal.pone.0197846

Martins, T., Valentim, A. M., Pereira, N., & Antunes, L. M. (2016). Anaesthesia and analgesia in laboratory adult zebrafish: A question of refinement. *Laboratory Animals*, *50*(6), 476–488. https://doi.org/10.1177/0023677216670686

Martins, T., Valentim, A. M., Pereira, N., & Antunes, L. M. (2019). Anaesthetics and analgesics used in adult fish for research: A review. *Laboratory Animals*, *53*(4), 325–341. https://doi.org/10.1177/0023677218815199

Matthews, J. L., Murphy, J. M., Carmichael, C., Yang, H., Tiersch, T., Westerfield, M., & Varga, Z. M. (2018). Changes to extender, cryoprotective medium, and in vitro fertilization improve zebrafish sperm cryopreservation. *Zebrafish*, *15*(3), 279–290. https://doi.org/10.1089/zeb.2017.1521

Mocho, J. P., & von Krogh, K. (2022). A FELASA working group survey on fish species used for research, methods of euthanasia, health monitoring, and biosecurity in Europe, North America, and Oceania. *Biology*, *11*(9), 1259. https://doi.org/10.3390/biology11091259

Mocho, J. P., Lang, F., Valentin, G., Bedu, S., McKimm, R., Ramos, J., Saavedra Torres, Y., Wheatley, S. E., Higgins, J., Millington, M. E., Lundegaard, P. R., Chamorro Valverde, R., Jenčič, V., & von Krogh, K. (2022). A multi-site assessment of anesthetic overdose, hypothermic shock, and electrical stunning as methods of euthanasia for zebrafish (Danio rerio) embryos and larvae. *Biology*, *11*(4), 546. https://doi.org/10.3390/biology11040546

Murugan, N. J., Vigran, H. J., Miller, K. A., Golding, A., Pham, Q. L., Sperry, M. M., Rasmussen-Ivey, C., Kane, A. W., Kaplan, D. L., & Levin, M. (2022). Acute multidrug delivery via a wearable bioreactor facilitates long-term limb regeneration and functional recovery in adult Xenopus laevis. *Science Advances*, *8*(4), eabj2164. https://doi.org/10.1126/sciadv.abj2164

Musk, G. C., Ezzy, B. J., Kenchington, L. M., Hopper, W. A., & Callahan, L. M. (2020). A comparison of buffered tricaine methanesulfonate (MS-222) and isoeugenol anesthesia for caudal fin clipping in zebrafish (Danio rerio). *Journal of the American Association for Laboratory Animal Science*, *59*(6), 732–736. https://doi.org/10.30802/AALAS-JAALAS-20-000027

Newby, N. C., Robinson, J. W., Vachon, P., Beaudry, F., & Stevens, E. D. (2008). Pharmacokinetics of morphine and its metabolites in freshwater rainbow trout (Oncorhynchus mykiss). *Journal of Veterinary Pharmacology and Therapeutics*, *31*(2), 117–127. https://doi.org/10.1111/j.1365-2885.2008.00939.x

Olt, J., Allen, C. E., & Marcotti, W. (2016). In vivo physiological recording from the lateral line of juvenile zebrafish. *Journal of Physiology*, *594*(19), 5427–5438. https://doi.org/10.1113/JP271794

Owen, J. P., & Kelsh, R. N. (2021). A suitable anaesthetic protocol for metamorphic zebrafish. *PloS One*. Public Library of Science ONE, *16*(3), e0246504. https://doi.org/10.1371/journal.pone.0246504

Philippe, C., Vergauwen, L., Huyghe, K., De Boeck, G., & Knapen, D. (2023). Chronic handling stress in zebrafish Danio rerio husbandry. *Journal of Fish Biology*, 103(2), 367–377. https://doi.org/10.1111/jfb.15453

Readman, G. D., Owen, S. F., Murrell, J. C., & Knowles, T. G. (2013). Do fish perceive anaesthetics as aversive?. *PloS One*, 8(9), e73773. https://doi.org/10.1371/journal.pone.0073773

Reilly, S. C., Quinn, J. P., Cossins, A. R., & Sneddon, L. U. (2008). Behavioural analysis of a nociceptive event in fish: Comparisons between three species demonstrate specific responses. *Applied Animal Behaviour Science, 114*(1–2), 248–259. https://doi.org/10.1016/j.applanim.2008.01.016

Rombough, P. J. (2007). Ontogenetic changes in the toxicity and efficacy of the anaesthetic MS222 (tricaine methanesulfonate) in zebrafish (Danio rerio) larvae. *Comparative Biochemistry and Physiology. Part A, Molecular and Integrative Physiology, 148*(2), 463–469. https://doi.org/10.1016/j.cbpa.2007.06.415

Ross, L. G., & Ross, B. (2008). *Anaesthetic and sedative techniques for aquatic animals* (3rd ed). Blackwell Publishing.

Rúa-Gómez, P. C., & Püttmann, W. (2012). Impact of wastewater treatment plant discharge of lidocaine, tramadol, venlafaxine and their metabolites on the quality of surface waters and groundwater. *Journal of Environmental Monitoring, 14*(5), 1391–1399. https://doi.org/10.1039/c2em10950f

Rúa-Gómez, P. C., & Püttmann, W. (2013). Degradation of lidocaine, tramadol, venlafaxine and the metabolites O-desmethyltramadol and O-desmethylvenlafaxine in surface waters. *Chemosphere, 90*(6), 1952–1959. https://doi.org/10.1016/j.chemosphere.2012.10.039

Sánchez-Vázquez, F. J., Terry, M. I., Felizardo, V. O., & Vera, L. M. (2011). Daily rhythms of toxicity and effectiveness of anesthetics (MS222 and eugenol) in zebrafish (Danio rerio). *Chronobiology International, 28*(2), 109–117. https://doi.org/10.3109/07420528.2010.538105

Schroeder, P. G., & Sneddon, L. U. (2017). Exploring the efficacy of immersion analgesics in zebrafish using an integrative approach. *Applied Animal Behaviour Science, 187*, 93–102. https://doi.org/10.1016/j.applanim.2016.12.003

Schroeder, P., Lloyd, R., McKimm, R., Metselaar, M., Navarro, J., O'Farrell, M., Readman, G. D., Speilberg, L., & Mocho, J. P. (2021). Anaesthesia of laboratory, aquaculture and ornamental fish: Proceedings of the first LASA-FVS Symposium. *Laboratory Animals, 55*(4), 317–328. https://doi.org/10.1177/0023677221998403

Shishis, S., Tsang, B., Ren, G. J., & Gerlai, R. (2023). Effects of different handling methods on the behavior of adult zebrafish. *Physiology and Behavior, 262*, 114106. https://doi.org/10.1016/j.physbeh.2023.114106

Sloman, K. A., Bouyoucos, I. A., Brooks, E. J., & Sneddon, L. U. (2019). Ethical considerations in fish research. *Journal of Fish Biology*, *94*(4), 556–577. https://doi.org/10.1111/jfb.13946

Sneddon, L. U. (2003). The evidence for pain in fish: The use of morphine as an analgesic. *Applied Animal Behaviour Science*, *83*(2), 153–162. https://doi.org/10.1016/S0168-1591(03)00113-8

Sneddon, L. U. (2012). Clinical anesthesia and analgesia in fish. *Journal of Exotic Pet Medicine*, *21*(1), 32–43. https://doi.org/10.1053/j.jepm.2011.11.009

Sneddon, L. U. (2015). Pain in aquatic animals. *Journal of Experimental Biology*, *218*(7), 967–976. https://doi.org/10.1242/jeb.088823

Sneddon, L. U., Braithwaite, V. A., & Gentle, M. J. (2003). Do fishes have nociceptors? Evidence for the evolution of a vertebrate sensory system. *Proceedings. Biological Sciences*, *270*(1520), 1115–1121. https://doi.org/10.1098/rspb.2003.2349

Sneddon, L. U., Elwood, R. W., Adamo, S. A., & Leach, M. C. (2014). Defining and assessing animal pain. *Animal Behaviour*, *97*, 201–212. https://doi.org/10.1016/j.anbehav.2014.09.007

Speedie, N., & Gerlai, R. (2008). Alarm substance induced behavioral responses in zebrafish (Danio rerio). *Behavioural Brain Research*, *188*(1), 168–177. https://doi.org/10.1016/j.bbr.2007.10.031

Steenbergen, P. J., & Bardine, N. (2014). Anti-nociceptive effects of buprenorphine in zebrafish larvae: An alternative for rodent models to study pain and nociception? *Applied Animal Behaviour Science*, *152*, 92–99. https://doi.org/10.1016/j.applanim.2013.12.001

Strichartz, G. R., Sanchez, V., Arthur, G. R., Chafetz, R., & Martin, D. (1990). Fundamental properties of local anesthetics. 2. Measured octanol: Buffer partition coefficients and pKa values of clinically used drugs. *Anesthesia and Analgesia*, *71*(2), 158–170. https://doi.org/10.1213/00000539-199008000-00008

Strykowski, J. L., & Schech, J. M. Effectiveness of recommended euthanasia methods in larval zebrafish (*Danio rerio*) (2015). *Journal of the American Association for Laboratory Animal Science*, *54*, 81–84.

Taylor, J. C., Dewberry, L. S., Totsch, S. K., Yessick, L. R., DeBerry, J. J., Watts, S. A., & Sorge, R. E. (2017). A novel zebrafish-based model of nociception. *Physiology and Behavior*, *174*, 83–88. https://doi.org/10.1016/j.physbeh.2017.03.009

Topic Popovic, N., Strunjak-Perovic, I., Coz-Rakovac, R., Barisic, J., Jadan, M., Persin Berakovic, A., & Sauerborn Klobucar, R. (2012).

Tricaine methane-sulfonate (MS-222) application in fish anaesthesia. *Journal of Applied Ichthyology, 28*(4), 553–564. https://doi.org/10.1111/j.1439-0426.2012.01950.x

Valentim, A. M., Félix, L. M., Carvalho, L., Diniz, E., & Antunes, L. M. (2016). A new anaesthetic protocol for adult zebrafish (Danio rerio): Propofol combined with lidocaine. *PloS One*. Public Library of Science ONE, *11*(1), e0147747. https://doi.org/10.1371/journal.pone.0147747

von Krogh, K., Higgins, J., Saavedra Torres, Y., & Mocho, J. P. (2021). Screening of anaesthetics in adult zebrafish (Danio rerio) for the induction of euthanasia by overdose. *Biology*, 10(11). https://doi.org/10.3390/biology10111133

Wallace, C. K., Bright, L. A., Marx, J. O., Andersen, R. P.; Mullins, M. C., & Carty, A. J. (2018). Effectiveness of rapid cooling as a method of euthanasia for young zebrafish (*Danio rerio*). *Journal of the American Association for Laboratory Animal Science, 57*, 58–63.

Westerfield, M. M. (1994). *The zebrafish book: A guide for the laboratory use of zebrafish Danio (Brachydanio) rerio*. University of Oregon Press.

Wilson, J. M., Bunte, R. M., & Carty, A. J. (2009). Evaluation of rapid cooling and tricaine methanesulfonate (MS222) as methods of euthanasia in zebrafish (*Danio rerio*). *Journal of the American Association for Laboratory Animal Science, 48*, 785–789.

Witeska, M., Dudyk, J., & Jarkiewicz, N. (2015). Haematological effects of 2-phenoxyethanol and etomidate in carp (Cyprinus carpio L.). *Veterinary Anaesthesia and Analgesia, 42*(5), 537–546. https://doi.org/10.1111/vaa.12242

Wong, D., von Keyserlingk, M. A. G., Richards, J. G., & Weary, D. M. (2014). Conditioned place avoidance of zebrafish (Danio rerio) to three chemicals used for euthanasia and anaesthesia. *PloS One*. Public Library of Science ONE, *9*(2), e88030. https://doi.org/10.1371/journal.pone.0088030

Wynd, B. M., Watson, C. J., Patil, K., Sanders, G. E., & Kwon, R. Y. (2017). A dynamic anesthesia system for long-term imaging in adult zebrafish. *Zebrafish, 14*(1), 1–7. https://doi.org/10.1089/zeb.2016.1289

Chapter 14

Severity assignment for EU and UK procedures on zebrafish

Jean-Philippe Mocho

Abstract

Under European legislation, a severity banding is assigned to the scientific procedures undergone by animals. A forecast of this is particularly useful in regard to ethical approval ahead of the project. It is also compulsory to report the severity reflecting the experience of each animal at the end of their use. It is important to distinguish that severity applies only to regulated procedures and not husbandry-related events. It is assigned to reflect the cumulative welfare impairment due to the series of procedures. Here we give examples of severity classification depicted through comparison tables and scenarios reflecting common zebrafish procedures. We cover cases of genetically altered colonies with and without harmful phenotypes, increased mortality, fin-clipping, toxicology assay, tumour growth, treatment with adverse effects and heart cryo-injury.

Keywords

zebrafish, *Danio rerio*, fish, severity, analgesia, harmful phenotype

Claire Allen and Jean-Philippe Mocho (eds) **Zebrafish**
DOI: 10.1079/9781800629431.0014,

Introduction

In Europe, animals undergoing procedures for scientific purposes are protected under legislation requiring the assignment of a severity grading to evaluate the amount of pain, suffering, distress and lasting harm that the animals suffer. Severity assessment is performed either prior to the ethical approval of the project, called prospective severity, or at the end of the use of the animal, called actual severity. Prospective severity is particularly useful for ethics committees and national authorities to weigh the harms inflicted on the animals versus the expected benefits to science enabled by the project. Prospective severity is sought for a group of fish belonging to a treatment or control group of an experiment. The actual severity is assessed retrospectively, it is assigned to each individual and it reflects the actual harms suffered by each fish due to its use in the experiment.

In all cases, severity is applied only to procedures that are regulated by the legislation; it does not apply to husbandry-related events (e.g. falling from net when tank cleaning), capture or transport of animals. Severity does not represent the well-being of the animal at the end of the experiment (i.e. the series of procedures). Rather, a severity band is assigned according to the accumulation of harm suffered by the fish during the experimental procedures and according to the worst severity suffered at any one time. The severity of an experiment depends on the length and intensity of the induced harm (including housing and environmental challenges and stress), the repetition of procedures (considering their frequencies and intervals), whether surgery was performed, the analgesic cover, the humane endpoints, the fate of the animal and the applied refinements and perioperative care. The severity band always reflects the worst severity suffered due to the experiment. Regulations define four severity bands: non-recovery, mild, moderate and severe. In the United Kingdom (UK), a fifth severity banding (named sub-threshold) is defined for the assignment of actual severity in the case of animals that underwent a regulated procedure but did not suffer any harm due to the experiment (UK Home Office, 2014). This is mostly useful in the case of genetically altered animals (GAA) with a non-harmful phenotype, a common occurrence in zebrafish lines.

In countries abiding by the Directive 2010/63/EU (European Union (EU), 2010), this concept is replaced by the requirement to perform a welfare assessment on new GAA colonies to potentially identify fish displaying a non-harmful phenotype. No severity assignment is thereafter required for fish with a non-harmful phenotype and not undergoing any regulated procedure. However, the use of these fish still needs to be reported every 5 years. It is to be noted that fin-clipping is a regulated procedure, so all fin-clipped fish will require severity assessment. The only fish not requiring severity classification therefore represent the small population of wild-type (WT) and GAA fish with non-harmful phenotypes genotyped by non-invasive means, such as microscopic identification of fluorescence at 3 dpf or similar (see Chapter 10 for more information). The European Commission has produced further guidance on the reporting of animal numbers, welfare assessments and compliance with other GAA breeding and maintenance matters. The guidance also explains when crosses of non-harmful phenotype lines are presumed to have automatically created new GAA lines with a non-harmful phenotype (European Commission, 2022).

Welfare assessments should be performed on a scientifically and statistically justified number of animals, and they should involve no fewer than 14 fish from at least two clutches. The welfare of fish of at least at two stages should be assessed: larval stage at the point of independent feeding; adult, sexually mature animals; and/or older animals where a later onset of disease is expected. The welfare assessment of a GAA colony should be repeated if there is a change in the animals' environmental conditions (including a change of establishment) and/or if new and persistent signs are detected. The welfare assessment should compare mortality rates and the prevalence of any other clinical sign (i.e. morbidity) between the GAA colony and the in-house WT colony of similar genetic background. There is a clear need, at least in larger zebrafish facilities, for a database to allow the recording and comparison of these data (see Chapter 7 for more information). Any comparison between GAA and WT colonies should statistically confirm that mortalities and morbidities are increased or within range. If the GAA colony suffers from a statistically significant increase in a mortality rate and/or in the occurrence of any clinical sign compared with the WT colony, the phenotype is harmful. In the absence of such a statistically significant increase, mortalities and morbidities in the GAA colony are deemed within range, no adverse effect is confirmed and the GAA may be considered as displaying a non-harmful phenotype (see Chapter 9 for more information). In the UK the absence of such an increase would potentially assess the GAA colony as showing sub-threshold severity. It is necessary to capture data at the GAA and WT population levels to report them at the individual level.

Further assistance with the assignment of severity banding can be found within Directive 2010/63/EU itself (European Union, 2010) and the European Commission's published guidance (European Commission, 2012). Norway's National Consensus Platform for the advancement of 'the 3 Rs' (replacement, reduction, refinement) in animal experiments (NORECOPA) published a useful guidance for prospective severity dedicated to fish procedures (Hawkins et al., 2011).

In this concluding chapter we aim to provide further help with a series of user-friendly tables giving examples of fish-specific severity. Remarkably, only one publication assesses severity according to refinements applied to procedures in fish. This shows a moderate severity for fin-clipping of zebrafish without perioperative analgesia, and a reduction to a mild severity when analgesia is administered (Deakin et al., 2019; see Chapter 13 for more information). In the absence of data from other publications, we base our examples on this publication. Another source of help is to develop score sheets to assess the well-being and suffering of the fish after each procedure and/or during the experiment. This can be inspired by welfare assessment in the European Commission guidance on GAA, or by published examples for zebrafish (European Commission, 2022; Martins et al., 2016).

Interestingly, there is often a common pattern among score sheets for aquatic animals, whether these are laboratory-bred freshwater or wild sea animals. One part is dedicated to physical appearance: whether the fish shows obvious lesions, is egg-bound or lacks muscle and fat (see Chapters 4 and 5 for more information). The latter can be scored using a body condition score (Carneiro et al., 2016; Clark et al., 2018; Lee et al., 2019; Mocho et al., 2022; Sanders et al., 2020; see Chapters 6 and 7 for more information). Currently there is noticeably no consensus on body condition scoring. The scoring system should ideally be adapted to experimental and local needs and practicalities. Similarly, some score sheets include evaluation of conditions unlikely to impact severity (e.g. lack of gill operculum cover). These conditions should be excluded when severity is assessed. Another section of score sheets is usually dedicated to behaviour monitoring. Indeed, changes in patterns such as feeding, swimming (e.g. buoyancy, staggered movements, loss of equilibrium, activity, speed), social interaction or fertility are valuable indicators of physiological and/or welfare disturbance (Martins et al., 2016). Score sheets should be adapted to each procedure because clinical impact varies with procedural adverse effects. The total score calculated in a score sheet triggers corrective measures when necessary, such as analgesic administration or euthanasia (see Chapter 13 for more information). These scores can also be directly related to the assignment of an actual severity band.

Table 14.1 summarizes severity band assignment. The threshold for a procedure to be regulated is defined as the introduction of a needle in accordance with good veterinary practice (European Union, 2010). This is not helpful in the case of zebrafish, because it is not translatable. In brief, if anaesthesia is involved the procedure is regulated. Similarly, if adverse effects are detectable at the tank side by simple observation of the fish, the procedure is regulated; the severity banding would then depend on the frequency and intensity of the adverse effects. As a simple rule, if a fish is anaesthetized and recovers from the anaesthesia, the minimum severity would be mild. If the fish did not recover from the anaesthesia, then regardless of what was performed on the anaesthetized fish, the severity of the procedure would remain non-recovery. If the anaesthetized fish underwent surgery and recovered, the minimum severity would be moderate. If the fish was then found dead and death could not be attributed to a cause independent of the procedure, the severity score would become severe. Finally, upper threshold would

Table 14.1 Severity bands. Sub-threshold: a severity band applicable only to actual severity assignment in the UK. Mild, non-recovery, moderate, severe: severity bands for prospective and actual severity in the EU and UK. Needle stick: the threshold of pain, suffering, distress or lasting harm for a procedure to require a severity assignment. Surgery: when surgery with recovery is performed, the severity is at least moderate. Mortality: when a procedure induces mortality, the severity is severe. Unalleviated suffering: procedures leading to unalleviated suffering are not authorized under normal circumstances.

Increasing severity
Sub-threshold (UK only)
< Needle stick
< Mild – Non-recovery
< Surgery
< Moderate
< Mortality
< Severe
< Unalleviated suffering

correspond to a fish undergoing unalleviated suffering (including endless inability to feed or move) and would not be authorized under normal circumstances.

Table 14.2 shows the actual severity assessment for WT and GAA, according to their genotype, phenotype, mortality and fate, for fish not undergoing any other procedure. Whilst euthanized animals are not included in the mortality comparison, it is essential to determine whether GAA mortality (i.e. the proportion of fish found dead) is significantly higher than that of the in-house WT colony from the same genetic background. If that were the case, any fish found dead in the GAA colony would be assigned to severe severity. For adequate comparison of mortality rates it is useful to focus on defined age/developmental stage categories and periods of time. Chapter 9 proposes statistical guidance for mortality rate comparison. Most interestingly, regulators do not define the statistical confidence required to confirm whether GAA colony mortality is within or beyond the corresponding range of the WT colony. The tables in the Chapter 9 clearly show that an increase in confidence in scoring

Table 14.2 Actual severity, according to genotype, phenotype, mortality and fate, for zebrafish not undergoing any other procedure. Mortality within range: mortality of the genetically altered zebrafish colony does not significantly exceed that of the in-house WT colony of the same genetic background. +: event has occurred. N/A: not applicable to WT zebrafish. –: event did not occur.

WT	GAA		Mortality within range	Fate		Severity
	Non-harmful	Harmful		Euthanized	Found dead	
+		N/A		Either way		Not reported
N/A	+	–	+	Either way		Sub-threshold (UK)/not reported (EU)
	–	+	+	Either way		Depends on displayed phenotype of at least mild severity
	–	+	–	+	–	Depends on displayed phenotype of at least mild severity
	–	+	–	–	+	Severe

requires a higher number of GAA found dead to determine a statistically significant difference in mortality rates. Such difference would trigger the assignment of a severe severity banding to all GAA found dead in the colony.

Table 14.3 focuses on the actual severity following anaesthesia and fin-clipping. Other procedures, such as a GAA suffering from a harmful phenotype, would impact the actual severity assignment at the end of the experiment. Because unexpected adverse effects and mortality remain a rare occurrence when fin-clipping, the key determinant factor here is the provision of analgesia allowing reduction in severity from moderate to mild (Deakin et al., 2019). However, if unexpected moderate adverse effects are detected, i.e. clinical signs beyond normal anaesthesia and fin-clipping recovery, it would be essential to euthanize the fish as soon as possible to alleviate unnecessary suffering and prevent the severity from becoming severe. If the

Table 14.3 Actual severity for anaesthesia and fin-clipping procedures only. If other procedures are performed, the worst severity prevails, e.g. according to harmful phenotype. Analgesia: analgesia is provided as a minimum after fin-clipping. Unexpected adverse effects: clinical signs of moderate severity, beyond normal anaesthesia and fin-clipping recovery. ASAP: as soon as possible. +: event has occurred. –: event did not occur. ±: event may or may not have occurred – does not affect the severity assessment.

Anaesthesia	Recovery	Fin-clipped	Analgesia	Unexpected adverse effects	Euthanised ASAP	Found dead soon after	Severity
+	–	–	–	–	–	–	Non-Recovery
+	+	–	–	–	–	–	Mild
+	+	+	+	–	–	–	Mild
+	+	+	–	–	–	–	Moderate
+	+	+	+/–	+	+	–	Moderate
+	+	+	+/–	+	–	–	Severe
+	+	+	+/–	+/–	–	+	Severe

fish was found dead soon after fin-clipping it would be difficult to attribute the death to a cause independent of the procedure, and severity would therefore be severe.

Table 14.4 combines Tables 14.2 and 14.3 in the context of a GAA that suffered moderately from a harmful phenotype and that was eventually fin-clipped. No other procedure was performed. In consequence, the actual severity at the end of the fish's life is never less than moderate. The table covers all eventualities regarding mortality, anaesthesia, recovery, fin-clipping, analgesia, unexpected moderate adverse effects and fate. The absence of the influence of analgesic provision on the actual severity in this case only highlights the limits of the severity assignment system: it does not reflect refinements and good practice. Obviously, even when severity is not reduced, we still ought to provide analgesia whenever useful to improve fish welfare. This is the most ethical choice and should be encouraged (see Chapter 13 for more information).

Tables 14.5–14.8 propose examples of severity assessment when fish undergo a selection of common procedures other than fin-clipping. These tables focus on the procedure and its outcome, regardless of the (non-)harmful phenotype of the fish or other procedures (e.g. fin-clipping) that the fish might otherwise undergo. These events would obviously have to be considered for the reported actual severity assignment at the end of the experiment (i.e. series of procedures or fish life). Table 14.5 focuses on the severity for fish undergoing a toxicology assay.

Tables 14.6 and 14.7 focus on p53 GAA homozygote fish (Berghmans et al., 2005). This harmful phenotype induces spontaneous tumours randomly. The appearance of a visible tumour is considered of moderate severity, or even severe if the tumour is particularly large and/or severely impairs fish welfare. To simplify these examples we will consider here that the appearance of a visible tumour induces only moderate severity. In Table 14.6, fish are treated whether a tumour is visible or not (control). The treatment may have some adverse effects, considered here as moderate and leading to animal euthanasia, for simplification. However, when fish display a tumour and suffer from treatment-induced adverse effects, the severity may become severe depending on the accumulation

Table 14.4 Actual severity for a genetically altered zebrafish that suffered moderately from a harmful phenotype. Severity is assigned according to the mortality, fate, anaesthesia recovery and fin-clipping (perioperative care and adverse effects). It is to be noted that the zebrafish did not undergo any other procedure. Adverse effects: unexpected clinical signs of moderate severity, beyond normal anaesthesia and fin-clipping recovery.

Mortality within range	Anaesthesia	Recovery	Fin-clipping	Analgesia	Immediately following fin-clipping			Fate		Severity
					Adverse effects	Euthanized ASAP	Found dead soon after	Euthanized	Found dead	
+	–	–	–	–	–	–	–	Either way		Moderate
–	–	–	–	–	–	–	–	+	–	Moderate
–	–	–	–	–	–	–	–	–	+	Severe
±	+	–	–	–	–	–	–	–	–	Moderate
+	+	+	–	–	–	–	–	Either way		Moderate
–	+	+	–	–	–	–	–	+	–	Moderate
–	+	+	–	–	–	–	–	–	+	Severe
+	+	+	+	±	–	–	–	Either way		Moderate
–	+	+	+	±	–	–	–	+	–	Moderate
–	+	+	+	±	–	–	–	–	+	Severe
±	+	+	+	±	+	+	–	–	–	Moderate
–	+	+	+	±	+	–	–	Either way		Severe
–	+	+	+	±	±	–	+	–	–	Severe

Table 14.5 Actual severity following a toxicology assay. Fish are immersed (or not) in a solution containing a pollutant. They display aversion for a few seconds but total immersion time is several hours. Other procedures potentially impacting severity are not considered. Adverse effects: clinical signs of moderate severity, beyond first aversion.

Immersion in pollutant	Adverse effects	Euthanized ASAP	Found dead during or soon after	Fish recover quickly from moderate adverse effects	Severity
–	–	–	–	–	N/A
+	–	–	–	–	Mild
+	+	+	–	–	Moderate
+	+	–	–	+	Moderate
+	+	–	–	–	Severe
+	±	–	+	–	Severe

of pain, suffering distress and lasting harm due to both welfare impairments.

Table 14.7 continues with the example of Table 14.6 but, in this case, the treatment induces only temporarily detectable adverse effects of mild severity. The treatment is therefore repeated. However, fish may be found dead as a consequence of the repetition. A severe severity is assigned to fish found dead.

Table 14.6 Actual severity for a p53 moderately harmful phenotype and treatment. Severity is assigned according to tumour growth and treatment, both of potentially moderate severity. Fish are euthanized if they display treatment-induced adverse effects. Other procedures potentially impacting the severity are not considered. Adverse effects: clinical signs due to treatment and of moderate severity.

Visible tumour	Treatment	Adverse effects	Severity
–	–	–	N/A
–	+	–	Mild
–	+	+	Moderate
+	–	–	Moderate
+	+	–	Moderate
+	+	+	Moderate or Severe

Table 14.7 Actual severity for a p53 moderately harmful phenotype and repeated treatment with mortality. Severity is assigned according to moderate tumour growth and mortality induced by repeated treatment. Other procedures potentially impacting the severity are not considered.

Visible tumour	Repeated treatment	Fish found dead	Severity
–	–	–	N/A
–	+	–	Mild
–	+	+	Severe
+	–	–	Moderate
+	+	–	Moderate
+	+	+	Severe

Table 14.8 Actual severity following heart cryo-injury. Severity is assigned according to anaesthesia recovery, analgesic provision, surgical adverse effects and fate. Other procedures potentially impacting the severity are not considered. Adverse effects: unexpected clinical signs of moderate severity, beyond normal anaesthesia and surgery recovery.

Anaesthesia recovery	Analgesia	Following recovery			Severity
		Adverse effects	Euthanized ASAP	Found dead	
–	±	N/A	N/A	N/A	Non-Recovery
+	+	–	–	–	Moderate
+	–	–	–	–	Severe
+	+	+	+	–	Moderate
+	–	+	+		Severe
+	±	+	–	–	Severe
+	±	+/–	–	+	Severe

Table 14.8 is based on an example of heart cryo-injury. In this case, all fish were operated on. The severity depends on anaesthetic recovery and analgesic provision. If unexpected moderate adverse effects were detected, i.e. clinical signs beyond normal anaesthesia and surgery recovery, it would be essential to euthanize the fish as soon as possible to alleviate unnecessary suffering and prevent the severity from becoming severe.

Conclusion

These tables are presented to simplify severity assignment for generic conditions, but we could not include all nuances. Severity assignments are sometimes complex and may require subjectivity to determine the severity of a specific adverse effect. This is not helped by the challenges faced by humans when assessing the welfare of aquatic species and early developmental stages. The lack of score sheets published by the zebrafish community is also a drawback. Indeed, score sheets are really useful tools that can add objectivity to a potentially biased assessment (Martins et al., 2016).

References

Berghmans, S., Murphey, R. D., Wienholds, E., Neuberg, D., Kutok, J. L., Fletcher, C. D., Morris, J. P., Liu, T. X., Schulte-Merker, S., Kanki, J. P., Plasterk, R., Zon, L. I., & Look, A. T. (2005). tp53 mutant zebrafish develop malignant peripheral nerve sheath tumors. *Proceedings of the National Academy of Sciences of the United States of America*, *102*(2), 407–412. https://doi.org/10.1073/pnas.0406252102

Carneiro, M. C., Henriques, C. M., Nabais, J., Ferreira, T., Carvalho, T., & Ferreira, M. G. (2016). Short telomeres in key tissues initiate local and systemic aging in zebrafish. *PLoS Genetics*. Public Library of Science Genetics, *12*(1), e1005798. https://doi.org/10.1371/journal.pgen.1005798

Clark, T. S., Pandolfo, L. M., Marshall, C. M., Mitra, A. K., & Schech, J. M. (2018). Body Condition Scoring for adult zebrafish (Danio rerio). *Journal of the American Association for Laboratory Animal Science*, *57*(6), 698–702. https://doi.org/10.30802/AALAS-JAALAS-18-000045

Deakin, A. G., Spencer, J. W., Cossins, A. R., Young, I. S., & Sneddon, L. U. (2019b). Welfare challenges influence the complexity of movement: Fractal analysis of behaviour in zebrafish. *Fishes*, *4*(1). https://doi.org/10.3390/fishes4010008

European Commission. (2012). Severity assessment framework. https://ec.europa.eu/environment/chemicals/lab_animals/pubs_guidance_en.htm Retrieved April 3, 2023

European Commission. (2022). Framework for the Genetically Altered Animals under Directive 2010/63/EU on the protection of animals used for scientific purposes. https://ec.europa.eu/environment/chemicals/lab_animals/pubs_guidance_en.htm Retrieved April 3, 2023

European Union. (2010). Directive 2010/63/EU of the European Parliament and of the Council; On the protection of animals used for scientific purposes. *Official Journal of the European Union, 276*, 33–79.

Hawkins, P., Dennison, N., Goodman, G., Hetherington, S., Llywelyn-Jones, S., Ryder, K., & Smith, A. J. (2011). Guidance on the severity classification of scientific procedures involving fish: Report of a Working Group appointed by the Norwegian Consensus-Platform for the Replacement, Reduction and Refinement of animal experiments (Norecopa). *Laboratory Animals, 45*(4), 219–224. https://doi.org/10.1258/la.2011.010181

Lee, C. J., Paull, G. C., & Tyler, C. R. (2019). Effects of environmental enrichment on survivorship, growth, sex ratio and behaviour in laboratory maintained zebrafish Danio rerio. *Journal of Fish Biology, 94*(1), 86–95. https://doi.org/10.1111/jfb.13865

Martins, T., Valentim, A. M., Pereira, N., & Antunes, L. M. (2016). Anaesthesia and analgesia in laboratory adult zebrafish: A question of refinement. *Laboratory Animals, 50*(6), 476–488. https://doi.org/10.1177/0023677216670686

Mocho, J. P., Collymore, C., Farmer, S. C., Leguay, E., Murray, K. N., & Pereira, N. (2022). FELASA-AALAS recommendations for monitoring and reporting of laboratory fish diseases and health status, with an emphasis on zebrafish (Danio rerio). *Comparative Medicine, 72*(3), 127–148. https://doi.org/10.30802/AALAS-CM-22-000034

Sanders, J. L., Monteiro, J. F., Martins, S., Certal, A. C., & Kent, M. L. (2020). The impact of Pseudoloma neurophilia infection on body condition of zebrafish. *Zebrafish, 17*(2), 139–146. https://doi.org/10.1089/zeb.2019.1812

UK Home Office. (2014). Advisory notes on recording and reporting the actual severity of regulated procedures. https://assets.publishing.service.gov.uk/government/uploads/system/uploads/attachment_data/file/810118/NotesActualSeverityReporting.pdf Retrieved April 3, 2023

Index

Printed and bound by CPI Group (UK) Ltd, Croydon, CR0 4YY
06/07/2026
02159837-0002